Lecture Notes in Artificial Intelligence

Edited by J. Siekmann

Subseries of Lecture Notes in Computer Science

418

K.H. Bläsius   U. Hedtstück
C.-R. Rollinger   (Eds.)

Sorts and Types in Artificial Intelligence

Workshop, Eringerfeld, FRG, April 24–26, 1989
Proceedings

Springer-Verlag

Berlin Heidelberg New York London Paris Tokyo Hong Kong

Editors

Karl Hans Bläsius
Fachhochschule Dortmund, Fachbereich Informatik
Sonnenstraße 96–100, D-4600 Dortmund, FRG

Ulrich Hedtstück
Claus-Rainer Rollinger
IBM Germany Scientific Center
Institute for Knowledge Based Systems
P.O. Box 80 08 80, D-7000 Stuttgart 80, FRG

CR Subject Classification (1987): I.2.3–4, F.4.1, I.2.7

ISBN 3-540-52337-5 Springer-Verlag Berlin Heidelberg New York
ISBN 0-387-52337-5 Springer-Verlag New York Berlin Heidelberg

© Springer-Verlag Berlin Heidelberg 1990
Printed in Germany

Printing and binding: Druckhaus Beltz, Hemsbach/Bergstr.
2145/3140-543210 – Printed on acid-free paper

Lecture Notes in Artificial Intelligence

Subseries of Lecture Notes in Computer Science
Edited by J. Siekmann

Lecture Notes in Computer Science

Edited by G. Goos and J. Hartmanis

Editorial

Artificial Intelligence has become a major discipline under the roof of Computer Science. This is also reflected by a growing number of titles devoted to this fast developing field to be published in our Lecture Notes in Computer Science. To make these volumes immediately visible we have decided to distinguish them by a special cover as Lecture Notes in Artificial Intelligence, constituting a subseries of the Lecture Notes in Computer Science. This subseries is edited by an Editorial Board of experts from all areas of AI, chaired by Jörg Siekmann, who are looking forward to consider further AI monographs and proceedings of high scientific quality for publication.

We hope that the constitution of this subseries will be well accepted by the audience of the Lecture Notes in Computer Science, and we feel confident that the subseries will be recognized as an outstanding opportunity for publication by authors and editors of the AI community.

Editors and publisher

Preface

The idea of organizing the workshop "Sorts and Types in Artificial Intelligence" which is documented by this book, arose in the pleasant and stimulating surroundings of the Institute for Knowledge Based Systems at IBM Germany. Due to a fortunate circumstance, several researchers investigating sorts and types met there and worked together in a very productive way. Putting our plan into action, we were lucky to gain the most competent researchers of this field in Germany as participants of the workshop.

We would like to thank all those attending the workshop, all our colleagues of the LILOG and PROTOS projects, and IBM Germany for the financial aid. In addition, we are grateful to Gabriele Kreutzner, who made the manuscript ready for the publisher.

K. H. Bläsius, U. Hedtstück, C.-R. Rollinger

Table of Contents

III. Sorts and Types in Natural Language (Understanding) Systems

Introduction

In the field of logic, the idea of partitioning the universe of a formal language by means of sorts has been investigated extensively since the work of J. Herbrand in the 1930s. Whereas the many-sorted logics analyzed at that time did not have any ordering structure on the set of sorts, in 1962 A. Oberschelp extended these investigations to the case that an ordering structure on the set of sorts is given. He provided a clear Tarski semantics for order-sorted logics and was, in our opinion, the first one to use the term "order-sorted". Deduction systems which take into account the sortal information contained in the sort hierarchy in a direct way (i.e. not by translating it into ordinary predicate logic and then using one of the well-known deduction systems for predicate logic) were not developed until the 1980s when C. Walther realized such a deduction system by extending J. A. Robinson's resolution calculus by order-sorted unification. In Walther's deduction system special algorithms are incorporated into the resolution mechanism in order to handle directly the sortal information contained in the sort hierarchy.

Current research activities in the field of order-sorted logics address, for instance, more complex ordering structures for sorts and their responsibility for the number of results of the unification process. Other actual investigations, particularly the works of A. Cohn and M. Schmidt-Schauß, deal with different kinds of sort restrictions for the terms used in an order-sorted logic.

Lately, order-sorted knowledge representation and order-sorted inference systems have gained importance for applications in the field of Artificial Intelligence. They offer the possibility to satisfy the requirements coming from research in Natural Language Understanding and Knowledge Representation to represent objects and common sense knowledge in a well-structured and clear way. Additionally, providing specific and efficient algorithms, the search spaces of problem solving processes can be drastically reduced, an effect which has been discovered in the field of Automated Theorem Proving.

One stream of these AI research activities emerged from the investigations in the field of semantic networks which lead to the knowledge representation language KL-ONE. Many current AI formalisms grew out of this formalism having as a common matter of concern a rich description language for sorts based on attributive concept descriptions. The set of all possible concepts (sorts) of such a language is partially ordered with respect to the so-called subsumption relationship. An essential aim pursued by the developers of languages of the KL-ONE family is to provide decidable and efficient algorithms for the information contained in the sort hierarchy, in particular an efficient subsumption test for pairs of concepts.

Parallel to the developments in AI, new representation formalisms and processing mechanisms have been developed in the field of Computational Linguistics. In the last ten years semantics based grammar formalisms like M. Kay's Functional Unification Grammar, S. Shieber's PATR-II formalism or H. Uszkoreit's Stuttgart Type Unification Formalism (STUF) have been developed in order to process natural language texts with the computer. The basic structures all these formalisms deal with are feature structures which also belong to the family of attributive concept descriptions. They are processed by a unification mechanism which is closely related to subsumption test algorithms for attributive concept descriptions.

The rich sort description language and the efficient sort algorithms of KL-ONE-like formalisms, the clear semantics as well as the well-known calculi for order-sorted logics and the ideas coming from unification based grammar formalisms were the motivation for the researchers of the LILOG project to develop the knowledge representation language L-LILOG which combines all these concepts within one common framework. LILOG is a project carried out (since 1986) by the Institute for Knowledge Based Systems (IKBS) at the Science Center of IBM Germany, Stuttgart, together with partners from the universities of Hamburg, Osnabrück, Saarbrücken, Stuttgart and Aachen. Its aim is to develop linguistic, logic, and AI methods as the basis for a text understanding system of German texts. In the same institute one subproject of the EUREKA project PROTOS is being undertaken with the ambition to enrich PROLOG by a powerful type concept. This situation led to a workshop held in Geseke, West Germany, in April 1989, sponsored and organized by IBM Germany. We brought together researchers from Logic, Theoretical Computer Science, Theorem Proving, Knowledge Representation, NL Systems, Linguistics, Logic Programming and Qualitative Reasoning in order to discuss the role of sorts and types in the various aspects of AI.

Very soon it became clear that also relative to sorts and types each discipline has its own problems, goals, formal concepts, special methods, and (sometimes) definitions. On the one hand, we had problems distinguishing clearly between sorts (in the sense of Knowledge Representation) and types (in the sense of Unification Grammar). On the other hand, it was problematic to bring together sorts (in the sense of Logic) and (data-) types (in the sense of programming languages). From a syntactic point of view, sorts were discussed as helpful syntactic constructs for describing natural languages while it was also claimed that sorts are cognitively adequate semantic entities.

These highly interesting and productive discussions raised a number of questions which we were unable to resolve. Perhaps a good way to sum up these problems is to present two statements concerning the question "What is a sort?" provided by Bernd Mahr and Arnold Oberschelp.

Bernd Mahr:

In his intuitionistic type theory, Per Martin-Löf distinguishes three different ways to explain what a judgement (the key concept of his theory) means:

1. What is a set?
2. What is it that we must know in order to have the right to judge something to be a set?
3. What does a judgement of the form "A is a set" mean?

The first is the ontological (ancient Greek), the second the epistemological (Descartes, Kant, ...), and the third the semantical (modern) way of posing essentially the same question.

If 'set' is replaced by 'sort', one observes that the simple question "What is a sort?" has more facets than just the ontological one and that it therefore cannot be answered in all its generality and independent from the particular field of consideration. Less difficult seems the question "What is sorting?" which only implicitly refers to the concept of sort. It turns the attention to pragmatics and purpose and may be answered in a concise way:

Sorting is a form of classification of objects in a descriptive language serving the purpose of restricting their use.

This definition binds the concept of sort to a descriptive language and explains the purpose of sorting: restricting the use of objects. Sorting is seen here as a technique to handle partiality. It may help clarification, ease of description, in reducing the search space, or to enhance verification. Properly used, it may form a tool for modeling and representation, and even be the object of discovery. Separating points from lines in geometry is a good example, and distinguishing scalars and vectors is another. From the technical point of view of description, defining roles in knowledge representation is not much different from the introduction of types in set theory which are used to avoid the well known anomalies.

Depending on the field of consideration, sorts may be called types, kinds, Merkmale, categories, or the like. It seems unnecessary and even useless to insist upon a general distinction between these names and to assume that each refers to a well distinguished concept. Sometimes technical reasons require the use of several names, as historical reasons and convention require sticking to a certain name, but in general no difference can be found.

Sorting has a long tradition as a technique in modeling and description. The nonterminals in triangular brackets in programming language syntax, like <statement>, <compound statement>, <if-statement>, and their interpretation as domains in denotational semantics is a prominent example. Polymorphic types in the programming language ML, sorts in algebraic specifications, the two types of nodes and places, in Petri-nets, the many categories in natural theory, or the nodes of a semantic network all define sorts, and are examples of sorting.

It seems appropriate to say that any descriptive language or formalism makes use of its own built-in or user-defined sorting concept. Whether it describes theories, specifications, programs, inference systems, graph structures, automata, or anything else, it involves notions of sort. Even untyped calculus, LISP, or plain PROLOG implicitly maintain sorting of their objects with respect to which they require consistency.

Concluding these remarks, the point of view taken here is that sorting is a technique in description. Only inside a field of consideration does it pose the ontological question "What is a sort?". Only with reference to a given description language does it raise the semantical question "What does a judgement of the form 'S is a sort' mean?". And only if we are given both, a field of consideration and a description language, is the epistemological question "What is it that we must know in order to have the right to judge something as a sort?" ready to be answered.

Arnold Oberschelp:

1. Sorts are sorts of individuals. One should not have too many sorts, just for some basic (natural?) kinds of individuals. Examples: points and lines (in plane geometry), vectors and scalars (in the theory of vector spaces), standard number sets $\mathbf{N}$, $\mathbf{Z}$, $\mathbf{Q}$, $\mathbf{R}$ (in arithmetic), physical objects, points in space and time, events, men, etc. (in other applications). One should not introduce sorts for nonzero reals, nonzero rationals, primes, odd numbers, spoons, cups, rivers, seagulls, women between 25 and 35, etc.

For any sort s there should be individual variables, say v_0^s, v_1^s, v_2^s, ... , and a sortal predicate D^s in the object language interpreted by the domain of sort s.

A sortal index s is not a symbol of the language but just an index to symbols (e.g. variables) of the language.

2. Let $\forall v \in A\ \phi(v)$ be $\forall v\ (v \in A \rightarrow \phi(v))$ (relativized universal quantifier) and $\exists v \in A\ \phi(v)$ be $\exists v\ (v \in A \wedge \phi(v))$ (relativized existential quantifier).

It is well known that a quantifier using a variable of sort s can be replaced by a quantifier using a universal individual variable relativized to the sortal predicate of sort s. This is not an argument showing that sorts should be eliminated altogether. But it shows that one does not need sorts in abundance since relativized quantifiers suffice. However, the demand for lots of predicates comes up which is not fulfilled by ordinary predicate logic (see 4).

In logic the unrestricted quantifier is the basic form and relativized quantifiers are defined. But perhaps it is more natural to have the relativized quantifiers as the basic form and to treat the variable as being declared for the scope of the quantifier to have domain A. However, this does not work for free variables.

3. Logical types can be considered as sorts with an internal syntactic structure having constraints for the domains. E.g. a functional type $(s_1, \ldots, s_n, s)$ can be considered as a sort made up of sorts $s_1, \ldots, s_n, s$ fulfilling the axiom:

$$D^{(s_1,\ldots,s_n,s)} = \{u \mid u : D^{s_1} \times \ldots \times D^{s_n} \rightarrow D^s\}.$$

4. The demand for lots of predicates can be fulfilled by an extension of predicate logic incorporating the abstraction operator $\{v|\phi\}$. This yields a name for any definable predicate and allows boolean operations $\cap$, $\cup$, $-$ on these predicates which sometimes are already called "sorts".

This book, which is divided into three parts, provides the reader with more detailed treatments of the problem of defining "sorts". The first part discusses sorts and types in Logic, Theorem Proving, and Logic Programming. Starting with "Order-Sorted Predicate Logic", A. Oberschelp presents the S-logic as the most general many-sorted predicate logic from a logical point of view. "Many-Sorted Inferences in Automated Theorem Proving" by C. Walther provides a short account of the influences a sort concept has on the calculation of an automated theorem proving system. In "Tableau Calculus for Order-Sorted Logic", P.H. Schmitt and W. Wernecke suggest a sound and complete calculus for order-sorted logic on the tableau method. In "A Calculus for Order-Sorted Predicate Logic with Sort Literals", U. Hedtstück and P. H. Schmitt present a calculus for a logic which takes into account sort information which is represented in declarations and by means of sort literals that are unary predicate symbols in the predicate logic language. With "Types, Modules, and Data Bases in the Logic Programming Language PROTOS-L", C. Beierle gives a survey on many-sorted, order-sorted, and polymorphic type concepts for logic programming and introduces the logic programming language PROTOS-L that has a polymorphic type concept and a module concept which allows for the integration of external data bases.

The second part deals with sorts and types in Knowledge Representation including Qualitative Reasoning. B. Nebel and G. Smolka start with "Representation and Reasoning with Attributive Descriptions" in order to survey terminological representation languages and feature-based unification grammars pointing out the similarities and differences between these two families of attributive description formalisms. U. Pletat and K. von Luck introduce the knowledge representation language L-LILOG in the article "Knowledge Representation in LILOG". The aspects discussed here focus on the sort concept and its significance for structuring knowledge bases. K. H. Bläsius, C.-R. Rollinger, and J.H. Siekmann then present the "Structure and Control of the Inference System for L-LILOG". B. Owsnicki-Klewe gives "A General Characterization of Term Description Languages" that is a short characterization of knowledge representation systems based on the KL-ONE paradigm. At the end of this chapter, W. Dilger and H. Voß point out that "Sorts in Qualitative Reasoning" are an integral part of QR models. They can be used to guide the construction of composite models from primitive ones and to specify models.

The third and last part considers sorts and types in Natural Language (Understanding) Systems. K. Eberle presents in "Eventualities in a Natural Language Understandig System" an approach that permits the sorting of eventualities according to a hierarchy of primitive sorts, the sorting of eventualities according to feature values, the discrimination of sub-events according to different dimensions, and the formalization of conditions constraining the construction of super-events from sub-events. For the treatment of (especially plural) NPs, J. Allgayer and C. Reddig-Siekmann developed a representational and inferential framework described in "What KL-ONE Lookalikes Need to Cope with Natural Language". Last, but not least, "Functor-Argument Structures for the Meaning of Natural Language Sentences and their Formal Interpretation" by B. Mahr and C. Umbach proposes a method of associating model-theoretic semantics with FAS expressions on the basis of signatures including overloading operators and polymorphic types.

The aim of this book is to reflect the substantial research done in AI on sorts and types. The main contributions come from Knowledge Representation and Theorem Proving but the important impulses come from the "application areas", i.e. Natural Language Understanding Systems, Computational Linguistics, and Logic Programming. This is one reason why we not only did not, but also could not cover AI totally in our discussion of sorts and types. There should be contributions from the perspectives of Computer Vision, Machine Learning, Planning, and many more. We could learn from them new requirements that we have not yet been able to consider, and we would hopefully get new solutions.

It is our hope that this book intensifies the discussion on sorts and types in AI and thus leads to new insights about representing knowledge.

K. H. Bläsius, U. Hedtstück, C.-R. Rollinger

I. Sorts and Types in Logic, Theorem Proving and Logic Programming

Order Sorted Predicate Logic

Arnold Oberschelp
Abteilung Logik, Philosophisches Seminar, Universität Kiel
Olhausenstraße 40, D-2300 Kiel

Contents
1. One sorted predicate logic
2. Introducing many sorted logic
3. Strict many sorted logic
4. Introducing order sorted logic
5. S-logic
6. Term declarations
7. Removing sortal constraints

1. One Sorted Predicate Logic

This is the usual and well known form of predicate logic. As an example we consider the one sorted language with individual variables $v_0, v_1, v_2, \ldots$, an individual constant c, a two place relational constant Q and a two place functional constant f. We consider some **formulas** of this language:

$$\text{let } \phi_1 \text{ be} \quad \forall v_0(c = v_0 \vee c\,Q\,v_0)$$
$$\text{let } \phi_2 \text{ be} \quad v_0\,Q\,v_1 \longrightarrow \exists v_2(v_0\,Q\,v_2\,Q\,v_1)$$
$$\text{let } \phi_3 \text{ be} \quad \exists v_2(c \neq v_2 \wedge f(v_0,v_2) = v_1) \longrightarrow v_0\,Q\,v_1$$

For an interpretion one needs a domain and denotations for the constants. This is what is called a **structure**. Examples:

$$\mathcal{N} = \langle \mathbb{N}, 0, <_{\mathbb{N}}, +_{\mathbb{N}} \rangle$$
$$\mathcal{R} = \langle \mathbb{R}, 0, <_{\mathbb{R}}, +_{\mathbb{R}} \rangle$$

Formula ϕ_1 has no free variables and is a **sentence**. The variable v_0 has a bound occurrence in this sentence. For the interpretation one has to know the domain of possible values of v_0, and this is given by the structure. A sentence has a truth value if a structure is given. E.g. ϕ_1 is **true** in $\mathcal{N}$ and **false** in $\mathcal{R}$. Formulas ϕ_1, ϕ_2 have free occurrences of variables. The structure is not sufficient for an interpretation. One needs in addition an **assignment** of values to the free variables. E.g. ϕ_2 is **satisfied** in $\mathcal{N}$ by 3 (as value for v_0), 5 (as value for v_1) but not satisfied for 3, 4. In $\mathcal{R}$ this formula is satisfied for any assignment, ϕ_2 is **valid** in $\mathcal{R}$. Formula ϕ_3 is valid in $\mathcal{N}$ but not valid in $\mathcal{R}$.

We have seen: for an interpretation bound variables need a domain and free variables need values.

The notion of logical consequence is based on the semantics outlined above. By the Gödel completeness theorem the consequence relation can be characterized by a proof system using axioms and rules. The rules can be given in various ways. We mention the rule of substitution which will play a special role in subsequent sections.

$$\text{SUB}_1 \quad \frac{\phi_1(v),\ldots,\phi_n(v) \vdash \phi_0(v)}{\phi_1(a),\ldots,\phi_n(a) \vdash \phi_0(a)}, \text{ if } \phi_i(a) \text{ is got from } \phi_i(v) \text{ by substituting term } a \text{ for variable } v \ (i=0,\ldots,n)$$

According to this rule one may substitute in a proofline (consisting of a finite set of assumption formulas and an assertion formula) an individual term a for all free occurrences of a variable v, of course, with the usual precautions against collisions of free and bound variables.

Axioms for identity can be given in the following form with $a, b, c, a_1, b_1, \ldots, a_n, b_n$ being any individual terms:

Axiom of self identity:
$$\vdash a=a$$

Axiom of comparability:
$$a=c, b=c \vdash a=b$$

Congruence axioms for relations (Q n-place relation symbol):
$$a_1=b_1,\ldots, a_n=b_n \vdash (Q \ni a_1 \ldots a_n) \longleftrightarrow (Q \ni b_1 \ldots b_n)$$

Congruence axioms for functions (f n-place function symbol):
$$a_1=b_1,\ldots, a_n=b_n \vdash f(a_1,\ldots,a_n)=f(b_1,\ldots,b_n)$$

2. Introducing Many Sorted Logic

It is quite natural to have individuals of different sorts. In geometry one has points and lines and uses different sorts of letters as variables, say P, Q,... for points and g, h,... for lines. Predicates often apply only to individuals of certain sorts. E. g. it is appropriate of two lines but not of two points to say whether they are othogonal or not.

One could take men and numbers as sorts of individuals. Then the blank of "...is a prime" would apply to numbers whereas the blank of "the father of ..." would apply to men. Filling in the blanks wrongly, e. g. as in

> Caesar is a prime
> the father of 733333

would yield undesirable strings.

Having different sorts of individuals is very useful and I think that logical systems should incorporate this feature. Therefore we assume that for a many sorted logical language there is a set S of sortal indices, for short **sorts**. For any sort $s \in S$ there are the **individual variables of sort s**:

$$v_0^s, v_1^s, v_2^s, v_3^s, \ldots$$

If v is a variable let s(v) be its sort, thus $s(v_i^r) = r$ for $r \in S$.

A structure $\mathcal{A}$ has to provide for any sort $s \in S$ a nonempty set $D_{\mathcal{A}}^s$, the **domain of sort** s. An assignment is a function which assigns to variables of sort s individuals of sort s as value.

However, many sorted logical systems are usually not considered in logic since it is possible to get along with just one sort of variables, provided that one has **sortal predicates** D^s. A quantifier with a variable v^s of sort s is equivalent to a quantifier with a "universal" individual variable u relativized to D^s:

$$\forall v^s \, \phi(v^s) \text{ is logically equivalent to } \forall u(u \in D^s \longrightarrow \phi(u)), \text{ for short } \forall u \in D^s \, \phi(u)$$
$$\exists v^s \, \phi(v^s) \text{ is logically equivalent to } \exists u(u \in D^s \wedge \phi(u)), \text{ for short } \exists u \in D^s \, \phi(u)$$

It is certainly not adequate for any collection of individuals to introduce an extra sort together with its own set of variables. Often it will be appropriate to work with relativized quantifiers and universal variables. Natural language does not use variables at all and has only relativized quantifiers. But a certain number of sorts for some basic kinds of individuals is very useful.

Up to now we have only demanded that a many sorted language has its individual variables divided into severals sorts, that a structure provides a domain for each sort, and that assignments have to take their values out of these domains. Nothing has been said about the syntax of a many sorted language. However, one often associates with the use of different sorts also some **sortal contraints** in the grammar in order to exclude strings which do not respect the sortal structure from being wellformed. This can actually be done in different ways.

3. Strict Many Sorted Logic

As starting point we will sketch a many sorted logic with rather narrow sortal constraints which will be called "strict".

In strict logic to any individual constant there is assigned exactly one sort, to any argument place of a relation or function symbol there is assigned exactly one sort and for any function symbol there is exactly one target sort. This can be described by giving to any nonlogical constant K of the language a signature $\sigma(K)$ which is a tuple of sorts. For an individual constant c we have $\sigma(c) = \langle s \rangle$ for some sort s. For an n-place relation constant Q we have $\sigma(Q) = \langle s_1, \ldots, s_n \rangle$ for sorts $s_1, \ldots, s_n$ belonging to the arguments of the relation. For an n-place function constant f we have $\sigma(f) = \langle s_1, \ldots, s_n, s \rangle$ for n argument sorts and a sort for the value of the function. By induction any wellformed term a of the strict language gets exactly one sort s(a). Furthermore in predicative formulas the argument terms should fit into the signature of the relation symbol and equalities are only allowed between terms of the same sort.

In a strict many sorted structure $\mathcal{A}$ the interpretations of the constants have to respect the signatures. An individual constant of signature $\langle s \rangle$ has to be interpreted by an element of $D_{\mathcal{A}}^s$. A relation constant of signature $\langle s_1, \ldots, s_n \rangle$ has to be interpreted by a subset of $D_{\mathcal{A}}^{s_1} \times \ldots \times D_{\mathcal{A}}^{s_n}$. And a function constant of signature

$\langle s_1, \ldots, s_n, s \rangle$ has to be interpreted by a function which is exactly defined on $D^{s_1}_{\mathcal{A}} \times \ldots \times D^{s_n}_{\mathcal{A}}$ and has values in $D^s_{\mathcal{A}}$. Any individual term of sort s will then be evaluated by an element of $D^s_{\mathcal{A}}$.

For the substitution rule it is now necessary that for variables of sort s individual terms of sort s are to be substituted.

$$\text{SUB}_2 \quad \frac{\phi_1(v), \ldots, \phi_n(v) \;\vdash\; \phi_0(v)}{\phi_1(a), \ldots, \phi_n(a) \;\vdash\; \phi_0(a)} \;,\; \text{if } \phi_i(a) \text{ is got from } \phi_i(v) \text{ by substituting term } a \text{ for variable } v \;(i=0,\ldots,n) \text{ and } s(v)=s(a)$$

Actually, it is not necessary to add the constraint $s(v) = s(a)$ since otherwise the result of the substitution would not be wellformed.

The identity axioms can be taken from one sorted logic. Of course all terms and formulas occuring in these axioms should be now in the strict many sorted language. This means that sorts and signatures fit together.

This ends the sketch of strict many sorted logic which by now will be sufficiently clear to the reader. The first many sorted logical systems by J. Herbrand (1930) and A. Schmidt (1938, 1951, see [3], [4]) were of this kind. Strict many sorted logic fits nicely into many situations. Consider e.g. points and lines, or vectors and scalars, or places, times, men. These are situations where the individuals of different sorts are so to speak "side by side". However, strict many sorted logic is not adequate if the individuals are arranged hierarchically and some sorts are subordinated to other sorts in a natural way. This leads to what is now called order sorted logic.

4. Introducing Order Sorted Logic

The classical case is given by the standard number sets which are included in each other in the following way:

$$\mathbb{N} \subsetneq \mathbb{Z} \subsetneq \mathbb{Q} \subsetneq \mathbb{R} \subsetneq \mathbb{C} \,.$$

It is common practice in mathematics to use more or less explicitly different sorts of letters for the different kinds of numbers. However, a formal representation in a strict many sorted language turns out to be very awkward. This is because arithmetical terms are of several sorts simultaneously and arithmetical predicates and functions can take arguments of severals sorts. If e.g. we take the signature of the less-than symbol to be $\langle \mathbb{R}, \mathbb{R} \rangle$ then it is not possible to formulate the archimedian axiom (that to any real number there is a bigger natural number) in the normal and natural way:

$$\forall x_0 \, \exists n_0 \; x_0 < n_0 \,.$$

In order to adapt many sorted artihmetic to the strict grammar one would need many less-than symbols $<_{\mathbb{NN}}, <_{\mathbb{NZ}}, <_{\mathbb{NQ}}, <_{\mathbb{NR}}, <_{\mathbb{ZN}}, \ldots$ (with signatures evident from the subscripts). For the other arithmetical concepts one would also need many versions and in addition many (nonlogical) equality symbols $=_{\mathbb{NZ}}, =_{\mathbb{ZN}}, \ldots$ in order to express that all these versions essentially amount to the same. Another possibility would be to use sortal tranfer functions id_{NZ}, id_{NQ}, id_{NR},... (as in some "overstructured" programming languages) which, however, are nothing but identity functions.

This is complicated and unnatural. According to the inclusion of the sortal domains any natural number **is** also a real number. Therefore any expression denoting a natural number should be allowed at places where names of real number are admitted. However, this is not the case in strict many sorted logic. The paradigm of strict many sorted logic is geometry. Arithmetic has not been taken into account in the first systems of many sorted logic.

Also outside mathematics concepts are often ordered hierarchically. Think of taxonomical systems (setters are dogs, dogs are mammals, mammals are vertebrates etc.). When introducing sorts one will often be in a situation where arithmetic rather than geometry is the paradigm.

This caused me almost 30 years ago to design many sorted logical systems which are more general than strict many sorted logic (see [2]). There are essentially two systems, one of which is called S-logic and will be presented in the next section. The system called $\mathfrak{S}$-logic will not be considered here. It deals with resolving ambiguities which occur quite often in mathematics when the same symbol is used with different denotations in the same formula. This occurs e.g. in a formula speaking about vectors when the symbol + at some occurrences means addition of vectors and in other occurrences addition of scalars.

5. S-Logic

We assume as before that a set S of sortal indices is given. To any sort there is a set of variables. A sort is interpreted by a domain which contains also the values for assignments to variables of that sort.

In addition there is given a **partial ordering** on the sorts. If $s \leq r$ then any structure $\mathcal{A}$ has to satisfy $D_{\mathcal{A}}^s \subseteq D_{\mathcal{A}}^r$. In my original paper I used the reverse ordering. But in the meantime the direction indicated above has become the normal one.

In S-logic any nonlogical constant gets a signature which is a set of tuples of the kind occuring in strict many sorted logic. The signature $\sigma(c)$ of an individual constant c is a set of one-tuples $\langle s \rangle$, the signature $\sigma(Q)$ of an n-place relation constant Q is a set of n-tuples $\langle s_1,...,s_n \rangle$, the signature $\sigma(f)$ of an n-place function constant f is a set of n+1-tuples $\langle s_1,...,s_n,s \rangle$.

The signatures impose constraints on structures admitted in S-logic. If $\langle s \rangle \in s(c)$ then the individual $c_{\mathcal{A}}$ is to be an element of $D_{\mathcal{A}}^s$. If $\langle s_1,...,s_n,s \rangle \in S(f)$ then the function $f_{\mathcal{A}}$ is at least to be defined on $D_{\mathcal{A}}^{s_1} \times ... \times D_{\mathcal{A}}^{s_n}$ having there values in $D_{\mathcal{A}}^s$.

By induction a set S(a) is assigned to any individual term a. The definition of this sortal set is as follows:

For variables v^s: $S(v^s) =_{Df} \{r \mid s \leq r\}$

For individual constants c: $S(c) =_{Df} \{r \mid (\exists s \in \sigma(c)) \; s \leq r\}$

For compound terms:
$$S(f(a_1,\ldots,a_n)) =_{Df} \{r \mid (\exists s_1 \ldots s_n s)(\langle s_1,\ldots,s_n,s \rangle \in \sigma(f) \\ \wedge s_1 \in S(a_1) \wedge \ldots \wedge s_n \in S(a_n) \wedge s \leq r)\}$$

If the sortal set $S(a)$ of the term a is nonempty then term a is wellsorted.

Any equality between wellsorted terms is a wellsorted formula:

$\qquad a = b \qquad$ is a wellsorted formula iff $\quad S(a) \neq \emptyset \neq S(b)$

Predicative formulas are wellsorted, if the terms fit into the signature:

$$(Q \ni a_1 \ldots a_n) \text{ is a wellsorted formula iff } (\exists s_1 \ldots s_n)(\langle s_1,\ldots,s_n \rangle \in \sigma(Q) \wedge s_1 \in S(a_1) \\ \wedge \ldots \wedge s_n \in S(a_n))$$

The set $S(a)$ contains the sorts such that the term a may be substituted for variables of these sorts. The substitution rule for S-logic is:

$$\text{SUB}_3 \quad \frac{\phi_1(v),\ldots,\phi_n(v) \vdash \phi_0(v)}{\phi_1(a),\ldots,\phi_n(a) \vdash \phi_0(a)}, \text{ if } \phi_i(a) \text{ is got from } \phi_i(v) \text{ by substituting term } a \text{ for variable } v \; (i = 0,\ldots,n) \text{ and } s(v) \in S(a)$$

In the identity axioms all terms and formulas have to be wellsorted.

S-logic is now sketched in sufficient detail. It is a generalization of strict many sorted logic which is got (except for the more general equalities in S-logic) by taking the partial ordering on the sorts to be the identity relation and taking all signatures to be singletons. The sortal set $S(a)$ for any term is either empty or a singleton in this case.

We consider an example. In an arithmetical language we may have (among others) sorts $\mathbb{N}$ and $\mathbb{R}$ with the partial ordering $\mathbb{N} \leq \mathbb{R}$, individual constants $0,1$ with signature $\{\langle \mathbb{N} \rangle\}$, π with signature $\{\langle \mathbb{R} \rangle\}$, two place relation constants $<$, $\mid$ (divisibility predicate) with signatures $\{\langle \mathbb{R},\mathbb{R} \rangle\}$ and $\{\langle \mathbb{N},\mathbb{N} \rangle\}$ resp., two place function constants $+$, $\cdot$ with signatures $\{\langle \mathbb{R},\mathbb{R},\mathbb{R} \rangle, \langle \mathbb{N},\mathbb{N},\mathbb{N} \rangle\}$, etc.

Then we have e.g. $S(0) = \{\mathbb{N},\mathbb{R}\}$. Therefore the term 0 may be substituted for variables of both sorts. But $S(\pi) = \{\mathbb{R}\}$. Therefore the term π may not be substituted for variables of sort $\mathbb{N}$. If n, m are variables of sort $\mathbb{N}$ and x, y are variables of sort $\mathbb{R}$ then we have the sortal sets $S(n+m) = \{\mathbb{N},\mathbb{R}\}$, $S(n+x) = S(x+y) = \{\mathbb{R}\}$. This means that the term $n+m$ may be substituted for variables of both sorts but the terms $n+x$ and $x+y$ may be substituted for variables of sort $\mathbb{R}$ only. In the argument places of the less-than predicate any well sorted terms may be substituted. The argument places of the divisibility predicate accept only terms having $\mathbb{N}$ in their sortal set.

6. Term Declarations

Order sorted logic has been used by Ch. Walther (see [6]) as a tool for automatic theorem proving in order to reduce the search space. Additional structure has been added, e.g. that the sorts form a lattice with a top sort (universal sort) and bottom sort (with empty domain) and a slightly different language using declarations is used. A recent dissertation by M. Schmidt-Schauß (see [5]) made further generalizations not covered by S-logic. Thus S-logic is by no means the most general many sorted predicate logic. This will be illustrated by an example (given by Schmidt-Schauß).

We assume two sorts NAT (for natural numbers) and EVEN (for even numbers), an individual constant 0 (for the number zero) and a one place function constant s (for the successor function).

The even natural numbers are contained in the natural numbers. This is expressed by a subsort declaration:

EVEN ⊏ NAT

Zero is an even number. This is expressed by a constant declaration:

0: EVEN

The successor function is a one place function from natural numbers into natural numbers. This is expressed by a function declaration:

$s(x_{NAT}):NAT$ for short $s:NAT \longrightarrow NAT$

All this is (up to a different notation) the same as it would be in S-logic. But how the successor function is related to the subsort cannot be expressed by a tuple in a signature of S-logic. The successor of a successor of an even number is even. This is expressed by a term declaration:

$s(s(x_{EVEN})):EVEN$

A **term declaration** has the general form:

a : s

The term a may be a compound term. If the term is of the form $f(v_1,\ldots,v_n)$ (with different variables $v_1,\ldots,v_n$), then we have a function declaration which corresponds to a tuple of sorts in a signature of S-logic.

A term declaration in a direct and shortcut way makes a term wellsorted. Subterms of the term need not to be wellsorted (but are so in well behaved cases). The term usually contains variables. Using wellsorted substitutions and exploiting the subsort relation yields further wellsorted terms.

It is beyond the scope of this paper (and beyond the competence of this author) to give an account of the use of order sorted logic in automated reasoning and AI.

7. Removing Sortal Constraints

An expression may be interpretable in a reasonable way without being wellsorted. Instead of classifying "Caesar is a prime" as ungrammatical and nonsensical it can also be classified as grammatically wellformed but simply false. Likewise one can say that it is false that a point is orthogonal to another point (since only a line can be orthogonal to another object, another line). It is possible to admit non-wellsorted expressions as nevertheless grammatically wellformed. Expressions which are not wellsorted may be unusual or uninteresting. But this is rather a matter of pragmatics than of semantics.

One might fear that undesirable consequences come up if nonwellsorted formulas are admitted since they are not altogether false. The negation of a sentence which is nonwellsorted and false is also nonwellsorted but true. For example, it is true:

Caesar is not a prime.

But this does not mean that Caesar has a proper divisor. In order that one cannot "prove" this, one has to prevent the substitution of the term "Caesar" for a variable of sort $\mathbb{N}$. A suitable constraint on the admitted substitutions is all that is needed for this.

If one rules out nonwellsorted expressions as ungrammatical then the sortal constraints of S-logic automatically take care of this. But out of the whole sortal apparatus only the relation $s(v) \in S(a)$ between variables v and terms a is needed in order to formulate the substitution rule SUB_3. This relation was called by me $\Sigma(v,a)$ and characterized by some conditions in [2] when the logical system Σ- logic was set up. In Σ-logic terms and formulas can be formed without respecting sortal constraints (just as in one sorted predicate logic). The substitution rule is SUB_3 with $\Sigma(v,a)$ restricting substitutions. Any nonwellsorted expressions can occur in the identity axioms thus allowing one to deduce further nonwellsorted formulas from nonwellsorted formulas. Nevertheless Σ-logic is a conservative extension of S-logic and thus equivalent to S-logic.

S-logic is not the most general many sorted logic. Term declarations are mentioned in section 6. Maybe more general order sorted logics having a more liberal grammar than S-logic can also be condensed into some suitable relation Σ of substitutability. The most general many sorted logic would be obtained if sortal constraints are skipped altogether. Then any term may be substituted for any variable and the relation of substitutability would become trivial. However, the substitution rule SUB_3 is no longer correct then (we can indeed sentences derive fron true sentences that Caesar has a proper divisor). Thus it seems that some sortal constraints - or equivalently some restrictions for substitutions - are necessary for a many sorted logic.

However, it is possibile to have variables of several sorts and remove sortal constraints in the grammar completely and in addition allow arbitrary substitutions. One simply has to add to the conclusion of the substitution rule a further assumption, that the substituted term denotes an individual of the right sort. This will be called the **sortal premise**. When adding the sortal premise arbitrary substitutions are admitted.

The sortal premise can be given by the formula:

$$\exists v^s \, v^s = a$$

Using sortal predicates D^s the sortal premise can be given in the form:

$$a \in D^s$$

The many sorted substitution rule for languages without sortal constraints in the grammar (and sortal predicates) is thus:

$$SUB_4 \quad \frac{\phi_1(v^s),\ldots,\phi_n(v^s) \ \vdash\ \phi_0(v^s)}{a \in D^s, \phi_1(a),\ldots,\phi_n(a) \ \vdash\ \phi_0(a)}, \quad \text{if } \phi_i(a) \text{ is got from } \phi_i(v^s) \text{ by substituting term } a \text{ for variable } v^s \ (i=0,\ldots,n)$$

We assume now that we have a rather rich language allowing us to define sortal predicates which therefore need not be taken as nonlogical constants. We furthermore assume that in the language we can express what is encoded in the signatures of S-logic. The role of signatures and sortal constraints can then simply be played by extra axioms which are to be added to the theory at hand.

A logical system with an abstraction operator $\{\ldots|\ldots\}$ which allows us to formulate the logical theory of relations and functions is suitable for this purpose. Several logical systems of this kind have been set up in [1]. The simplest of these system is called elementary logic LE (keeping class terms and individual terms syntactically apart), more general is class theoretical logic LC (with general terms and allowing classes to be individuals), the most general system is "Ausdruckslogik" LA (treating formulas as "boolean terms" and making no syntactical difference between formulas and terms).

Sortal predicates are defined by an abstraction term using a trivial condition, say $v_o^s = v_0^s$.

$$D^s =_{Df} \{v_0^s \mid v_0^s = v_0^s\}$$

Signatures of S-logic can be translated into axioms which have to be added in order to replace the sortal constraints of S-logic. One has to take:

$c \in D^s$,	iff	$\langle s \rangle \in \sigma(c)$
$Q \subseteq D^{s_1} \times \ldots \times D^{s_n}$,	iff	$\langle s_1, \ldots, s_n \rangle \in \sigma(Q)$
$(\forall v_1^{s_1} \ldots v_n^{s_n}) f(v_1^{s_1}, \ldots, v_n^{s_n}) \in D^s$,	iff	$\langle s_1, \ldots, s_n, s \rangle \in \sigma(f)$

Inclusions and boolean relations between sorts can easily be expressed, e. g. :

$$D^{s_1} \subseteq D^{s_2} , \qquad D^{s_1} = D^{s_2} \cap D^{s_3} , \text{ etc.}$$

Other many sorted logics which are more general than S-logic can be similarly translated into the general language. A term declaration a:s would be expressed by:

$$\forall \ldots (a \in D^s)$$

with $\forall \ldots$ denoting the universal closure.

The presentation of the "rich" many sorted logical systems is a subject of its own and cannot be given in this paper. The substitution rule SUB_4 allows not only to remove sortal constraints from many sorted predicate logic, it also allows (as indicated above) to fuse individual terms and class terms (into general terms) and even general terms and formulas (into expressions, "Ausdrücke").

The field of more or less general many sorted logical systems with sortal constraints in the grammar has lost some of its interest for me. When teaching logic I usually only present the extreme positions of the spectrum indicated above, namely strict many sorted logic which I then call LP (predicate logic), and the fairly general systems LC or LA with an abstraction operator (see [1]) which go beyond predicate logic but still stay inside the field of first order logic.

References

[1]   J.-M. Glubrecht, A. Oberschelp, G. Todt
        Klassenlogik
        BI, Mannheim, Wien, Zürich 1983

[2]   A. Oberschelp
        Untersuchungen zur mehrsortigen Quantorenlogik
        Math. Annalen 145, S. 297-333 (1962)

[3]   A. Schmidt
        Über deduktive Theorien mit mehreren Sorten von Grunddingen
        Math. Annalen 123, S. 485-506 (1938)

[4]   A. Schmidt
        Die Zulässigkeit der Behandlung mehrsortiger Theorien mittels der üblichen
        einsortigen Prädikatenlogik
        Math. Annalen 123, S. 187-200

[5]   M. Schmidt-Schauß
        Computational Aspects of an Order-Sorted Logic with Term Declarations
        Lecture Notes in Artificial Intelligence, LNCS vol. 395, Springer-Verlag 1989

[6]   C. Walther
        A Many-Sorted Calculus Based on Resolution and Paramodulation
        Proc. 8th IJCAI, Karlsruhe 1983

Many-Sorted Inferences
in
Automated Theorem Proving

Christoph Walther

Institut für Logik, Komplexität
und Deduktionssysteme
Universität Karlsruhe

Contents

Preface

The use of sorts is a well established technique in computer science. For instance, most programming languages have a sort concept (better known as *types* in this area) to improve the detection of coding errors and to increase computational efficiency by shifting certain calculations from compile time to run time.

In this article we present a short account of the influences a sort concept has on the calculations of an *automated theorem proving* system. We shall assume familiarity for our presentation with the basic notions of formal logic as well as resolution and paramodulation based theorem proving, as provided, for instance in [Chang and Lee 1973, Loveland 1978, Walther 1987(a)]. We intend to present the main ideas and concepts of sorts in automated theorem proving, without hampering readability by a thorough formal investigation, which can be found elsewhere [Walther 1987].

Our initial motivation for investigating many-sorted theorem proving was to take advantage of a representation of mathematical truth, which is (a bit) closer to the human style of representing mathematics as compared to plain first-order logic. But as time passed, it was recognized that the advantages of many-sorted representation and reasoning are by no means restricted to mathematical domains, but are useful also in other areas of artificial intelligence as, for instance, *natural language processing* and *knowledge representation*. This resulted in further developments and also in alternative approaches to many-sorted reasoning as presented in the other articles of this volume.

Abstract

A brief account of the basic ideas and advantages of many-sorted first-order logic is given. Based on this survey, a many-sorted version of a resolution calculus is proposed. The advantages of such a calculus and the problems related to its definition are illustrated with several examples. We also discuss the problems, which arise when extending this calculus with equality reasoning and present a many-sorted version of the paramodulation rule. We show how the structure of a sort hierarchy influences the inferences of our many-sorted calculus and discuss the ways to state certain axioms of a theorem proving problem by an adequate definition of a sort hierarchy. We conclude with a brief survey on related work.

1   MANY-SORTED LOGIC

1.1   INTRODUCTION

A successful method humans use in reasoning is to assume a structured universe of discourse. We assume that the universe is divided into certain subuniverses, instead of having a single one. Certain relations exist between pairs of these subuniverses - they may be disjoint, they may have non-empty intersections or one may be completely contained in another. We also assume that certain objects belong to certain subuniverses and that certain mappings and relations are meaningful only for certain subuniverses.

In geometry, for instance, we may consider a universe of discourse which is divided into *points, lines, angles, triangles, rectangles, polygons* etc., where we assume, for instance, that points and angles have no objects in common, or that triangles as well as rectangles are also polygons. Here we may think of specific points p_1, p_2, p_3, or of some specific angles α and β. We may associate an *area* with each polygon, where we assume that each area is a *real number*, i.e. another subuniverse, which is disjoint with each of the other ones. Of course, our implicit understanding of areas prevents us from talking of the area of a point or of an angle, and also from assuming that the area of a polygon can be a triangle.

Thinking of a structured universe helps humans to draw logical consequences efficiently. It also helps to avoid wasting our time with meaningless or wrong conclusions. Given, for instance, that the area of each polygon is a *positive* real number, we implicitly and immediately assume that the same holds true for triangles as well as for rectangles. But we will never conclude that a triangle is parallel to itself (which is meaningless) from the fact the each line is parallel to itself. We also do not assume that the sum of all angles in a polygon is 180° (which is wrong), despite the fact that this holds true for each triangle.

We think of structured universes in almost every branch of mathematics, and therefore it is not a mere accident that almost all mathematical textbooks are written in a language which reflects the specific (sub)universes of discourse, albeit often very implicitly.

This paper is concerned with a mechanization of this aspect of human reasoning. Our starting point is some first-order calculus, which defines how we obtain theorems from given hypotheses by pure syntactic reasoning. This calculus is extended to a *many-sorted* calculus, i.e. a calculus which allows us to make explicit all the implicit assumptions and notions discussed above. The development of a many-sorted (*mehrsortig*) version of some given (sound and complete) first-order one-sorted or unsorted (*einsortig, unsortiert*) calculus is well known in the area of formal logic [Oberschelp 1962, Enderton 1972]: the first-order language has to be extended to a many-sorted language, the semantics for this language have to be defined, and the rules of inference have to be modified accordingly. As we shall see, it is possible in this case to shift all the semantic and implicit argumentation to the syntactic and explicit level of formal first-order reasoning.

In this paper we first discuss the advantages of many-sorted reasoning and then we apply the ideas to the resolution calculus extended by the paramodulation rule. The many-sorted calculus thus obtained is a true generalization of the unsorted system, i.e. our many-sorted resolution calculus is identical to its unsorted counterpart, if the set of sort symbols is a singleton (yielding a one-sorted calculus).

The advantages of such a calculus and the problems related to its definition are illustrated with several examples. The basic ideas of many-sorted resolution based theorem proving are easy to understand and an implementation is obtained without severe difficulties. Certain deductions are decidable by the many-sorted unification procedure. The practical application of this calculus in automated theorem proving leads to a drastic reduction of the search space and to shorter refutations of smaller sets of shorter clauses. Without sorts, a theorem prover has to search for these deductions instead, giving rise to overhead and deadends.

1.2 THE MANY-SORTED LANGUAGE

Our subuniverses of discourse are given names, called *sort symbols*, where the set of all sort symbols is partially ordered by the *subsort order* relation, thus expressing the inclusion relations which hold between the subuniverses under consideration. We may use, for instance, *angle, triangle, polygon* etc. as sort symbols, where we define triangle as a *subsort* of polygon. A set of sort symbols with the subsort order imposed on it is called a *sort hierarchy*.

Variable and function symbols (of our given calculus) are associated with a sort symbol, called the *rangesort*. The *sort of a term* is determined by the rangesort of its outermost symbol. This is a syntactic formulation of the fact that the (semantic) object represented by a given term is a member of the subuniverse represented by the sort of that term.

We also restrict the domains of functions and predicate symbols to terms of certain sorts (one for each argument position), called the *domainsorts* for the respective argument positions. This syntactic requirement expresses the fact that not *each* term is meaningful as an argument to *each* function or predicate symbol. In geometry, for instance, we only want to have terms of sort *line* as arguments of a predicate *parallel* and we demand that only terms of sort *polygon* are used as arguments of the function *area*. The definitions of the range- and domainsorts for all the variable, function and predicate symbols under consideration are collected in a so-called (many-sorted) *signature*.

Consequently we demand that only those well-formed formulas of our given calculus are allowed in its many-sorted version, which are *well-sorted* (*sortenrecht*). In the construction of these well-sorted formulas we define that each variable and each constant is a well-sorted term and we allow for each argument position of a function or predicate symbol only well-sorted terms of the *domainsort* or of a *subsort* of this domainsort stipulated for the argument position of the respective function or predicate symbol. Hence, for instance, we can apply the function *area* to terms of sort *polygon* but also to terms of sort *triangle* or *rectangle*, provided both sorts are subsorts of *polygon*. But we never use a term of sort *angle* as an argument of *area*.

1.3 SEMANTICS OF THE MANY-SORTED LANGUAGE

Having defined the many-sorted language, i.e. the set of all well-formed and well-sorted formulas of our many-sorted calculus, we need a notion of truth for these formulas. Of course, the semantic notions have to mirror our assumptions of sorts as names of subuniverses and of the subsort order as an inclusion relation.

The necessary extensions of the semantic notions are straightforward: as intended, we demand that the *domain* of an interpretation is divided into certain *subdomains*. An interpretation associates each sort symbol with a certain subdomain, such that a subdomain is included in another one (in the set-theoretical sense), if the sort symbol designating the included subdomain is a subsort of the sort symbol designating the other one.

Furthermore, an interpretation associates each function symbol with a mapping and each predicate symbol with a relation defined on the subdomains given by the interpretation for the respective domainsort symbols and yielding always an element of the subdomain associated with the rangesort in the case of a function symbol.

Finally, each *variable assignment* assigns a variable of a certain sort to an element of the subdomain associated with that sort. The remaining semantic notions, e.g. the semantics of the connectives and quantifiers, are the same as in the unsorted case. We let $\models_\Sigma \Phi$ denote the validity of a well-sorted formula Φ and use HYP $\models_\Sigma \Phi$ to indicate that Φ is semantically implied by the set of hypotheses HYP, whose members are well-sorted formulas.

1.4   MANY-SORTED CALCULI

The inference rules of the many-sorted calculus correspond to the inference rules of the given calculus, but with the restriction that only well-sorted formulas can be deduced by an application of the restricted inference rules. Starting with well-sorted formulas this guarantees that only well-sorted formulas are derived in a deduction of the many-sorted calculus. We let $\vdash_\Sigma \Phi$ denote that Φ is a theorem of the many-sorted calculus and we write HYP $\vdash_\Sigma \Phi$ to indicate that there is a deduction of Φ from the set of hypotheses HYP in the many-sorted calculus.

Obviously we are only interested in a many-sorted calculus which is sound and complete, i.e. we have to choose our modifications of the given inference rules such that

(1) $\models_\Sigma \Phi$ iff $\vdash_\Sigma \Phi$

is guaranteed for each well-sorted formula Φ. Having defined a many-sorted calculus which satisfies (1), we have succeeded in shifting the implicit and semantic notions associated with our thinking of a structured universe of discourse, to a purely syntactic level of formal reasoning.

1.5   EFFICIENCY OF MANY-SORTED REASONING

But what have we gained and what are the advantages of using this new calculus? The advantages can be seen if we compare the deductive capabilities of both calculi, i.e. if we use the given unsorted calculus instead of its many-sorted version to prove theorems.

To facilitate a comparison between both calculi, we need a means to express the sort of a term in the unsorted calculus: we use sort symbols as unary predicate symbols. Now we can *explicitly* represent the relations between the function symbols and the sort symbols as well as the subsort order by a set A^Σ of so-called *sort axioms (Sortenaxiome)*, i.e. a set of first-order (unsorted) formulas. For instance,

$$\forall x \, polygon(x) \implies real\_number(area(x))$$

is a sort axiom expressing that *area* maps *polygons* into *real numbers*, and

$$\forall x \, triangle(x) \implies polygon(x)$$

is the sort axiom for our assumption that *triangles* are *polygons*.

For a well-sorted formula Φ like, for instance,

$$\forall x{:}line \; \exists y{:}line \; parallel(xy)$$

the *relativization* Φ^Σ (*Sortenbeschränkung, Relativierung*) of Φ is the unsorted version of Φ, as e.g.,

$$\forall x \, line(x) \implies (\, \exists y \, line(y) \wedge parallel(xy) \,)$$

where sort symbols are used as unary predicate symbols to express the sorts of the variables. Given the notions of *sort axioms* and *relativizations*, the relation between a first-order calculus and its many-sorted version can now be described by the following diagram:

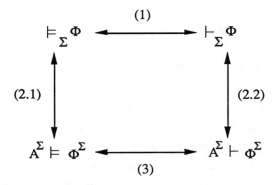

Figure 1.5.1 *Unsorted and many-sorted logic*

The equivalences (1) and (3) are given by the soundness and completeness of both calculi, where $\models$ and $\vdash$ denote semantic implication and formal deduction in the unsorted calculus. The equivalences (2.1) and (2.2) form the so-called *Sort-Theorem*. Equivalence (2.1) is the *model-theoretic part* and (2.2) is the *proof-theoretic part* of the Sort-Theorem.

Now we can see what advantages we have in using a many-sorted calculus: the proof-theoretic part (2.2) of the Sort-Theorem shows that a well-sorted formula Φ is a theorem of the many-sorted calculus iff its relativization Φ^Σ, which is more *complex* than Φ, can be deduced in the unsorted calculus from the set of sort axioms A^Σ as *additional hypotheses*. Hence we obtain a *shorter deduction* with *smaller formulas* from *a smaller set of hypotheses*, when proving $\vdash_\Sigma \Phi$ instead of $A^\Sigma \vdash \Phi^\Sigma$. The reason is that *deductions about sorts*, which are performed explicitly in the unsorted calculus (using sort axioms and relativizations), are *built into the inference mechanism* in a many-sorted calculus.

But we can also see that in an unsorted calculus a lot of *useless* formulas can be deduced, which do not have a many-sorted counterpart, because these many-sorted formulas would be meaningless or wrong. Given, for instance, the axiom

$$\forall \, x{:}line \ parallel(xx)$$

and a term α of sort angle, we cannot deduce

$$parallel(\alpha\alpha)$$

in the many-sorted calculus, because this formula is not well-sorted and is therefore meaningless. But in the unsorted calculus we may obtain from the relativization

$$\forall x \ line(x) \Rightarrow parallel(xx)$$

by universal instantiation the formula

$$line(\alpha) \Rightarrow parallel(\alpha\alpha)$$

which is true but useless, since $line(\alpha)$ does not hold. Also it is impossible to deduce from the given axiom

$$\forall \, x{:}triangle \ angle\_sum(x) \equiv 180°$$

for any polygon *p* the well-sorted but *wrong* formula

$$\text{angle\_sum(p)} \equiv 180°$$

whereas in the unsorted calculus we do not have any difficulties inferring the useless truth

$$\text{triangle(p)} \;\Rightarrow\; \text{angle\_sum(p)} \equiv 180°$$

from the relativized axiom

$$\forall x \; \text{triangle(x)} \;\Rightarrow\; \text{angle\_sum(x)} \equiv 180°.$$

These observations lead to the conclusion that a many-sorted calculus is a good foundation for automated theorem proving, because it is a syntactical device to draw logical consequences *efficiently* and to avoid wasting time with the derivation of *useless* conclusions.

2 MANY-SORTED THEOREM PROVING

2.1 INTRODUCTION

Given a first-order calculus and its many-sorted version, the Sort-Theorem tells us that every truth which can be deduced in the many-sorted calculus can also be deduced in the unsorted system, cf. Figure 1.5.1. But conversely, there are theorems of the unsorted calculus which do not have many-sorted counterparts, cf. Section 1.5. However, this does not entail an incompleteness property of the many-sorted system, because these many-sorted non-theorems are meaningless (wrt. the semantics of many-sorted logic). Hence both calculi are equivalent regarding soundness and completeness, and this is the reason why many-sorted logic has received scarce attention in the field of formal logic.

Since logicans use a formal calculus to study the foundations, the nature, and the limitations of formal reasoning, there is a good reason to choose as simple a calculus as possible. Investigations are much simplified if all features which do not represent a fundamental idea (like, for instance, variables and quantifiers constitute the fundamental difference between propositional and first-order logic) are excluded from the formal system.

The situation changes drastically, however, if we are concerned with *automated theorem proving*. Here a formal calculus is a device and a practical tool to find and compute proofs and theorems. Now efficiency is a concern (which it is not in formal logic) and therefore the kind of a calculus is of considerable interest, because the performance of a theorem proving system depends directly on the performance of the calculus it implements.

The resolution calculus of *Robinson* [Robinson 1965] is a first-order calculus which has been proved to be useful and successful in automated theorem proving. This calculus was augmented with the *paramodulation* rule by *Wos* and *Robinson* [Wos and Robinson 1973], i.e. an inference rule designed to implement equality reasoning efficiently. The outstanding features of this calculus are its *deduction incompleteness* and the *principle of most generality*, which are the reasons for the efficiency of the resolution calculus[1].

The resolution calculus is deduction incomplete, because not all *valid* formulas, i.e. valid *clauses*, can be deduced. As a consequence a theorem prover based on this calculus does not deduce all the truths from a set of formulas which hold *independently* from the set. This feature counts for efficiency, because it helps to keep the search space small.

But it is also impossible to deduce all the (non-valid) formulas from a given set of formulas which are semantically implied by the set. Again, this feature helps to keep the search space small and therefore helps to obtain an efficient theorem proving system.

However, for each non-valid formula Φ which is semantically implied by a set of formulas, a formula Φ' can be deduced in the resolution calculus such that Φ' semantically implies Φ. This feature, known as the *principle of most generality* (because Φ' is more general than Φ), entails the *refutation completeness* of the resolution calculus.

Our intention here is to extend the resolution calculus to a many-sorted calculus such that all the useful features of the given system, viz. soundness, deduction incompleteness, the principle of most generality, and of course the refutation completeness, are maintained.

---

[1] Here and subsequently we mean a calculus consisting of resolution, factorization *and paramodulation* as the only rules of inference, when we talk about the "resolution calculus".

2.2 MANY-SORTED RESOLVENTS AND FACTORS

Having defined a many-sorted language as in Section 1.2, the definition of a *many-sorted clause language* is straightforward: a *well-sorted atom* is a well-sorted atomic formula, a *well-sorted literal* is a well-sorted atom or its negation and a *well-sorted clause* is a finite set of well-sorted literals.

The many-sorted versions of the resolution and factorization rule have to be defined such that only well-sorted clauses can be deduced from well-sorted clauses by the many-sorted inference rules. Since the inference rules of the resolution calculus are based on the key concept of *unification*, we have to focus our attention on the many-sorted version of this notion.

For the many-sorted resolution calculus we demand for all *substitutions* that (1) variables are only replaced by *well-sorted* terms and that (2) the sort of a term is always *equal* to or a *subsort* of the sort of the variable it replaces. Those substitutions are called *well-sorted* and it is easily seen that an application of a well-sorted substitution to a well-sorted term (well-sorted literal or well-sorted clause) always yields a well-sorted term (well-sorted literal or well-sorted clause).

Now many-sorted resolution and factorization is defined as in the unsorted case, but with the proviso that only *well-sorted clauses* and *well-sorted most general unifiers* are used when resolvents and factors of the many-sorted resolution calculus are computed. Let us illustrate the many-sorted resolution calculus and also its advantages by a simple example (extracted from *Schubert's Steamroller Problem* which is discussed in [Walther 1984]) and consider the following statements:

Example 2.2.1 (i) Birds and snails are animals and (ii) there are some of each of them. (iii) Also there are some plants. (iv) Every animal likes to eat all the animals which do not like to eat all plants. (v) Birds do not like to eat snails. (vi) Therefore there is an animal that likes to eat all plants.

The problem is to show that statement (vi) is true, whenever statements (i) - (v) are assumed to be true. We start with a many-sorted axiomatization of the problem:

Our sort hierarchy consists of sorts B, S, A, and P denoting birds, snails, animals, and plants, respectively. We define B and also S as subsorts of A, thus obtaining an axiomatization of statement (i). We further assume, that there are constants b of Sort B, s of sort S and p of sort P and so an axiomatization of statements (ii) and (iii) is also obtained. Although these constants are not used in a proof, we do need them for theoretical reasons. By the presence of these constants, our sorts cannot be 'empty', i.e. the subdomains associated with the sorts by any interpretation cannot be empty, thus guaranteeing the sound-ness of our derivations, cf. [Walther 1987]. Figure 2.2.1 is a diagram of the sort hierar-chy under consideration.

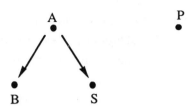

Figure 2.2.1   *The sort hierarchy of Example 2.2.1*

Using L(xy) as an abbreviation for 'x likes to eat y', the remaining statements are axio-matized by the well-sorted clauses shown in Figure 2.2.2:

(1) $\{L(a_1a_2), L(a_2p_1)\}$ (2) $\{\neg L(b_1s_1)\}$ (3) $\{\neg L(a_3h(a_3))\}$

Figure 2.2.2   *A many-sorted axiomatization for Example 2.2.1*

Here h is a skolem function with domainsort A and rangesort P (mapping an animal to a plant it refuses to eat), which is necessitated by the existential quantification in the nega-ted conclusion (vi). The subscripted lower case letters, e.g. a_1, a_2, p_1, ... are universally quantified variables of the sorts denoted by the corresponding upper case letter, e.g. A, P, ...

We can resolve upon the literals 1(1) and 2(1) in Figure 2.2.2 using the well-sorted most general unifier $\{a_1\leftarrow b_1, a_2\leftarrow s_1\}$ (but not $\{b_1\leftarrow a_1, s_1\leftarrow a_2\}$, because this most gene-ral unifier violates the above condition (2) of well-sortedness). However, there is no such resolvent upon the literals 1(2) and 2(1), since there is no subsort relation between S and P and as a consequence the variables p_1 and s_1 of sort P and S are not unifiable under sorts.

From the well-sorted clause set of Figure 2.2.2, we obtain a solution of the problem by the following many-sorted refutation:

(R1) $\{L(s_1\,p_1)\}$; many-sorted resolvent of 1(1) and 2(1)

(R2) □　　　　　; many-sorted resolvent of R1(1) and 3(1)

Here we use the well-sorted most general unifier $\{a_1 \leftarrow b_1, a_2 \leftarrow s_1\}$ for the computation of R1 and the well-sorted most general unifier $\{a_3 \leftarrow s_1, p_1 \leftarrow h(s_1)\}$ to obtain R2.

Now let us consider an unsorted formulation of the problem: Using the predicates

$B(x)$ - x is a bird,	$S(x)$ - x is a snail,
$A(x)$ - x is an animal,	$P(x)$ - x is a plant

as abbreviations, we obtain the set of clauses shown in Figure 2.2.3 as an unsorted clause axiomatization of the problem.

(1) $\{\neg A(x), \neg A(y), \neg P(z), L(xy), L(yz)\}$　(2) $\{\neg B(x), \neg S(y), \neg L(xy)\}$

(3) $\{\neg A(x), \neg L(xh(x))\}$

(4) $\{\neg B(x), A(x)\}$　　　　　　　　(5) $\{\neg S(x), A(x)\}$

(6) $\{B(b)\}$　　　　　　　　　　　　(7) $\{S(s)\}$

(8) $\{P(p)\}$　　　　　　　　　　　　(9) $\{\neg A(x), P(h(x))\}$

Figure 2.2.3　*An unsorted axiomatization for Example 2.2.1*

Here b, s and p are skolem constants necessitated by the existential quantifications in statements (ii) and (iii), and h is the skolem function as in Figure 2.2.2. Clauses (3) and (9) stem from the negation of statement (vi).

If we compare both axiomatizations, the advantages of the many-sorted approach become apparent: we only need 2 resolution steps for a many-sorted refutation, compared to 7 steps in the unsorted case. Also the search space is reduced very much in the many-sorted case compared to the unsorted one, because we have only 3 clauses with 4 literals, compared to 9 clauses with 19 literals. The reason is that sort information has to be represented explicitly in the unsorted case by *sort literals*, i.e. literals whose predicate letters are sort symbols. Using sort literals, we replace the given many-sorted clauses by their *relativizations* and extend the resulting clause set by the clausal representations for the *sort axioms* to obtain an unsorted clausal representation of the problem.

Note that the clauses (4) and (5) in Figure 2.2.3 are the sort axioms for the subsort order relation, clauses (6), (7) and (8) are the sort axioms stating the sorts of the constants and clause (9) is the sort axiom describing the domain- and rangesort of the skolem function h. Clauses (1) - (3) in Figure 2.2.3 are the relativizations of the corresponding clauses in Figure 2.2.2, which are obtained by extending a well-sorted clause with all the sort literals of form $\neg Q(x)$, where x is a variable of sort Q in the given well-sorted clause.

The resulting search space is further reduced by the *constraints imposed on many-sorted unification* [Walther 1986]. For instance, we can compute the resolvent $R=\{\neg A(x), \neg A(y), \neg P(z), \neg B(y), \neg S(z), L(xy)\}$ upon the literals 1(5) and 2(3) in Figure 2.2.3 giving rise to the computation of several *pure clauses*: $\neg S(z)$ can only be resolved upon 7(1), yielding the pure clause $\{\neg A(x), \neg A(y), \neg P(s), \neg B(y), L(xy)\}$ (because $\neg P(s)$ cannot be resolved upon). Also $\neg P(z)$ in R can only resolved upon the literals 8(1) and 9(2) yielding the pure literals $\neg S(p)$ and $\neg S(h(w))$ respectively.

In the many-sorted case these *deadends are impossible*: the resolution step upon the literals 1(2) and 2(1) in Figure 2.2.2, which corresponds to the computation of R, is blocked, because the variables s_1 and p_1 have no *well-sorted* unifier.

As a result the size of the initial *search space* is reduced to 2 many-sorted resolvents and no many-sorted factors, compared to an initial search space of 18 resolvents and 2 factors in the unsorted case. This example reveals the advantages of the many-sorted resolution calculus compared to the unsorted one: we obtain a *shorter* refutation of *shorter* clauses from a *smaller* set of clauses and also a *reduction* of the search space.

Note also that the *Herbrand Universe* H for the clause set of Figure 2.2.3 (as well as for the original Steamroller Example) is infinite, since we have {h(b), h(h(b)), h(h(h(b))), ...} ⊆ H because of the function symbol h. But for the many-sorted clause set of Figure 2.2.2 (as well as for the many-sorted version of the Steamroller Example), we obtain a *finite* Herbrand Universe $H_\Sigma = \{b, s, p, h(b), h(s)\}$, and consequently unsatisfiability (under sorts) is *decidable* for these examples. The reason is, that in the many-sorted case nested applications of the function symbol h are impossible, because the rangesort P of h neither equals nor is a subsort of the domainsort A of h.

The many-sorted resolution calculus is sound, because only well-sorted most general unifiers are used. But unfortunately this calculus, as defined so far, is (refutation) *complete* if and only if the sort hierarchy under consideration forms a *forest structure*, i.e. a collection of trees (like the sort hierarchy in Figure 2.2.1). The reason is, that the *Unification Theorem* [Robinson 1965] can be generalized to the many-sorted case only with this restriction:

Theorem 2.2.1 [Walther 1986]

A sort hierarchy is a forest structure

iff

for each set of well-sorted atoms D: if D has a well-sorted unifier, then D has a well-sorted most general unifier.

2.3 NON-FOREST SORT HIERARCHIES

As a consequence of Theorem 2.2.1, the many-sorted resolution calculus, as defined so far, is (refutation) incomplete if non-forest structured sort hierarchies are used. We shall illustrate this with another example:

Example 2.3.1 (i) Reptiles and birds are animals, which breathe by lungs and lay eggs. (ii) There are reptiles as well as birds. (iii) Each animal with lungs and wings has feathers. (iv) All animals without wings are poikilothermic, provided they lay eggs. (v) There is an animal which has feathers and lays eggs, or there is a poikilotherm, which breathes by lungs.

We give a many-sorted axiomatization expressing that statement (v) is true whenever statement (i) - (iv) are assumed to be true: our sort hierarchy consists of sorts R, B, L and E denoting reptiles, birds, animals with lungs and animals which lay eggs, respectively. We define R and B as subsorts of L and also of E, thus obtaining a sortal representation of statement (i). We can do so because of statement (ii), which guarantees that neither R nor B are empty sorts (and by the subsort order L and E consequently are also non-empty). Figure 2.3.1 is a diagram of this sort hierarchy, which obviously is not a forest structure.

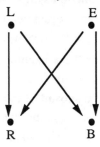

Figure 2.3.1 *The sort hierarchy of Example 2.3.1*

Using the predicates

W(x) - animal x has wings   F(x) - animal x has feathers
P(x) - animal x is poikilothermic

as abbreviations, we obtain the following many-sorted clausal representation of the problem:

(1) $\{\neg W(l_1), F(l_1)\}$ (2) $\{W(e_1), P(e_1)\}$

(3) $\{\neg F(e_2)\}$ (4) $\{\neg P(l_2)\}$

Figure 2.3.2 *A many-sorted axiomatization for Example 2.3.1*

Clauses (1) and (2) are the formal representations of statements (iii) and (iv), and clauses (3) and (4) stem from the negation of statement (v). As in the previous example, the subscripted lower case letters l_i and e_i (i=1,2) are universally quantified variables of the sorts denoted by the corresponding upper case letter L and E.

Now, assume that r_1 and b_1 are variables of sort R and B respectively. Then $\{l_1, e_1\}$ has well-sorted unifiers, e.g. $\{l_1 \leftarrow r_1, e_1 \leftarrow r_1\}$ and also $\{l_1 \leftarrow b_1, e_1 \leftarrow b_1\}$. But $\{l_1, e_1\}$ has no *well-sorted most general unifier*, because neither of the most general unifiers $\{l_1 \leftarrow e_1\}$ and $\{e_1 \leftarrow l_1\}$ is well-sorted. As a consequence, the many-sorted resolution calculus - as defined so far - is refutation incomplete, because we cannot compute a many-sorted resolvent from the set of clauses in Figure 2.3.2, although this set is unsatisfiable.

We shall overcome this problem by using a technique borrowed from *unification under equational theories* [Plotkin 1972, Fages and Huet 1983, Siekmann 1984]: We use *complete and minimal sets* of well-sorted unifiers instead of most general unifiers when we compute many-sorted resolvents, factors and paramodulants. The existence of these sets is guaranteed by the following theorem:

Theorem 2.3.1 [Walther 1986]
If a set D of well-sorted atoms has a well-sorted unifier, then D has a complete and minimal set of well-sorted unifiers.

In the above example, for instance, $\{\{l_1 \leftarrow r_1, e_1 \leftarrow r_1\}, \{l_1 \leftarrow b_1, e_1 \leftarrow b_1\}\}$ is a minimal and complete set of well-sorted unifiers for $\{l_1, e_1\}$. The minimality of such a set guarantees that the modified many-sorted calculus still satisfies the principle of most generality and the property of completeness entails the refutation completeness of the calculus. Hence we obtain:

Theorem 2.3.2 [Walther 1987]
A set S of well-sorted clauses is unsatisfiable under sorts iff the empty clause can be derived from S using the inference rules of many-sorted resolution and many-sorted factorization.

With this extension, we obtain the following many-sorted refutation from the clause set in Figure 2.3.2:

(R1) $\{\neg W(r_1)\}$; many-sorted resolvent of 1(2) and 3(1)

(R2) $\{P(r_1)\}$; many-sorted resolvent of 2(1) and R1(1)

(R3) □   ; many-sorted resolvent of 4(1) and R2(1)

We use the first element of the minimal and complete set of well-sorted unifiers $\{\{l_1 \leftarrow r_1, e_2 \leftarrow r_1\}, \{l_1 \leftarrow b_1, e_2 \leftarrow b_1\}\}$ for the computation of R1, and the sets $\{\{e_1 \leftarrow r_1\}\}$ and $\{\{l_2 \leftarrow r_1\}\}$ for the computation of R2 and R3, respectively.

As indicated, our definitions allow us to establish a correspondence between the semantic notions for the resolution calculus and its many-sorted version with the model-theoretic part of the Sort-Theorem, cf. Figure 1.5.1:

Theorem 2.3.3 [Walther 1987]
A set S of well-sorted clauses is unsatisfiable under sorts if and only if the union of the set of sort axioms with the set of relativized clauses of S is unsatisfiable.

Hence with Theorem 2.3.2 the deductive notions of both calculi can be related with the *proof-theoretic* part of the Sort-Theorem:

Corollary 2.3.4

The empty clause can be derived from a set S of well-sorted clauses in the many-sorted resolution calculus using the inference rules of many-sorted resolution and factorization if and only if the empty clause can be derived in the resolution calculus from the union of the set of sort axioms with the set of relativized clauses of S using the inference rules of resolution and factorization.

2.4 MEET-SEMILATTICES AS SORT HIERARCHIES

However, since a complete and minimal set of unifiers can be arbitrarily large (although always finite, if the set of sort symbols is finite) it is possible to have an arbitrarily large number of many-sorted resolvents (or factors) for a given pair of clauses and literals resolved upon (or a given subset of a clause). Assume, for instance, that sorts s_1 and s_2 have 1000000 common but incomparable subsorts and that the variables x and y are of sort s_1 and sort s_2 respectively. Then we find that each complete and minimal set of well-sorted unifiers of $\{x,y\}$ has 1000000 members, and as a consequence a clause like $\{P(x),P(y)\}$ has 1000000 independent many-sorted factors.

If we ignore for the moment that this example is a rather artificial one, we may consider this effect as a possible disadvantage of the many-sorted resolution calculus, because the search space of an automated theorem prover can be swamped with a huge number of clauses by a single inference step. But if we use the unsorted calculus instead, we have 2000000 sort axioms in this example, each of which has to be used to compute the 1000000 relativizations of all the many-sorted factors. Hence the size of the search space would be even larger in the resolution calculus.

But there is also a simple technical remedy to this kind of problem: if we embed the given sort hierarchy into a *meet-semilattice* (i.e. each pair of sort symbols has an infimum) by inventing additional sorts, we find that each complete and minimal set of well-sorted unifiers is always a *singleton*.

Theorem 2.4.1 [Walther 1986]
If a sort-hierarchy is a meet-semilattice, then each complete and minimal set of well-sorted unifiers is a singleton.

For the sort hierarchy of Figure 2.3.1, for instance, we obtain the meet-semilattice by invention of a sort X, representing the union of reptiles and birds, and a sort Ø. Figure

2.4.1 is a diagram of the extended sort hierarchy.

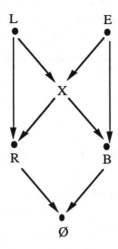

Figure 2.4.1 *A meet semi-lattice for the sort hierarchy of Example 2.3.1*

Now $\{l_1,e_1\}$ has the singleton $\{\{l_1\leftarrow x_1,e_1\leftarrow x_1\}\}$ as a complete and minimal set of well-sorted unifiers. The sort Ø is needed here only for formal reasons, i.e. to obtain a meet-semilattice, if there are sorts in the given hierarchy without common subsorts, like R and B. Since we assume that there are no terms of sort Ø, terms without common subsorts never have a well-sorted unifier.

But we prefer partially ordered sets of sort symbols instead of meet-semilattices in our definition of the many-sorted resolution calculus, because many-sorted axiomatizations tend to be unnatural or unintuitive (like the union X of reptiles and birds) if only meet-semilattices can be used to define a sort hierarchy.

2.5 MANY-SORTED PARAMODULATION

Paramodulation is an inference rule which integrates equality reasoning into the resolution calculus [Wos and Robinson 1973]. But before we define a many-sorted version of this inference rule, we have to stipulate the domainsorts for the equality predicate ≡. When concerned with equality reasoning, we assume that each sort hierarchy has a least upper bound, say any, and we define any as the domainsort for both argument positions of ≡. Hence for all well-sorted terms q and r, q≡r is a well-sorted atom.

But this definition seems to ignore the information given by the sorts of the terms, because terms of arbitrary sort can be in the equality relation. One feels that it is more natural for a many-sorted logic to have the sorts of equal terms related somehow by the subsort order relation.

Given the equality $q \equiv r$, it seems meaningful to have that the sort of q is a subsort of, or equal to, the sort of r (or vice versa). A more liberal restriction would be to demand that the sorts of q and r have at least one common subsort, because the subdomains associated with the sorts of the terms must have a non-empty intersection, unless the equality is unsatisfiable.

However, those sortal restrictions for the equality predicate entail (implicitly) a strong assumption on the semantics of many-sorted logic in general and the sort hierarchies under consideration in particular. We suppose in this case, that if sort s_1 is not a subsort of sort s_2, the subdomain associated with s_1 by any interpretation is never included in the subdomain associated with s_2, and that the subdomains associated with s_1 and s_2 always have an empty intersection if s_1 and s_2 do not have a common subsort. Also the interpretation of a term t cannot be an element of the subdomain associated with a sort symbol s, provided the sort of t is not a subsort of s. Hence, as opposed to ordinary first-order logic, we assume certain propositions to be false by *default*, unless the contrary is stated explicitly.

We dispense with sortal restrictions on the domainsorts of the equality predicate, because many-sorted logic with default assumptions would not be compatible with unsorted logic. Also, this would require that sort hierarchies and signatures are complete in the sense that they mirror all inclusion and membership relations on the syntactic level which may hold on the semantic level.

As opposed to many-sorted resolution and factorization, we do not necessarily obtain a well-sorted clause as the result of a paramodulation step when we restrict ourselves to using *well-sorted* unifiers. Whenever the sorts of the terms in an equality disagree, it may happen that a term in a literal is replaced by a term whose sort is neither a subsort of, nor equal to, the domainsort of the specific argument position it occupies. Of course, we cannot allow such deductions of ill-sorted clauses in our calculus. Hence, in the case of paramodulation (and in contrast to resolution and factorization), the derivations of the many-sorted system are influenced also by the *domainsorts* stipulated for the function and predicate symbols.

A straightforward idea to overcome this problem is to define a many-sorted paramodulant as a paramodulant generated from well-sorted clauses (using a member of a minimal and complete set of well-sorted unifiers), which in addition is a *well-sorted clause*.

Since the empty substitution is the only unifier in the ground case (and, of course, this substitution is well-sorted), this restriction is effective already in the ground case. But for the ground case completeness is maintained, because we can always assemble the 'right' equations by paramodulating into equalities:

Example 2.5.1 Let S={ {P(b)}, {¬P(b')}, {b≡a}, {a≡b'} } be a set of well-sorted clauses such that the rangesort of the constant a is sort A, the rangesort of the constants b and b' as well as the domainsort of the predicate P is sort B, and A is not a subsort of B. Then neither {P(a)} nor {¬P(a)} is a well-sorted clause, i.e. these clauses cannot be obtained from S by many-sorted paramodulation. But with the many-sorted paramodulant {b≡b'} of {b≡a} and {a≡b'}, we obtain the many-sorted paramodulants {P(b')} and {¬P(b)}, each of which leads to a many-sorted refutation of S. ◆

But unfortunately the above definition of a many-sorted paramodulant does not guarantee the refutation completeness of our many-sorted calculus in the general case, as can be verified with the following example:

Example 2.5.2 (i) Reptiles and birds are animals, which breathe by lungs and lay eggs. (ii) There are reptiles as well as birds. (iii) There is a bird which hatches out the eggs it lays. (iv) There is a reptile or a bird which does not hatch out its eggs. (v) There are at least two animals which breathe by lungs.

The problem is to show that statement (v) is true, whenever statements (i) - (iv) are assumed to be true. We use H(x) as an abbreviation for 'animal x hatches out its eggs', where we define sort E as the domainsort for the unary predicate H, because 'hatch out eggs' is a meaningful characterization only for animals which lay eggs. With the same sort hierarchy as given for Example 2.3.1 (cf. Figure 2.3.1), we obtain the following many-sorted clausal representation of the problem:

(1) {H(b')}   (2) {¬H(b),¬H(r)}   (3) {l_1≡l_2}

Figure 2.5.1 *A many-sorted axiomatization for Example 2.5.2*

Clauses (1) and (2) are the formal representations of statements (iii) and (iv), and clause (3) stems from the negation of statement (v). The symbols b', b and r are skolem constants of sort B and R respectively, which are necessitated by the existential quantifications in statements (iii) and (iv). As in the previous example, l_1 and l_2 are universally quantified variables of the sort L.

From the clause set in Figure 2.5.1, 6 paramodulants can be computed in the first generation, viz. $\{H(l_i)\}$, $\{\neg H(l_i), \neg H(r)\}$ and $\{\neg H(b), \neg H(l_i)\}$ (where i=1,2). But each of these paramodulants is ill-sorted, because L (i.e. the sort of l_i) is neither a subsort of nor equal to E (i.e. the domainsort of H) and therefore $H(l_i)$ is not a well-sorted atom. Since there are no many-sorted resolvents and factors, a many-sorted refutation does not exist for the above clause set.

As can be seen from this example, our calculus would be (refutation) incomplete if a many-sorted paramodulant were defined only as a paramodulant which is a well-sorted clause. But which necessary and meaningful conclusions should be drawn from the clauses in Figure 2.5.1, which cannot be obtained with the given restriction?

We infer from our sort hierarchy and from the semantics of many-sorted logic, that everything which is true for animals with lungs, must hold for birds and also for reptiles. Hence we know that the clauses

(3.1) $\{l_1 \equiv b_1\}$ (3.2) $\{l_1 \equiv r_1\}$

must be true, provided clause (3) is true (where b_1 and r_1 are variables of sort B and R respectively). Replacing clause (3) by the clauses (3.1) and (3.2) in the above set, we obtain the following many-sorted refutation from the modified clause set:

(P1) $\{H(b_1)\}$; many-sorted paramodulant of 1(1) and 3.1(1)

(P2) $\{H(r_1)\}$; many-sorted paramodulant of 1(1) and 3.2(1)

(R3) $\{\neg H(r)\}$; many-sorted resolvent of P1(1) and 2(1)

(R4) $\square$; many-sorted resolvent of P2(1) and R3(1)

To compute this deduction, we use the well-sorted unifiers $\{l_1 \leftarrow b'\}$, $\{l_1 \leftarrow b'\}$, $\{b_1 \leftarrow b\}$ and $\{r_1 \leftarrow r\}$, each of which is the only member of the respective complete and minimal set of unifiers. Formally, we obtain clauses (3.1) and (3.2) by replacing the variable l_2 in clause (3) by the variables b_1 and r_1 respectively. Such a variable replacement is sound and yields a well-sorted clause, provided a variable is replaced by a variable the sort of which is a subsort of the sort of the replaced variable. We call the application of a sound variable replacement a *coercion*, similar to the coercion operations in programming languages with sorts (there often called *types*). If we allow coercions as additional inference rules, we are now able to infer all well-sorted clauses which are necessary for a many-sorted refutation with paramodulation.

However, an *unrestricted* usage of coercions would destroy the principle of most generality. Assume, for instance, that we extend our given sort hierarchy by sorts S and D, standing for snakes and ducks, such that S is a subsort of R and D is a subsort of B. Consequently, S and D are also subsorts of L, and we obtain by unrestricted coercion

$$(3.3) \quad \{l_1 \equiv s_1\} \qquad\qquad (3.4) \quad \{l_1 \equiv d_1\}$$

(where s_1 and d_1 are variables of sort S and D respectively). But obviously, we do not need clauses (3.3) and (3.4) for a many-sorted refutation of the above problem, because we can use clauses (3.1) and (3.2). Hence we infer more well-sorted *instances* of clause (3) than we need to obtain a solution.

Moreover, we want to have coercions only when they are *necessary* for a refutation, i.e. to enable the computation of a well-sorted paramodulant which is necessary for a solution. Assume, for instance, we want to prove, that 'there are at least two animals which lay eggs' from the statements (i) - (iv) of Example 2.5.2. Then we obtain

$$(3') \quad \{e_1 \equiv e_2\}$$

instead of clause (3), and each of the 6 well-sorted paramodulants $\{H(e_i)\}$, $\{\neg H(e_i), \neg H(r)\}$ and $\{\neg H(b), \neg H(e_i)\}$ (i=1,2) leads to a many-sorted refutation of the clauses (1), (2) and (3') without any use of coercions.

Technically, we map the problem of finding most general and necessary coercions into a *many-sorted unification problem*: whenever a term t is replaced by a term r in an unrestricted paramodulation step, we create a fresh variable z of the same sort as the domainsort of the argument position the term r occupies in the paramodulant P. Then we use each member of a minimal and complete set of well-sorted unifiers for $\{r, z\}$ as a *coercion substitution*, which applied to P yields a paramodulant of the many-sorted calculus (and obviously the result is always a well-sorted clause).

Given the clause set of Figure 2.5.1, we paramodulate clauses 1(1) and 3(1), using the well-sorted unifier $\{l_1 \leftarrow b'\}$, and obtain $\{H(l_2)\}$. Letting e_1 denote a fresh variable of sort E (i.e. the domainsort of H), $\{\{l_2 \leftarrow b_1, e_1 \leftarrow b_1\}, \{l_2 \leftarrow r_1, e_1 \leftarrow r_1\}\}$ is a minimal and complete set of well-sorted unifiers of $\{l_2, e_1\}$. The application of both coercion substitutions to $\{H(l_2)\}$ now yields the many-sorted paramodulants $\{H(b_1)\}$ and $\{H(r_1)\}$ of

1(1) and 3(1).

With this definition of the many-sorted paramodulation rule, we obtain a refutation complete many-sorted version of a resolution calculus with paramodulation, and we can prove the following theorem:

Theorem 2.5.1   [Walther 1987]
A set S of well-sorted clauses is equality-unsatisfiable under sorts iff the empty clause can be derived from S using the inference rules of many-sorted resolution, many-sorted factorization and many-sorted paramodulation.

From the clause set of Figure 2.5.1, for instance, the following refutation in our many-sorted calculus is obtained :

(P1) $\{H(b_1)\}$; many-sorted paramodulant of 1(1) and 3(1)

(P2) $\{H(r_1)\}$; many-sorted paramodulant of 1(1) and 3(1)

(R3) $\{\neg H(r)\}$; many-sorted resolvent of P1(1) and 2(1)

(R4) $\square$;   many-sorted resolvent of P2(1) and R3(1) .

2.6   MANY-SORTED AXIOMATIZATIONS

The principal idea of many-sorted logic is to represent certain truths by defining a sort hierarchy and a many-sorted signature, instead of having these truths expressed *explicitly* by first-order formulas. Hence it is possible to incorporate certain deductions *into the inference machinery* of a calculus, instead of performing them explicitly by inference steps. In case of the many-sorted resolution calculus, this incorporation is realized (mainly) in the modified unification procedure, cf. [Walther 1986, Walther 1987].

But which kinds of truths can be represented by a sort hierarchy and a many-sorted signature? We call statements of naive set theory, which can be defined in terms of the subsort order and of the signature, *sortal statements*. Figure 2.6.1 is a collection of sortal statements and their equivalent representations by the subsort order and the signature:

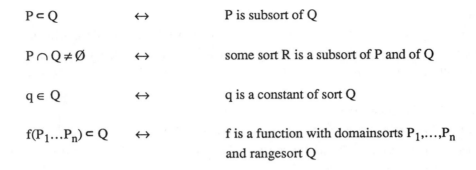

$P \subseteq Q$	$\leftrightarrow$	P is subsort of Q
$P \cap Q \neq \emptyset$	$\leftrightarrow$	some sort R is a subsort of P and of Q
$q \in Q$	$\leftrightarrow$	q is a constant of sort Q
$f(P_1 \ldots P_n) \subseteq Q$	$\leftrightarrow$	f is a function with domainsorts $P_1, \ldots, P_n$ and rangesort Q

Figure 2.6.1 *Sortal statements and their equivalent representations by the sort hierarchy and the signature*

Obviously, the performance of a many-sorted logic increases with an increasing number of sortal statements in an axiomatization, because the number of explicit inferences decreases if we do not use sort literals. But unfortunately, we cannot express a *negative sortal statement*, i.e. the negation of a sortal statement, like $P \not\subseteq Q$ or $P \cap Q = \emptyset$, as a sortal statement. Consequently, we cannot represent each unary predicate symbol in an unsorted clause set by a sort symbol. Let us consider the following example:

Example 2.6.1 [Chang and Lee 1973] (i) Some patients like all doctors. (ii) No patient likes any quack. (iii) Therefore, no doctor is a quack.

We can represent the set of patients by a sort P. But if we represent the set of doctors and also the set of quacks by sorts, we are unable to axiomatize the conclusion (iii), because we cannot state that $D \cap Q = \emptyset$. Hence we have to choose sort D, say, to represent the set of doctors, and with a predicate Q(x) for 'x is a quack' we can symbolize (iii) as $\forall x : D \; \neg Q(x)$.

However, there is a simple trick to have *both* doctors and quacks as sorts. Since we are interested in resolution based theorem proving, it suffices to show that the conjunction of statements (i) and (ii) with the *negation* of statement (iii) is unsatisfiable. But the negation of (iii), i.e. 'there is a doctor who is a quack', can be represented within the sort hierarchy: we let X denote a sort symbol standing for the intersection of doctors and quacks, i.e. X is a subsort of D and of Q. Because of X we have $D \cap Q \neq \emptyset$, i.e. a sortal statement for the negation of statement (iii). Figure 2.6.2 is a diagram of this sort hierarchy.

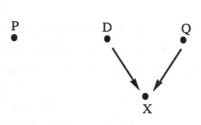

Figure 2.6.2 *The sort hierarchy for Example 2.6.1*

Using L(xy) for 'x likes y', p as a constant of sort P, and d_1, p_1, q_1 and x_1 as variables of sort P, D, Q and X respectively, we obtain the following many-sorted clausal representation for Example 2.6.1:

(1) $\{L(p\ d_1)\}$ (2) $\{\neg L(p_1\ q_1)\}$

With the well-sorted unifier from the minimal and complete set $\{\{p_1 \leftarrow p,\ d_1 \leftarrow x_1,\ q_1 \leftarrow x_1\}\}$, we derive the empty clause with a single many-sorted resolution step. Note that without D and Q having a common subsort X, the variables d_1 and q_1 would not have a well-sorted unifier. For the corresponding unsorted axiomatization, we obtain the empty clause with 4 resolution steps from a set of 5 clauses with 8 literals, cf. [Chang and Lee 1973].

Obviously, we can use this method whenever the theorem to be shown can be represented as a disjunction of negative sortal statements. But it fails if we have a negative sortal statement in the *axiomset* of a problem. For problems of this kind, we take advantage of the fact that sortal statements can also be expressed by many-sorted first-order equations:

$P \subset Q$	$\leftrightarrow$	$\forall x{:}P\ \exists y{:}Q\ x{\equiv}y$
$P \cap Q \neq \emptyset$	$\leftrightarrow$	$\exists x{:}P\ \exists y{:}Q\ x{\equiv}y$
$q \in Q$	$\leftrightarrow$	$\exists y{:}Q\ q{\equiv}y$
$f(P_1...P_n) \subset Q$	$\leftrightarrow$	$\forall x_1,...,x_n{:}P\ \exists y{:}Q\ f(x_1...x_n){\equiv}y$

Figure 2.6.3 *Sortal statements and their equivalent representations by many-sorted equations*

Since we have these equations in our object language, we are also able to represent negative sortal statements, like $\forall x{:}P\ \forall y{:}Q\ \neg x{\equiv}y$ stands for $P{\cap}Q{=}\emptyset$, without resorting to the use of sort literals. Let us illustrate the advantage of these axiomatizations with another example:

Example 2.6.2 [Chang and Lee 1973] (i) No used-car dealer buys a used car for his family. (ii) Some people who buy used cars for their families are absolutely dishonest. (iii) Conclude that some absolutely dishonest people are not used-car dealers.

We let B, D, U and X be sorts standing for people, who <u>b</u>uy used cars, absolutely <u>d</u>ishonest people and <u>u</u>sed-car dealers. We define X as a subsort of B and of D, i.e. $B{\cap}D{\neq}\emptyset$, and thus an axiomatization of statement (ii) by a sortal statement is obtained. We also define D as a subsort of U, i.e. $D{\subset}U$, and obtain an axiomatization of the negated statement (iii), viz. 'all absolutely dishonest people are used-car dealers', by a sortal statement. Figure 2.6.4 is a diagram of this sort hierarchy.

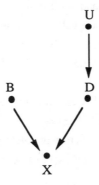

Figure 2.6.4 *A sort hierarchy for Example 2.6.2*

The remaining statement (i) is a negative sortal statement, viz. $U{\cap}B{=}\emptyset$, and is represented explicitly by the many-sorted unit clause

(1) $\{\neg u_1{\equiv}b_1\}$

(compared to 4 clauses with 6 literals in an unsorted axiomatization). From (1), we obtain the empty clause by resolution with the *reflexivity clause* $\{z_1{\equiv}z_1\}$, using the well-sorted unifier $\{z_1 \leftarrow x_1, u_1 \leftarrow x_1, b_1 \leftarrow x_1\}$, (where the sort of the variable z_1 is any). Note that without U and B having a common subsort X, the variables u_1 and b_1 of sort U and B would not have a well-sorted unifier.

However, we use many-sorted equations not only for the representation of negative sortal statements. Since it may become necessary to verify sortal statements if the given sort hierarchy becomes very large or complex, we also need a means to represent *non-negative* sortal statements explicitly. Again, many-sorted equalities can be used to axiomatize sortal statements without sort literals:

Example 2.6.3 (i) Every man is mortal. (ii) Socrates is a man. (iii) Therefore Socrates is mortal.

Defining sort MAN as a subsort of sort MORTAL, and Socrates as a constant of sort MAN, we obtain an axiomatization of statements (i) and (ii) by sortal statements only. Hence we are unable to give an axiomatization for statement (iii) without sort literals, unless we use many-sorted equalities. For the example above, statement (iii) is represented as $\exists x{:}MORTAL\ socrates{\equiv}x$, which after negation yields the clause:

(1) $\{\neg socrates{\equiv}m_1\}$,

where m_1 is a variable of sort MORTAL. From (1), we obtain the empty clause by resolution with the reflexivity clause $\{z_1 \equiv z_1\}$, using the well-sorted unifier $\{z_1 \leftarrow socrates, m_1 \leftarrow socrates)$.

3   *CONCLUSION*

An increase of efficiency of the many-sorted calculus presented here can be obtained by extending this calculus with *polymorphic* function symbols. This allows the sort of a term $f(t_1...t_n)$ to vary with the sorts of its arguments $t_1,...,t_n$ and consequently more deductions are shifted to the many-sorted unification algorithm, which otherwise have to be performed explicitly [Schmidt-Schauss 1985a]. Also efficiency can be increased by combining many-sorted unification with *unification under equational theories* [Schmidt-Schauss 1986]. To benefit from many-sorted resolution also for problems stated in an unsorted axiomatization, an algorithm has been developed which translates unsorted clause sets into many-sorted ones, where certain unary predicates from the unsorted formulation are used as sort symbols in the many-sorted version [Schmidt-Schauss 1985b].

However, we do not claim that the calculus proposed here is the only approach to utilize many-sorted reasoning for resolution based theorem proving - see, for instance, *Cohn's* proposal of a many-sorted resolution logic [Cohn 1987].

The advantages of many-sorted reasoning were early recognized within the field of automated theorem proving, e.g. [Hayes 1971, Henschen 1972], and several theorem proving programs have been based on some kind of many-sorted calculus, e.g. [Weyhrauch 1977, Champeaux 1978]. More recently many-sorted reasoning has also found attention in other fields of automated deduction, as e.g. rewrite systems [Cunningham and Dick 1985, Goguen et al. 1985], parsing with Horn-clause grammars and knowledge retrieval [Frisch 1985 a Frisch 1985 b] or logic programming [Ait-Kacki and Nasr 1986, Huber and Varsek 1987].

REFERENCES

AIT-KACKI, H. and NASR, R. *LOGIN: A Logic Programming Language with Built-in Inheritance.* J. Logic Programming, vol 3, no 3, 1986

CHAMPEAUX, D. De *A Theorem Prover Dating a Semantic Network.* Proceedings of the AISB/GI Conference, Hamburg, 1978

CHANG, C.-L. and LEE, R. C.-T. *Symbolic Logic and Mechanical Theorem Proving.* Academic Press, 1973

COHN, A. G. *A More Expressive Formulation of Many-Sorted Logic.* J. Automated Reasoning, vol 3, no 2, 1987

CUNNINGHAM, R. J. and DICK, A. J. J. *Rewrite systems on a lattice of types.* Acta Informatica, vol 22, no 2, 1985

ENDERTON, H. *A Mathematical Introduction to Logic.* Academic Press, Orlando 1972

FAGES, F. and HUET, D. *Complete Sets of Unifiers and Matchers in Equational Theories.* Proceedings of the 8th CAAP, Lecture Notes in Computer Science 159, L'Aquila, 1983

FRISCH, A. M. *An Investigation into Inference with Restricted Quantification and a Taxonomic Representation.* SIGART Newsletter 91, 1985 (a)

FRISCH, A. M. *Parsing with Restricted Quantification.* Proc. of the AISB Conference, 1985 (b)

GOGUEN, J. A., JOUANNAUD, J.-P. and MESEGUER, J. *Operational Semantics for Order-Sorted Algebra.* Proceedings of the 12th ICALP, Lecture Notes in Computer Science 194, Springer-Verlag, 1985

HAYES, P. *A Logic of Actions.* Machine Intelligence 6, 1971

HENSCHEN, L. J. *N-Sorted Logic for Automatic Theorem Proving in Higher Order Logic.* Proceedings of the ACM Conference, Boston, 1972

HUBER, M. and VARSEK, I. *Extended Prolog for Order-Sorted Resolution.* Proceedings of the 1987 Symposium on Logic Programming, San Francisco, 1987

LOVELAND, D. W. *Automated Theorem Proving: A Logical Basis.* North-Holland Publishing Company, 1978

OBERSCHELP, A *Untersuchungen zur mehrsortigen Quantorenlogik.* Mathematische Annalen 145, 1962

PLOTKIN, G. *Building-in Equational Theories.* Machine Intelligence 7, 1972

ROBINSON, J. A. *A Machine-Oriented Logic Based on the Resolution Principle.* J. ACM, vol 12, no 1, 1965, also in: *Automation of Reasoning - Classical Papers on Computational Logic*, vol 1. J. Siekmann and G. Wrightson, Eds., Springer-Verlag, 1983

SCHMIDT-SCHAUSS, M. *A Many-Sorted Calculus with Polymorphic Functions based on Resolution and Paramodulation.* Proceedings of the 9th International Joint Conference on Artificial Intelligence (IJCAI.85), Los Angeles, 1985 (a)

SCHMIDT-SCHAUSS, M. *Mechanical Generation of Sorts in Clause Sets.* MEMO SEKI-85-VI-KL, Fachbereich Informatik, Universität Kaiserslautern, 1985 (b)

SCHMIDT-SCHAUSS, M. *Unification in Many-Sorted Equational Theories.* Proceedings of the 8th International Conference on Automated Deduction (CADE-86), Lecture Notes in Computer Science 230, Springer-Verlag, 1986

SIEKMANN, J. *Universal Unification.* Proceedings of the 7th International Conference on Automated Deduction (CADE-84), Lecture Notes in Computer Science 170, Springer-Verlag, 1984

WALTHER, C. *The Markgraf Karl Refutation Procedure: PLL - A First-Order Language for an Automated Theorem Prover.* Interner Bericht 35/82, Institut für Informatik I, Universität Karlsruhe, 1982

WALTHER, C. *A Mechanical Solution of Schubert's Steamroller by Many-Sorted Resolution.* Proceedings of the 4th National Conference on Artificial Intelligence (AAAI-84), Austin, 1984, revised version in: Artificial Intelligence, vol 26, no2, 1985

WALTHER, C. *A Classification of Many-Sorted Unification Problems.* Proceedings of the 8th International Conference on Automated Deduction (CADE-86), Lecture Notes in Computer Science 230, Springer-Verlag, 1986, revised version appeared as: *Many-Sorted Unification*, J. ACM, vol 35, no 1, 1988

WALTHER, C. *A Many-Sorted Calculus Based on Resolution and Paramodulation.* Research Notes in Artificial Intelligence, Pitman, London, and Morgan Kaufmann, Los Altos, 1987

WALTHER, C. *Automatisches Beweisen.* In: *Künstliche Intelligenz - Theoretische Grundlagen und Anwendungsfelder*, Th. Christaller, H.-W. Hein und M. M. Richter (Hrsg.), Frühjahrsschulen Dassel 1985 und 1986, Informatik Fachberichte 159, Springer-Verlag, 1987 (a)

WEYHRAUCH, R. W. *FOL: A Proof Checker for First-Order-Logic.* MEMO AIM-235.1, Stanford Artificial Intelligence Laboratory, Stanford University, 1977

WOS. L. and ROBINSON, G. *Maximal Models and Refutation Completeness: Semidecision Procedures in Automatic Theorem Proving.* In: *Wordproblems.* W: W. Boone, F. B. Cannonito and R. C. Lyndon, Eds., North-Holland, 1973, also in: *Automation of Reasoning - Classical Papers on Computational Logic*, vol 2. J. Siekmann and G. Wrightson, Eds., Springer-Verlag, 1983

Tableau Calculus for Order Sorted Logic

P.H. Schmitt
Universität Karlsruhe
Institut für Logik, Komplexität und
Deduktionssysteme
Postfach 6980
D-7500 Karlsruhe 1
EARN/BITNET: PSCHMITT at DKAUNI0I

W.Wernecke
IBM Deutschland GmbH
Scientific Center
Institute for Knowledge Based Systems
Knowledge and Language Processing 1
Wilckensstr. 1a
D-6900 Heidelberg 1
Phone + 49-6221-404-309
EARN/BITNET: WERNECKE at DHDIBM1

Abstract

In this paper we discuss a calculus for order sorted logic, based on an extension of the tableau method. We first present the background of our investigation by specifying the underlying representation language and its interpretation. Then we introduce the tableau calculus for order sorted logic and show the completeness of the method. Finally we discuss some issues related to the implementation of the presented calculus.

Introduction

Many-sorted logic, first invented for formal investigations of mathematical concepts, is now being used in many areas of computer science, for example in algebraic specifications of abstract data types ([Eh85]), in relational data base theory ([Ga78]) and in logic programming ([My84]). In the area of Artificial Intelligence, many-sorted languages are e.g. used for the conceptual modelling of application domains in knowledge based systems ([LILOG88; We88]).

For many of these applications it is quite natural to use order sorted logic (a many sorted logic with a partial ordering on the set of sorts), because very often the domain under consideration can be partitioned into hierarchically organized concepts. As an example one might consider the representation of taxonomic information in knowledge based systems. Compared to ordinary (one sorted) predicate logic, one of the advantages of using many sorted languages is based on the fact that it is more adequate to denote subsets of a domain by distinguished sort names, than by using unary predicate symbols. In some sense this is only 'syntactic sugar', because using many sorted logic does not increase the expressive power of ordinary (one sorted) first order predicate logic ([Ob62]). The benefit from the syntactic point of view is that formulae in many sorted languages are more compact and thus more readable.

From the computational point of view, however, order sorted calculi offer advantages. The essential point is that sorted variables restrict the domain of possible interpretations. For theorem prover based on unification, this is of great importance, because this leads to more efficient processing.

In this paper we will discuss a calculus for order sorted logic, based on an extension of the (analytic) tableau method. We will first present (chapter 1) the background of our investigation by specifying the underlying representation language and its interpretation. In chapter 2 we introduce the tableau calculus for order sorted logic and show the completeness of the method. In chapter 3 we finally discuss some issues related to the implementation of the presented calculus.

Related work

Mathematical investigations of many sorted logic dates back to J. Herbrand ([He]).

Many important results were obtained by the work of A. Oberschelp ([Ob62]), who proved the equivalence of many sorted logic to one sorted logic and gave soundness and completeness results for a calculus of Montague and Henkin.

In the area of automated deduction, investigations of many sorted logic have been carried out e.g. by [Wa83; Co87; Fr].

More recently the development of representation formalism which combine sorted logic with frame based formalism has been considered ([Ai86]).

Extensions of the programming language Prolog by type concepts (e.g. [My84]), also relates to the area discussed in this paper.

1   The language and its interpretation

The purpose of this chapter is to introduce the underlying representation language. This is done by specifying the vocabulary of the language and the construction rules for well-formed expressions. Finally we outline how these expressions are interpreted. The notion of satisfaction is defined by providing a semantic definition of truth in the style of Tarski.

In the following for a set S , S^* denotes the set of all finite strings of elements of S including the empty string e. Pow(S) denotes the set of all subsets of S.

Syntax

Definition 1.1 :

A **many sorted language** L_S consist of the following :

- a nonempty set S of sort names

- for every s in S, a countably infinite set X_s of variable symbols (elements in X_s are called variables of sort s, and $X = \bigcup \{X_s \mid s \in S\}$)

- an $S^* \times S$ indexed family F of sets of function symbols

- an S^* indexed family P of sets of predicate symbols

- the propositional operators : $\wedge, \vee, \neg, \rightarrow$

- the quantifiers $\forall, \exists$

In the following we will consider many sorted languages with additional internal structure in the set of sorts.

Definition 1.2 :

A **sort hierarchy** Σ_S is a pair $\Sigma_S = (S, \leq)$, where S is partially ordered by $\leq$.

For a language L_S and a sort hierarchy Σ_S, the set T_Σ of all well-formed terms over Σ is defined inductively for every s in S :

$$T_{0,s} = \bigcup \{X_{s'} \mid s' \leq s\} \cup \bigcup \{F_{e,s'} \mid s' \leq s, e \in S^*\}$$

$$T_{n+1,s} = T_{n,s} \cup \{f(t_1, \dots, t_m) \mid f \in F_{(s_1, \dots, s_m, s')}, t_i \in T_{n,s_i} \text{ for } s' \leq s, m \geq 1\}$$

$$T_s = \bigcup \{T_{n,s} \mid n \in \mathbb{N}\}$$

$$T_\Sigma = \bigcup \{T_s \mid s \in S\}$$

Observe, that for every term t, there is a minimal sort s, such that t belongs to T_s. This s is called the sort of t, sort(t) = s.

The set **AT** of **atomic formulas** consist of expressions of the form:

- $p(t_1, \dots, t_n)$, for p in $P_{(s_1, \dots, s_n)}$ and t_i in T_{s_i}

The set **Fo** of first order **formulas** consist of :

- all atomic formulas

- expressions of the form $\phi_1 \wedge \phi_2$, $\phi_1 \vee \phi_2$, $\phi_1 \rightarrow \phi_2$, and $\neg \phi_1$, for formulas ϕ_1, ϕ_2

- expressions of the form $\forall x \phi, \exists x \phi$, for variables x and formulas ϕ

Semantics

For a many sorted language L_S a **structure** A_S (an L_S-interpretation) consist of

- an S indexed family of nonempty sets A_s
  ($U = \bigcup \{A_s \mid s \in S\}$ is called the universe of A_s)

- an $S^* \times S$ indexed family f of functions $f_{us}: F_{us} \to [A_u \to A_s]$
  (for $u = s_{i_1} \dots s_{i_n}$ and $A_u = A_{s_{i_1}} \times \dots \times A_{s_{i_n}}$)

- an S^* indexed family p of functions $p_u: P_u \to Pow(A_u)$.

The well-formed expressions of the language L_S are now given an interpretation $I = I_{AS}$ dependent on a structure A and an interpretation of the variables, where we require that, for variables x of sort s, $I(x)$ is an element of A_s
For terms t of form $f(t_1, \dots, t_n)$ we set $I(f(t_1, \dots, t_n)) = I(f)(I(t_1), \dots, I(t_n))$

The extension of I to all other formulas is carried out as in the case of usual predicate logic and will be omitted here.

The fact that a structure A_S satisfies a sentence ϕ is denoted by $A_S \models \phi$. As usual we call a sentence ϕ a logical consequence of a set F of sentences, in symbols $F \vdash \phi$, if for every structure A_s , such that $A_s \models \psi$ for every ψ in F, also $A_s \models \phi$ holds. F is called inconsistent, if there is no A_s, such that $A_s \models \psi$ for all ψ in F.

2   The calculus

In this chapter we will present a calculus for order sorted predicate logic, which is based on the tableau method ([Sm68]). For presentation purposes our formulation of the method as a calculus for many sorted logic is a modification of the classical calculus as presented in [Sm68; St87]. The actual implementation is a variation of that method, as it make use of a more efficient computation rule, which can be regarded as an extension of Prolog to full first order predicate logic [Sd85].

The tableau method

Informally a tableau is a binary tree with nodes labeled by formulas. For the construction of a tableau we distinguish four classes of formulas: α-formulas, β-formulas, γ-formulas and δ-formulas. The construction rules for a tableau associate with every formula (normally two) other formulas. The correspondence is given in the following four tables.

α- Formulas

α	α_1	α_2
$\phi \wedge \psi$	ϕ	ψ
$\neg(\phi \vee \psi)$	$\neg\phi$	$\neg\psi$
$\neg(\phi \rightarrow \psi)$	ϕ	$\neg\psi$
$\neg\neg\phi$	ϕ	-

β - Formulas

β	β_1	β_2
$\neg(\phi \wedge \psi)$	$\neg\phi$	$\neg\psi$
$\phi \vee \psi$	ϕ	ψ
$\phi \rightarrow \psi$	$\neg\phi$	ψ

γ - Formulas

γ	γ_1
$\forall x_s \phi(x_s)$	$\phi(x_s)$
$\neg \exists x_s \phi(x_s)$	$\neg\phi(x_s)$

δ - Formulas

δ	δ_1
$\neg \forall x_s \phi(x_s)$	$\neg\phi(x_s)$
$\exists x_s \phi(x_s)$	$\phi(x_s)$

A branch B of a tableau is called **closed**, if for a formula ϕ, both ϕ and $\neg\phi$ occur in B.

A branch B that is not closed, is called **open**. A tableau can only be extended by extending an open branch.

For a set of sentences F the basic notion is that of **a tableau for F** .

The construction of a tableau for F is now defined by the following steps

A tree consisting only of the root node labeled by an arbitrary formula taken from F is a tableau for F.

Let T be an already constructed tableau for F , B an open branch and ϕ a formula occuring as label somewhere in B.

1. If ϕ is an α-formula, the tableau T is extended by placing the new node n_1 below the leaf of branch B and by placing another new node n_2 below n_1, labeling n_i by α_i (If ϕ is a α-formula by the last row in the α-table, then only one node labeled by α_1 is added).

2. If ϕ is a β-formula, T is extended by continuing B at its leaf with two branches, one containing a new node labelled β_1 , and another one containing a new node labeled β_2 .

3. If ϕ is a γ-formula, with (bounded) variable x of sort s, then T is extended by one new node at the leaf of B labeled by $\gamma_1(t)$, where t is a ground term of T_s.

4. If ϕ is a δ-formula, T is extended by one new node at the leaf of B labeled by $\delta_1(c_s)$, where c_s is a new constant symbol of sort s, not occuring in the tableau so far.

5. The tableau T may be extended by adding a new node at the leaf of branch B labeled by an arbitrary formula from F.

The first four steps are called applications of the α-,β-,γ- or δ- rule respectively, while step five is called an application of the premise-rule.

To guarantee completeness of the method, we need to formulate an exhaustion condition for a branch.

A branch B of a tableau for S is called **exhausted**, if either :

- B contains all formulas from F as labels, or

- if an α-formula occurs as a label, then both α_1,α_2 occur as labels, or

- if a β-formula occurs as a label, then either β_1 or β_2 occurs, or

- if a γ-formula occurs as a label in B, then also $\gamma_1(t)$ occurs for every Σ-well-formed ground term t of the appropriate sort, that can be built by using constant and function symbols appearing in B, or

- if a δ-formula occurs as a label in B, then also $\delta_1(c_s)$ occurs for at least one term of sort s .

A tableau is **complete**, if all its branches are either closed or exhausted.

We note that if no γ-formula ever occurs as a label in a branch B, then the branch will be closed or exhausted after finitely many steps. If any γ-formula is present, then fulfilling the exhaustion condition may require infinitely many steps. In this case some kind of bookkeeping of already used formulas is required to ensure that no formula will be neglected in constructing a tableau. We will suppose such a fair procedure, which guarantes that if the process of extending a branch does not terminate, the resulting infinite branch will be exhausted (Smullyan's procedure for constructing a systematic tableau is an example of such a fair procedure).

We are now able to state the fundamental result :

Theorem 2.1 :

A set of (closed) formulae F is inconsistent iff there is a closed tableau for F.

The two implications of the theorem are referred to as the **soundness** (implication from right to left) and the **completeness** (implication from left to right) of the calculus.

To prove **soundness**, we have to show that the existence of a closed tableau for F implies the inconsistency of F.

It should be obvious by looking at the definition of the rules for constructing a tableau, that the rules preserve satisfiability. To be concrete we call a tableau T for F satisfiable, if there is at least one branch B in T such that $F \cup F_B$ is satisfiable, where F_B is the set of formulas occuring as labels in B. Now its easy to see, that for a satisfiable tableau T_1 , the tableau T_2 resulting from T_1 by applying one of the α-,β-,γ-,δ-rules or the premise-rule is also satisfiable. By induction it follows that for a satisfiable set F, at least one branch in the corresponding tableau is also satisfiable (note in particular, that the induction step involving an application of the δ-rule only works when the introduced constant symbol does not occur earlier on the branch).

Now, if T is a closed tableau for F, then for no branch B, F_B can be satisfiable, i.e. T is not satisfiable and therefore by the contraposition of the implication that has just been proved, F is inconsistent.

To prove **completeness**, we have to show that for every inconsistent set F, there is a closed tableau for F. We proceed by contraposition, i.e. we show that if there is no closed tableau for F then F is satisfiable.

We first note that if there is no closed tableau for F, then there is an exhausted open branch B, such that every formula in F occurs on B. This is ensured by the fairness condition for the tableau construction. The missing argument for the completion of the proof is the content of the following lemma:

Lemma (Hintikka) :

Every exhausted open branch B is satisfiable.

Proof:

The model we will construct consists of the (Herbrand) universe H of all Σ-wellformed ground terms that can be built from function symbols and constants appearing in B. The subset H_s, for sort symbol s, consists of all ground terms t, such that sort(t) = s.

An atomic ground formula $p(t_1, \dots , t_n)$ gets the truth value 'True', if $p(t_1, \dots , t_n)$ occurs as label in B, and 'False' otherwise. Note that not both $p(t_1, \dots , t_n)$ and $\neg p(t_1, \dots , t_n)$ can occur in B. We will now show by induction on the degree of formulas (number of occurrences of logical operator and quantifier) that every formula occurrying as a label in B is satisfiable under the given interpretation. Let ϕ be any formula in B and d its degree. The case d = 0 is obvious.

Now let d > 0 and suppose every formula in B of lower degree is true, then the following cases are consequences of the exhaustion condition:

- If ϕ is an α-formula, then α_1 and α_2 occur as labels and are of lower degree, hence both are true, hence ϕ is true.

- If ϕ is a β-formula, then β_1 or β_2 is in B and hence β_1 or β_2 is true by induction hypothesis, hence ϕ is true.

- If ϕ is a γ-formula, then for **every** Σ-well-formed ground term t of appropriate sort, $\gamma_1(t)$ is true, which implies the satisfiablity of ϕ.

- If ϕ is a δ-formula, then at least one $\delta_1(c_s)$ is true, hence ϕ must be true.

□

Corollary 2.2. :

Let F be a set of closed formulas, ϕ a single closed formula. Then $F \vdash \phi$ iff there is a closed tableau for $F \cup \{\neg \phi\}$.
□

3   Computational aspects

In this chapter we will discuss computational aspects related to an efficient realization of the calculus presented above.

Definition :
For an order sorted language L with sort hierarchy Σ a substitution $\theta = \{x_1/t_1, \dots, x_n/t_n\}$ is called a Σ-**substitution** (or well-sorted substitution), if $\Sigma \models sort(t_i) \leq sort(x_i)$

Later we will need the following technical lemma, which states that Σ-substitution behaves well:

Lemma :

If θ and τ are Σ-substitutions, then so is $\theta \circ \tau$

□

We first observe that from a computational point of view the formulation of the γ-rule is very inefficient. Undirected substitution of ground terms is obviously not the best way to check for complementary literals. A modification of the γ-rule resulting in a replacement of the universally quantified variable by a free variable of the same sort will allow the use of **unification** for checking the closure condition for a branch. Modifying the γ-rule also require a modification of the δ-rule. The modification is as follows:

- If ϕ is a γ-formula, with (bounded) variable x of sort s, then T is extended by one new node at the leaf of B labeled by $\gamma_1(y_s)$, where y_s is a new variable symbol of sort s.

- If ϕ is a δ-formula, including the free variables $x_1, \ldots, x_n$ with $sort(x_i) = s_i$, then T is extended by one new node at the leaf of B labeled by $\delta_1(f(x_1, \ldots, x_n))$, where f is a new function symbol in $F_{(s_1, \ldots, s_n, s)}$, not occuring in the tableau so far.

We then call a branch B of a tableau **closed** if there are formulas ϕ and $\neg\psi$ in B and a Σ-substitution θ such that $\theta(\phi) = \theta(\psi)$.

The proof of the correctness of the tableau proof-procedure thus modified is analogous to the correctness proof given before, since we may assume without loss of generality that the given closed tableau contains no more free variables.

To prove completeness of the modified version, we first need a modified definition of an exhausted branch. Only the clause for γ-formulas needs to be changed:

- If a γ-formula occurs as a label in B, then $\gamma_1(y_i)$ occurs for infinitely many variables $y_i \in X_s$.

Now let B be an exhausted, open branch in the modified tableau and θ be a Σ- substitution from the set of all variables y_i occuring in B, such that

- for every term t of sort s', there is a variable y of sort s, with $s' \leq s$, such that $\theta(y) = t$.

Let B_θ be the branch arising from B by changing all labels ϕ in B to $\theta(\phi)$. Then B_θ is an exhausted open branch in the original sense and satisfiability of B_θ follows as before.

As an example showing that even the modified γ-rule may have to be applied more than once, consider the unsatisfiable set

$$F = \{\forall x(p(x) \rightarrow p(f(x))), p(a), \neg p(f(f(a)))\}$$

A closed tableau for F may look like this:

$$\neg p(f(f(a)))$$

$$p(a))$$

$$\forall x(p(x) \rightarrow p(f(x)))$$

$$p(y) \rightarrow p(f(y))$$

$\neg p(y)$	$p(f(a))$
closed by	$p(z) \rightarrow p(f(z))$
$\{y/a\}$	

$\neg p(z)$	$p(f(f(a)))$
closed by	closed
$\{z/f(a)\}$	

In the case of unsorted first order logic, it is well known that if a unifier exists then there exists a most general unifier (mgu) such that each other unifier is an instance of the mgu. For order sorted logic however this is in general not true. Consider the following example. Let $t_1 \equiv x_{s_1}$ and $t_2 \equiv y_{s_2}$. And let the partial order of $S = \{s_0, s_1, s_2, s_3, s_4, s_5\}$ be as in the following diagram

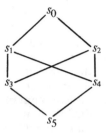

In this situation, there are two unifiers, namely z_{s_3} and z_{s_4}, and neither is an instance of the other.

It turns out that if one excludes situations as in the above diagram by the requirement that for every subset of S there exists a greatest lower bound, then for any Σ-unifiable termset a unique most general unifier exists.

Theorem ([Wa88])

If Σ is a meet semilattice, then for Σ-unifiable termsets there always exists a most general Σ- unifier.
□

For sort hierarchies $\Sigma_S = (S, \leq)$ which are meet semilattices, the algorithm computing the most general Σ-unifier for Σ-well-formed terms, requires in its essential step the computation of greatest lower bounds for sort symbols (see e.g. [Wa83]). Looking at the unifiability problem as the problem of solving certain kinds of equations, the corresponding solution sets consists of equations of the form $x_{s_1} = t_{s_2}$, which are solvable either if $t \in X_{s_2}$ and $\mathrm{glb}(s_1, s_2) \neq \perp$, or if x does not occur in t and $s_2 \leq s_1$.

The requirement that Σ has to be a meet semilattice is not too restrictive, as the following theorem ([McN36]) shows:

Theorem
Every partial ordering $\Sigma = (S, \leq)$ can be extended to a lattice $\Sigma^* = (S^*, \leq)$ with $S \subseteq S^*$, and existing infima and suprema in S are preserved.

The proof of the above theorem is constructive in the sense that one can extract an algorithm which computes a corresponding lattice structure for a given partial ordering.

In the following we will now briefly discuss an alternative approach, which deals with an implementation of our concepts in Prolog.
Using Prolog as the implementation language it is natural to use the built in (Robinson) unification as far as possible. It turns out that one can code information concerning the order structure into appropriate data structure, such that under some restrictions, sorted unification reduces to (Robinson) unification. We restrict our considerations to order structures which can be represented in form of a tree. In other words, sorts s_1 and s_2 do not have common subsorts unless

$s_1 \leq s_2$ or $s_2 \leq s_1$. For the coding of the order structure we use incomplete date structures as provided by Prolog ([St86]).

For an example let us consider the following sort hierarchy

For every sort s_i in the tree the representation is such that starting from the root s_1 the whole branch from s_1 to s_i forms an incomplete data structure $[s_1, \dots , s_i \mid X]$, which is a list whose head contains $s_1, \dots , s_i$ as elements and whose tail is the Prolog variable X.

One can easily verify that with this representation, order sorted unification reduces to Robinson unification.

Summary and open problems

We have presented a sound and complete calculus for order sorted logic based on the tableau method.
In addition we discussed implementation issues related to an efficient realization of our concepts.

However there is an unsatisfactory situation that arises when one consider other deductive calculi for (variants) of order sorted logic. Various variants of sorted logic (e.g. with more expressive sort structure) are discussed in the literature. In introducing a new logic , one has to prove e.g. soundness and completeness results. Doing this from scratch over and over again is unnecessary, if these results may be reconstructed from general results, when the underlying principles are known.

To underscore the current lack of a remedy for this situation, we state the following

Research Problem:

What is an adequate general framework for establishing results (e.g. soundness and completeness) for variants of many sorted predicate logic ?

References

[Ai86]   H. Ait-Kaci, R. Nasr. *LOGIN: A Logic Programming Language with Built-In Inheritance.* J. of Logic Programming 3, pp. 185-215 ,1986.

[Co87]   A.G. Cohn. *A More Expressive Formulation of Many-Sorted Logic.* Journal of Automated Reasoning 3,2, pp.113-200,1987.

[Eh85]   H. Ehrig and B. Mahr. *Fundamentals of Algebraic Specification 1.* Springer-Verlag, 1985.

[Fr]   A. Frisch. *A General Framework for Sorted Deduction.* Fundamental Results on Hybrid Reasoning (Draft)

[Ga78]   H. Gallaire and J. Minker (Ed.). *Logic and Databases.* Plenum Press, New York, 1978.

[He]   J. Herbrand. *Logical Writings.* Edited by W. Goldfarb, D. Reidel Publishing Company.

[LILOG88] C. Beierle et al. *The Knowledge Representation Language L_{LILOG}.* LILOG Report 41, IBM Germany, Stuttgart, 1988.

[McN36]   A. M. MacNeille. *Partially ordered sets.* Transactions of the American Math. Society, Vol. 42, pp.416-460.

[My84]   A. Mycroft, R.A. O'Keefe. *A Polymorphic Type System for Prolog.* Artifcial Intelligence 23,3, pp.295-307, 1984.

[Ob62]   A. Oberschelp. *Untersuchungen zur mehrsortigen Quantorenlogik.* Mathematische Annalen 145, pp.297-333, 1962.

[St87]   P.H. Schmitt. *The THOT Theorem Prover.* IBM Germany, Heidelberg Scientific Center, TR 87.09.007, 1987

[Sd85]   W. Schönfeld. *Prolog extensions based on tableau calculus.* Proc. 9th Int. Conf. Artificial Intelligence, Los Angeles, Vol.2 pp.730-732, 1985.

[Sm68]   R.M. Smullyan. *First-order logic.* Ergebnisse der Mathematik Bd. 43, Springer-Verlag, 1968.

[St86]   L. Sterling, E. Shapiro *The Art of Prolog.* Advanced Programming Techniques.

[Wa83]   C. Walther. *A many sorted calculus based on resolution and paramodulation.* Proc. of the 8th Int. Conf. Artificial Intelligence, Karlsruhe, 1983.

[Wa88]   C. Walther. *Many sorted unification.* JACM Vol. 35, No.1, pp. 1-17, 1988.

[We88]   W. Wernecke et al. *Das wissensbasierte System KEYSTONE.* Informatik Forsch. Entw. (1988) 3, 153-163.

A Calculus for Order-Sorted Predicate Logic with Sort Literals

U. Hedtstück
IBM Germany Scientific Center
Institute for Knowledge Based Systems
Project LILOG
POBox 80 0880
D-7000 Stuttgart 80
hdstueck at ds0lilog.bitnet

P. H. Schmitt
Institute for Logic, Complexity, and Deduction Systems
University of Karlsruhe
POBox 6980
D-7500 Karlsruhe 1

1 Introduction

In the LILOG project the knowledge representation formalism L_{LILOG} was developed for natural language understanding [Beierle *et al.*, 1988]. L_{LILOG} is a logical representation formalism which is based on first order order-sorted predicate logic [Walther, 1987] incorporating a representation formalism for taxonomical information, based on ideas coming from the KL-ONE family of representation formalisms [Brachman and Schmolze, 1985].

In L_{LILOG} there are two principal possibilities to represent sort information. The first one is within the framework of order-sorted logic. Sorts are declared for terms by restricting variables to sorts and attaching domain sorts and range sorts to the function symbols. In addition, the argument positions of predicate symbols are restricted by sorts. The set of sorts is partially ordered by the subsort relation.

In [Walther, 1987] a resolution based reasoning system for order-sorted predicate logic was presented together with necessary and sufficient conditions for the sort hierarchy which guarantee soundness and completeness of the reasoning system.

The second representation principle for sort information in L_{LILOG} is to use sorts as unary predicate symbols in the predicate logic language by means of so-called *sort literals*. This allows for explicit reasoning about sort information contained in the sort hierarchy. Since this may eventually alter the set of possible models, special reasoning mechanisms have to be provided.

In this paper we want to present a calculus for a logic which takes into account sort information which is represented both in the declarations and by means of sort literals. The inference rules for this calculus depend strongly on the way the sort hierarchy is interpreted. In [Cohn, 1987] Cohn presented the LLAMA system which is based on such a logic where the sort hierarchy is interpreted in a *lattice theoretic* way. This means essentially that greatest lower bounds are interpreted by intersections. In our logic the sort hierarchy is interpreted *order theoretically*, where the only restriction for interpretations is the stipulation that subsort relationships be interpreted by subset relationships.

After introducing the formalism for our logic we present a resolution based reasoning system. In order to take into account during unification the sort information expressed by means of sort literals we allow substitutions which lead to terms being not well-sorted. Additional sort constraints guarantee well-sortedness implicitly. This is an idea which was presented by A. Oberschelp in [Glubrecht *et al.*, 1983], and which is also used in Cohn's LLAMA system [Cohn, 1987]. The rest of this paper is devoted to the proof that our reasoning system is sound and complete.

2 The Formalism

2.1 Syntax

In this section we define the formalism for our logic. For the definition of the basic notions we follow [Beierle *et al.*, To appear].

Definition 1 (Signature): An (order-sorted) *signature* $\Sigma = (\mathbf{S}, \mathbf{P}, \mathbf{F})$ consists of

1. a partially ordered set of sorts $(\mathbf{S}, \leq)$ with least element $\bot$ and greatest element $\top$. $(\mathbf{S}, \leq)$ is called the *sort hierarchy*. We require that there are no infinite increasing chains in $(\mathbf{S}, \leq)$.

2. an $\mathbf{S}^* \times \mathbf{S}$-indexed family of sets of function symbols $(\mathbf{F}_{w,S})_{w \in \mathbf{S}^*, S \in \mathbf{S}}$.

   For $f \in \mathbf{F}_{w,S}$ we write $f : S_1 \ldots S_n \to S$ where $w = S_1 \ldots S_n$.

   w is called the *arity* and S the *coarity* or *target sort* of f. The elements of w are the *argument sorts* of f.

3. an $\mathbf{S}^*$-indexed family of sets of predicate symbols $(\mathbf{P}_w)_{w \in \mathbf{S}^*}$.

   For $p \in \mathbf{P}_w$ we write $p : S_1 \ldots S_n$ where $w = S_1 \ldots S_n$

We assume that the sets of function and predicate symbols are disjoint. The least element $\bot$ may not occur in the arities and coarities of $\mathbf{F}$ and $\mathbf{P}$. Moreover, in order to avoid problems with *empty sorts* ([Goguen and Meseguer, 1986]) we assume that for every sort (except $\bot$) there is at least one ground term of this sort (see the definition of Σ-terms).

Σ is a signature with *sort predicates* if for every sort S in $\mathbf{S}$ there is a unary predicate, also denoted by S, with argument sort $\top$, i.e. $S \in \mathbf{P}_\top$.

Definition 2 (Well-Sorted Terms): Given a signature $\Sigma = (\mathbf{S}, \mathbf{P}, \mathbf{F})$, a family of *sorted variables* over Σ is an $\mathbf{S}$-indexed family V of variables $V = \{V_S \mid S \in \mathbf{S}\}$ where $V_\bot = \emptyset$. The family of *well-sorted terms* T_{ws} over Σ and V is the least $\mathbf{S}$-indexed family of sets such that

1. $v_S \in T_{ws}$ for every $v_S \in V_S$; the sort of v_S is S.

2. $c \in T_{ws}$ for every $c \in \mathbf{F}_{\epsilon,S}$ where ϵ is the empty string of sorts; the sort of c is S.

3. if $t_1, \ldots, t_n$ are terms in T_{ws} of sort $S'_1, \ldots, S'_n$, respectively, then $f(t_1, \ldots, t_n) \in (T_{ws})_S$ if $f \in \mathbf{F}_{S_1 \ldots S_n, S}$ and $S'_i \leq S_i$ for $i \in \{1, \ldots, n\}$; the sort of $f(t_1, \ldots, t_n)$ is S.

For any well-sorted term t we represent its sort (given by the previous definition) by the function $sort(t)$. A *well-sorted ground term* is a well-sorted term in which no variable occurs.

Let T_{us}, the set of *unsorted terms*, be the set of terms which is built using the set of sorted variables V and the function symbols of Σ without taking into account the sort restrictions. Clearly, T_{us} is a superset of T_{ws}.

Definition 3 (Well-Sorted Formulas): The set Fml_{ws} of *well-sorted formulas* over Σ and V is defined by:

1. Let $t_1, \ldots, t_n$ be well-sorted terms of sort $S_1', \ldots, S_n'$, respectively, and let $p \in \mathbf{P}_{S_1 \ldots S_n}$ with $S_i' \leq S_i$ for $i \in \{1, \ldots, n\}$. Then $p(t_1, \ldots, t_n)$ is in Fml_{ws}.

2. If F is a well sorted formula, so is $\neg F$.

3. If F_1 and F_2 are well-sorted formulas, so are $F_1 \wedge F_2$, $F_1 \vee F_2$, and $F_1 \rightarrow F_2$.

4. If x is a sorted variable of sort S and F is a well-sorted formula, so are $\forall x{:}S\ F$ and $\exists x{:}S\ F$.

If, in the above definition, we do not take into account the sort restrictions, then we get a superset Fml_{us} of formulas which we call the set of *unsorted formulas*.

We use brackets in the usual way, or omit them according to the usual preference rules. The respective notions for *atomic formulas*, *literals* and *clauses* are defined in a straightforward manner.

If K_1, K_2 are two clauses, we call K_2 a *subclause* of K_1 and write $K_2 \subseteq K_1$ or $K_1 \supseteq K_2$, if every literal in K_2 also occurs in K_1.

If Σ is a signature with sort predicates, then the literals, in which the predicate symbol is a sort, are called *sort literals* (in [Cohn, 1987] the term *characteristic literal* is used).

2.2 Semantics

We provide our formalism with a set theoretic semantics, which means that sorts are interpreted by subsets of a given universe. The bottom sort $\perp$ is interpreted by the empty set, the top sort $\top$ is interpreted by the nonempty universe. Any other sort is interpreted by a nonempty subset of the universe. The subsort relationship is interpreted by the respective subset relationship. It is then clear how the sort declarations for function and predicate symbols restrict the possible interpretations for function and predicate symbols (for more details see [Beierle *et al.*, To appear]).

Let Σ be a signature with sort predicates. For a sort S and a model A, let A_S be the set which interprets the sort S in A, and let S_A be the unary relation on the universe of A which interprets the sort predicate S (recall that we denote a sort and its sort predicate by the same symbol). Then we assume that the set of elements a of A, for which $S_A(a)$ holds, is exactly the set A_S.

We are mainly interested in well-sorted formulas, but at intermediate stages unsorted formulas also arise. How do we interpret these?

We use the same class of models as just introduced and the same definitions without any changes. The interpretation f_A of a unary function f of target sort S and argument sort S' is a partial function on the universe of A with domain $A_{S'}$. For $a \notin A_{S'}$, $f_A(a)$ is not defined. If c is a constant symbol of sort S_1, $S' \not\leq S_1$, then $f(c)$ is not well-sorted. But $(f(c))_A$ may still be defined in A if $c_A \in A_{S'}$.

For emphasis let us repeat the truth definition for an atomic formula $p(t)$ in the model A. If p is a unary predicate symbol of sort S, then the interpretation p_A of p in A is of course a subset of A_S, $p_A \subseteq A_S$. Let us further assume that t is a ground term of sort S' with $S' \not\leq S$. Therefore $p(t)$ is not well-sorted. If t_A is the interpretation of t in A, the truth definition reads:

$$A \models p(t) \quad iff \quad t_A \in p_A.$$

In case $A \models p(t)$, then, of course, $A \models S(t)$, and also $t_A \in A_S$; in particular, t_A will be defined.

2.3   Interpretations of a Sort Hierarchy

Whereas the sort hierarchy of a signature is considered to be static, the sort information contained in the sort hierarchy may be interpreted in different ways.

One way is to interpret different sorts of the sort hierarchy by different sets in a model, the subsort relationships by corresponding subset relationships, and greatest lower bounds (in case they exist) by intersections. This is what we would like to call a *lattice theoretic interpretation* of a sort hierarchy.

If we do not require to interpret different sorts by different sets and greatest lower bounds by intersections, then we speak of an *order theoretic interpretation* of a sort hierarchy. In such an interpretation subsorts still have to be interpreted by corresponding subsets, but different sorts may be interpreted by the same set, and greatest lower bounds may be interpreted by subsets of the corresponding intersection in a model.

Example: Consider the simple sort hierarchy

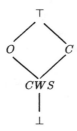

where

$$
\begin{aligned}
C &= \text{sort of all churches} \\
O &= \text{sort of all old buildings} \\
CWS &= \text{sort of all churches worth seeing}
\end{aligned}
$$

Then the model with the universe

$$A = \{St.\ Peter,\ St.\ Mary,\ St.\ Catherine,\ town_hall,\ theatre\}$$

and sort interpretation

$$
\begin{aligned}
A_C &= \{St.\ Peter,\ St.\ Mary,\ St.\ Catherine\} \\
A_O &= \{St.\ Peter,\ St.\ Catherine,\ town_hall\} \\
A_{CWS} &= \{St.\ Peter\}
\end{aligned}
$$

is perfectly admissible according to our definitions. But $A_{CWS} \subset A_O \cap A_C$ would violate a lattice theoretic interpretation.

Whereas sort information expressed by means of sort literals has no influence on the lattice theoretic interpretation of a sort hierarchy, this is not true if a sort hierarchy is interpreted order theoretically.

For example, if two given sorts are incomparable in the sort hierarchy, then, under an order theoretic interpretation, the set of models is restricted by axioms, which imply that the two sorts either are equal, or one of them is a subsort of the other. If the sort hierarchy is interpreted lattice theoretically, then such formulas obviously lead to inconsistencies, since greatest lower bounds in the static sort hierarchy never change, whereas the intersections in a model may be changed.

2.4 Substitutions

A *well-sorted substitution* σ is a mapping from a subset $dom(\sigma)$ of the set of all sorted variables V into the set of well-sorted terms T_{ws}.

We assume throughout this paper that $dom(\sigma)$ is finite. For brevity, we mostly speak of *substitutions* instead of well-sorted substitutions.

The following three classes of substitutions will be of interest in our context:

$$
\begin{aligned}
Subst &= \text{the set of all substitutions} \\
OSubst &= \text{the set of all substitutions } \sigma, \\
&\quad \text{such that for all } x \in dom(\sigma), \\
&\quad sort(\sigma(x)) \le sort(x) \\
Subst_\top &= \text{the set of all substitutions } \sigma, \\
&\quad \text{such that for all } x \in dom(\sigma), \\
&\quad \text{for which } \sigma(x) \text{ is a variable,} \\
&\quad sort(\sigma(x)) = \top.
\end{aligned}
$$

The substitutions of $OSubst$ are called *order-sorted substitutions* in this paper.

For substitutions σ, ϱ, by $\sigma\varrho$ we denote the concatenation of substitutions, i.e., $\sigma\varrho(x) = \sigma(\varrho(x))$.

For an arbitrary substitution σ the deviation from being order-sorted is compensated by the (well-sorted) formula

$$
SL(\sigma) = \bigwedge\{S_i(\sigma(x)) : x \in dom(\sigma) \text{ and } sort(\sigma(x)) \not\le sort(x) = S_i\}
$$

3 Inference Rules

In this section we present a resolution-based deduction system for our order-sorted logic with sort literals. The deduction system is designed under the assumption that the sort hierarchy is interpreted order theoretically.

We shall not consider the treatment of equality, nor the factorization problem, which in ordinary resolution has to be taken into account for completeness, and for which in our calculus a corresponding inference rule would be necessary, too.

The first inference rule, called the *extended order-sorted resolution rule* or *EOS resolution rule* allows us to resolve two literals, although the sort hierarchy does not contain enough information to guarantee the compatibility of the respective argument sorts. For a unifying substitution the deviation of being order-sorted is expressed by a conjunction of sort literals as described in the previous section. By this, the proof for compatibility is left to the reasoning mechanism which operates on the predicative level.

EOS Resolution Rule: Suppose we have the two clauses $L_1 \vee A$ and $L_2 \vee B$ where L_1 and L_2 are two literals with the same predicate symbol and with different signs, and A and B are disjunctions of literals. If $\sigma \in Subst_\top$ is such that $\sigma(L_1)$ and $\sigma(L_2)$ are complementary literals, then the two clauses may be resolved yielding the resolvent $\sigma(A) \vee \sigma(B) \vee \neg SL(\sigma)$.

For the inference rules we use a schematic representation

EOS Resolution Rule:

$$
\frac{L_1 \vee A, L_2 \vee B}{\sigma(A \vee B) \vee \neg SL(\sigma)}
$$

Allowing for substitutions of $Subst_T$, the terms occuring during a proof are not always well-sorted. But if we generate for any pair x and t, where $sort(x) = S_1$, the sort literal $S_1(t)$, we force any evaluation of the term t, that leads to a successful refutation, to be of the sort S_1. This implicitly guarantees well-sortedness. If the conjunction $SL(\sigma)$ of generated sort literals is the empty conjunction then σ is an order-sorted substitution. In this case our inference rule coincides with order-sorted resolution (cf. [Walther, 1987]).

Remark: The EOS resolution rule may be viewed as partial theory unification w.r.t. the theory which is given by the sort hierarchy and by the set of all formulas of a given knowledge base which are connected to a formula containing at least one sort literal [Stickel, 1985].

Now we formulate the *subsort resolution rule* (which is a generalization of the subsort resolution rule described in [Bollinger *et al.*, 1988]) as follows:

Let A and B be disjunctions of literals and $S_1(t_1)$, $S_2(t_2)$ be two sort literals. Then subsort resolution is given by

Subsort Resolution Rule:

$$\frac{\neg S_1(t_1) \vee A, S_2(t_2) \vee B}{\sigma(A \vee B) \vee \neg SL(\sigma)}$$

where $\sigma \in Subst_T$ such that $\sigma(t_1) = \sigma(t_2)$ and $S_2 \leq S_1$

In order to detect inconsistencies within a sort literal, which may be recognized neither by the EOS resolution rule nor by the sort resolution rule, we need the following *elimination rule*:

Elimination Rule:

$$\frac{\neg S_1(t) \vee A}{\sigma(A) \vee \neg SL(\sigma)}$$

where $\sigma \in Subst_T$ and $sort(\sigma(t)) \leq S_1$.

Remark: Sort literals disappear either by the elimination rule, or by EOS resolution, if the sorts used as predicate symbols are identical, or by subsort resolution. Of course, by means of the conjunction $SL(\sigma)$, sort literals are newly introduced.

4   Soundness and Completeness of the Inference Rules

In this section we present proofs for the soundness and completeness of our calculus. As usual we prove the propositional case and extend this result to the predicate logic case by means of a lifting lemma.

In case the substitution σ involved in the inference rules of the last section is empty we call the corresponding version the *PL-case (propositional logic case)* of the rule. These are of the form:

$$\mathbf{PL - EOS} : \quad \frac{L \vee A, \neg L \vee B}{A \vee B}$$

$$\textbf{PL} - \textbf{SUBS} : \quad \frac{\neg S_1(t) \vee A,\, S_2(t) \vee B}{A \vee B}$$
$$\text{where } S_2 \leq S_1.$$

$$\textbf{PL} - \textbf{E} : \quad \frac{\neg S_1(t) \vee A}{A}$$
$$\text{where } sort(t) \leq S_1.$$

The following lemma provides a crucial step towards the appropriate version of the lifting lemma.

Lemma 1: Let $\varrho \in Subst_\top$, $\sigma \in Subst$, then $\neg SL(\sigma\varrho)$ is a subclause of $\neg\sigma(SL(\varrho)) \vee \neg SL(\sigma)$.

Proof: Let $\neg S_i(\sigma\varrho(x))$ be a literal in $\neg SL(\sigma\varrho)$. By definition of $SL(\sigma\varrho)$ we must have $sort(\sigma\varrho)) \not\leq sort(x) = S_i$. If $sort(\varrho(x)) \not\leq sort(x)$ happens to be true, we get $notS_i(\varrho(x))$ as a literal in $\neg SL(\varrho)$, and therefore $\neg S_i(\sigma\varrho(x)) = \neg\sigma(S_i(\varrho(x)))$ as a literal in $\neg\sigma(SL(\varrho))$.

Now we consider the case $sort(\varrho(x)) \leq sort(x)$. Then $\varrho(x)$ has to be a variable, since otherwise $sort(\varrho(x)) = sort(\sigma\varrho(x))$ would contradict the fact that $sort(\sigma\varrho(x)) \not\leq sort(x)$.

Furthermore we must have $\varrho(x) = x$. Otherwise $sort(\varrho(x)) = \top$ has to be true, since $\varrho \in Subst_\top$. But this would entail $sort(x) = \top$ making $sort(\sigma\varrho(x)) \not\leq sort(x)$ impossible. From $\varrho(x) = x$ we obtain $sort(\sigma\varrho(x)) = sort(\sigma(x)) \not\leq sort(x)$, and thus $\neg S_i(\sigma\varrho(x)) = \neg S_i(\sigma(x))$ occurs in $\neg SL(\sigma)$. $\quad\square$

To prove completeness of the proposed inference system we try to follow the usual procedure to reduce the problem first to the propositional, i.e. quantifierfree case, and then solve the propositional case.

The usual way to obtain ground clauses by application of ground substitutions does not work immediately. We need the following concept.

For $\sigma \in Subst$ and a clause A we denote by $[\sigma]A$ the clause $\sigma(A) \vee \neg SL(\sigma)$. For a set D of clauses $[Subst]D$ denotes the set $\{[\sigma]A : \sigma \in Subst,\ A \in D\}$.

For a set D of clauses we define

$$
\begin{aligned}
Subst_{SL}(D) &= \{\sigma(K) : \text{for } K \in D, \\
&\qquad \sigma \in Subst_\top, \text{ such that } \sigma(K) \text{ is variablefree}\} \\
OSubst(D) &= \{\sigma(K) : \text{for } K \in D, \\
&\qquad \sigma \in OSubst \text{ and } \sigma(K) \text{ is variablefree}\}
\end{aligned}
$$

We will prove a slightly stronger completeness result, in that the application of the elimination rule, which infers $\sigma(A) \vee \neg SL(\sigma)$ from $\neg S_1(t) \vee A$, is permitted only if, in addition to the condition $sort(\sigma(t)) \leq S_1$, the term $\sigma(t)$ is not a variable. We call this the *restricted elimination rule*.

Now, continuing the above definitions, we define

$$
\begin{aligned}
I(D) \;=\; &\text{the smallest set of clauses containing } D, \text{ and which is} \\
&\text{closed under the EOS-, SUBS- and the restricted E-rule}
\end{aligned}
$$

Lemma 2:

1. For every clause R in $I(D)$ and every ground substitution σ there is a subclause R_1 of $[\sigma](R)$, such that $R_1 \in I([Subst](D))$.

2. If M is a model for all formulas in $I([Subst](D))$, then M is also a model for all formulas in $OSubst(I(D))$.

Proof:

(1) We proceed by induction on the definition of $I(D)$.

If $R \in D$, then $R_1 = [\sigma](R) \in [Subst](D) \subseteq I([Subst](D))$. For the induction step three cases have to be considered.

Case 1: R is an EOS-resolvent of $K_1 = L_1 \vee A$ and $K_2 = \neg L_2 \vee B$ in $I(D)$.

For some $\varrho \in Subst_T$ we thus have $R = \varrho(A \vee B) \vee \neg SL(\varrho)$ and $\varrho(L_1) = \varrho(L_2)$. By induction hypothesis there are subclauses K_1' of $[\sigma\varrho](K_1)$, K_2' of $[\sigma\varrho](K_2)$, such that K_1', $K_2' \in I([Subst](D))$. We try to apply the EOS-rule on the pair K_1', K_2'. Since K_1' is only a subclause of $[\sigma\varrho](K_1)$ it might not contain the literal $\sigma\varrho(L_1)$, i.e., K_1' is a subclause of $\sigma\varrho(A) \vee \neg SL(\sigma\varrho)$. But then, by Lemma 1, K_1' is already a subclause of $[\sigma](R) = \sigma\varrho(A \vee B) \vee \neg\sigma(SL(\varrho)) \vee \neg SL(\sigma)$.

Analogously, K_2' is already a subclause of $[\sigma](R)$ if it does not contain the literal $\neg\sigma\varrho(L_2)$.

If both cases just considered do not hold, then we may use the PL-EOS-rule to obtain the resolvent $R_1 \in I([Subst](D))$ from K_1' and K_2'. R_1 will be of the form $\sigma\varrho(R_0) \vee \neg SL(\sigma\varrho)$, where R_0 is a subclause of $A \vee B$. By Lemma 1 R_1 is a subclause of $[\sigma](R)$.

Case 2: R is a SUBS-resolvent of $K_1 = \neg S_1(t_1) \vee A$ and $K_2 = S_2(t_2) \vee B$ in $I(D)$.

For some $\varrho \in Subst_T$ we thus have $R = \varrho(A \vee B) \vee \neg SL(\varrho)$, $\varrho(t_1) = \varrho(t_2)$ and $S_2 \leq S_1$. By induction hypothesis there are subclauses K_1' of $[\sigma\varrho](K_1)$, K_2' of $[\sigma\varrho](K_2)$, such that K_1', $K_2' \in I([Subst](D))$. It can be seen from this point that the argument parallels exactly that of case 1.

Case 3: R is a restricted-E-resolvent of $K = \neg S_1(t) \vee A$ in $I(D)$.

There is thus some $\varrho \in Subst_T$ satisfying $R = \varrho(A) \vee \neg SL(\varrho)$ and $sort(\varrho(t)) \leq S_1$. By induction hypothesis there is a subclause K' of $[\sigma\varrho](K)$ in $I([Subst](D))$. If the literal $\neg\sigma\varrho(S_1(t))$ does not occur in K', then K' is a subclause of $\sigma\varrho(A) \vee \neg SL(\sigma\varrho)$, and therefore, by Lemma 1, also of $[\sigma](R) = \sigma\varrho(A) \vee \neg\sigma(SL(\varrho)) \vee \neg SL(\sigma)$. If $\neg\sigma\varrho(S_1(t))$ occurs in K', then we have $sort(\sigma\varrho(t)) = sort(\varrho(t))$, since $\varrho(t)$ is not a variable, and the restricted PL-E-rule is applicable on K' yielding R_1 in $I([Subst](D))$. R_1 will be a subclause of $\sigma\varrho(a) \vee \neg SL(\sigma\varrho)$, and thus, again by Lemma 1, a subclause of $[\sigma](R) = \sigma\varrho(A) \vee \neg\sigma(SL(\varrho)) \vee \neg SL(\sigma)$.

(2) Let $\sigma(K)$ be a formula in $OSubst(I(D))$. By part (1) there is a subclause K' of $[\sigma](K)$ in $I([Subst](D))$. Since σ is in $OSubst$ $[\sigma](K)$ equals $\sigma(K)$. Since K' is true in M this also holds for the superclause $\sigma(K)$ of K'. $\square$

Lemma 3:

1. For every clause R in $I([Subst](D))$ there is a clause $K \in I(D)$ and a substitution σ such that $[\sigma](K)$ is a subclause of R.

2. If the empty clause is not in $I(D)$, then it is also not in $I([Subst](D))$.

Proof:

(1) We proceed by induction on the definition of $I([Subst](D))$.

If $R \in [Subst](D)$, then $R = [\varrho](K) \vee \neg SL(\varrho)$ for some $K \in D$, and the claim is satisfied for $\sigma = \varrho$.

In the inductive step there are again three cases to be considered.

Case 1: R is an PL-EOS-resolvent of $R_1 = L_1 \vee A$ and $R_2 = \neg L_1 \vee B$ in $I([Subst](D))$

By induction hypothesis there are substitutions σ_1, σ_2 and K_1, $K_2 \in I(D)$, such that $[\sigma_1](K_1) \subseteq R_1$ and $[\sigma_2](K_2) \subseteq R_2$.

We may assume that K_1 and K_2 do not have free variables in common, and thus use one substitution σ instead of σ_1 and σ_2. If either L_1 is not a literal in $[\sigma](K_1)$, or $\neg L_2$ fails to occur in $[\sigma](K_1)$, then $[\sigma](K_1) \subseteq R$, or $[\sigma](K_2) \subseteq R$, respectively, and we are through. Otherwise, the EOS-rule can be applied to K_1, K_2 using σ as a unifying substitution. Let K be the resolvent. Then $K \in I(D)$, and $[id](K) = K \subseteq R$.

Case 2: R is a PL-SUBS-resolvent of $R_1 = \neg S_1(t) \vee A$ and $R_2 = S_2(t) \vee B$ in $I([Subst](D))$.

The proof exactly parallels that of case 1.

Case 3: R is a restricted PL-E-resolvent of $\neg S_1(t) \vee A$ in $I([Subst](D))$.

By induction hypothesis there is some $K_1 \in I(D)$ and a substitution σ such that $[\sigma](K_1) \subseteq \neg S_1(t) \vee A$. If $\neg S_1(t)$ does not occur in $[\sigma](K_1)$, then we have already $[\sigma](K_1) \subseteq R$. Otherwise, we use the restricted E-rule on K_1 with unifying substitution σ obtaining a resolvent $R_1 \in I(D)$ with $[id](R_1) = R_1 \subseteq R$.

(2) Assume for the sake of a contradiction that the empty clause $\square$ is in $I([Subst](D))$. By part (1) $[\sigma](K) \subseteq \square$ for some $K \in I(D)$, which forces K to be the empty clause. $\square$

Example: Let S_3 be a common subsort of S_1 and S_2, c a constant of sort S_3, and

$$D = \{\forall x : S_1 \; A(x), \forall y : S_2 \; \neg A(y), S_1(c), S_2(c)\}.$$

There are two ways to deduce the empty clause $\square$ from D.

First we compute the most general unifier σ of $A(x)$ and $A(y)$.

$$\sigma(x) = \sigma(y) = z, \; with \; sort(z) = \top \, .$$

The resolvent is $\neg S_1(z) \vee \neg S_2(z)$. By two resolution steps using $S_1(c)$ and $S_2(c)$ we arrive at $\square$.

In the second computation we use the substitution

$$\mu(x) = \mu(y) = c \, ,$$

which is well-sorted, and obtain $\square$ right away.

Theorem 4 (Completeness Theorem): A set D is satisfiable, iff the empty clause $\square$ is not contained in $I(D)$.

Proof: The soundness part of the theorem, i.e., the claim that satisfiability of D implies the satisfiability of $I(D)$, and therefore $\square \notin I(D)$, is easily proved. We give the details for the EOS-rule only.

Let A be a model for

$$\forall x_1 : S_1...\forall x_n : S_n(L_1 \vee A)$$
$$\forall x_1 : S_1...\forall x_n : S_n(L_2 \vee B)$$

and $\sigma \in Subst_T$ such that σ unifies the atoms of L_1 and L_2. We claim that A is also a model of

$$\forall z_1 : T_1...\forall z_k : T_k(\sigma(A \vee B) \vee \neg SL(\sigma))$$

where $z_1,\ldots,z_k$ are all variables in $\sigma(A \vee B)$.

Let $a_1,\ldots,a_k$ be arbitrary elements of A. If for some variable x of sort S in $dom(\sigma)$

$$A \models \neg S(\sigma(x))[a_1,\ldots,a_k]$$

then $A \models \neg SL(\sigma)[a_1,\ldots,a_k]$, and we are through. We may therefore assume for all i that

$$A \models S(\sigma(x_i))[a_1,\ldots,a_k]$$

is true.

Let $b_i = \sigma(x_i)_A[a_1,\ldots,a_k]$. Since L_1, L_2 are complementary either $A \models A[b_1,\ldots,b_n]$ or $A \models B[b_1,\ldots,b_n]$. Therefore $A \models (A \vee B)[b_1,\ldots,b_n]$ which is equivalent to $A \models \sigma(A \vee B)[a_1,\ldots,a_k]$.

For the remaining implication, we assume $\square \notin I(D)$, and construct a Herbrand model for D.

We will show that $OSubst(I(D))$ has a Herbrand model H. This implies that H is also a model of $I(D)$. This is true, because all free variables in $I(D)$ are implicitly universally quantified, a variable of sort S by the quantifier $\forall x{:}S$, and the only elements that in H may be assigned to x are variablefree terms of sort $\leq S$. This also explains why $OSubst(I(D))$ suffices.

By Lemma 2 (2) it suffices to construct H satisfying $I(Subst_{SL}(D))$. The Assumption $\square \notin I(D)$ implies, by Lemma 3 (2), $\square \notin I(Subst_{SL}(D))$.

To summarize, what remains to be done, for any given set M of quantifierfree clauses such that $\square \notin I(M)$, is to construct a Herbrand model H for $I(M)$. Note that in order to obtain $I(M)$ only the PL-cases of the inference rules are required.

To define H, we have to specify for each quantifierfree atomic formula A whether A is true in H or not.

Let $\{A_\alpha : 0 \leq \alpha < \gamma\}$ be some enumeration of all variablefree atomic formulas. As an important restriction we require that all atomic formulas of the form $S_i(t)$ appear before all formulas of the form $S_j(s)$ for $i < j$. To satisfy this restriction we have to admit enumerations up to ordinals γ strictly greater than ω.

We will define truth or falsity of A_n in H by induction on n. We let A_α be true in H, unless

1. there is a clause K in $I(M)$ containing only negative literals, $K = K_1 \vee \neg A_\alpha$. K_1 only contains literals $\neg A_\beta$ for $\beta < \alpha$, and $H \models \neg K_1$,

or

2. $A_\alpha = S_2(t)$, and there is a negative clause $K = K_1 \vee \neg S_1(t)$ in $I(M)$, such that $H \models \neg K_1$, and $S_2 < S_1$, and K_1 only contains literals of the form $\neg A_\beta$ for $\beta < \alpha$.

In particular, A_0 is set to be true in H, unless

1. $\neg A_0 \in I(M)$

or

2. $A_0 = S_2(t), \neg S_1(t) \in I(M)$, and $S_2 < S_1$.

This completes the definition of H.

We claim
(C1) $H \models I(M)$
(C2) If $sort(t) \leq S_1$, then $H \models S(t)$
(C3) If $S_2 \leq S_1$, then $H \models \forall x : \top(S_2(x) \to S_1(x))$.

Proof of (C1): Let $K \in I(M)$. We shall prove $H \models K$ by induction on the number of positive literals in K.

The induction base consists thus in the consideration of negative clauses $K \in I(M)$.

First we know that K is not the empty clause. Thus $K = \neg A_{\alpha_1} \vee \cdots \vee \neg A_{\alpha_k}$ with $\alpha_1 < \cdots < \alpha_k$. If H does not satisfy $\neg A_{\alpha_1} \vee \cdots \vee \neg A_{\alpha_{k-1}}$, then, by construction, $H \models \neg A_{\alpha_k}$.

Now, assume that for all clauses in $I(M)$ with less than n positive literals the claim (C1) is proved, and K contains n positive literals $K = K_1 \vee A_\alpha$.

If $H \models A_\alpha$, then $H \models K$, and we are through. If $H \models \neg A_\alpha$, then one of the above mentioned cases 1 or 2 have to be true.

Let us first consider case 1.

There is a negative clause $K' = K_1' \vee \neg A_\alpha \in I(M)$ with $H \models \neg K_1'$. Since $I(M)$ is closed under EOS-resolution also $K'' = K_1' \vee K_1 \in I(M)$. Since K'' contains less than n positive literals $H \models K''$ is true, and therefore $H \models K_1$. This yields again $H \models K$.

Now consider case 2.

There is a negative clause $K' = K_1' \vee \neg S_1(t), A_\alpha = S_2(t), S_2 < S_1$, and $H \models \neg K_1$. Since $I(M)$ is closed under subsort resolution, also $K'' = K_1' \vee K_1 \in I(M)$, and the argument proceeds as above.

Proof of (C2): Assume $sort(t) \leq S$ and $H \models \neg S(t)$. There are again the two cases 1 and 2 to be considered.

Let us consider case 2 first this time.

If $H \models \neg S(t)$, then by construction of H there is a clause $K = K_1 \vee \neg S_1(t) \in I(M)$, such that $S \leq S_1$ and $H \models \neg K_1$. By the E-rule also $K_1 \in I(M)$ which contradicts (C1). Thus $H \models S(t)$.

The same argument applies to case 1.

Proof of (C3): Assume $H \models S_2(t)$, with the aim of showing $H \models S_1(t)$.

$H \models S_1(t)$ may be wrong for the two reasons provided for in the construction of H.

Firstly, there may be a negative clause $K = K_1 \vee \neg S_1(t) \in I(M)$, such that $H \models \neg K_1$, and all literals in K_1 occur before $S_1(t)$ in the given enumeration. Since $S_2 \leq S_1$, it follows from the constraint put on the enumeration, that all literals also occur before $S_2(t)$, but this contradicts $H \models S_2(t)$.

Secondly, there may be a negative clause $K = K_1 \vee \neg S_3(t) \in I(M)$ with $S_1 < S_3$, with $H \models \neg K_1$, and all literals in K_1 occur before $S_1(t)$ in the given enumeration. Again, since all literals in K_1 occur before $S_2(t)$ in the enumeration $S_2(t)$ would not have been set true in the construction of H. Therefore, $H \models S_1(t)$ is true.

There is still one detail to be checked. We require that in a model the interpretation of a sort S coincides with the universe U_S of all elements of sort S. For the Herbrand model we have not specified so far what these universes should be. Given property (C2) we may define for each sort symbol S

the universe U_S to be just the interpretation of the sort S. $\quad\square$

Remark:

It follows from the proof of Theorem 4 that we may restrict the application of the EOS- and SUBS-rule to cases where one of the parent clauses is negative without loosing completeness.

References

[Beierle et al., To appear] C. Beierle, U. Hedtstück, U. Pletat, and J. Siekmann. An Order-Sorted Predicate Logic with Closely Coupled Taxonomic Information. To appear.

[Beierle et al., 1988] C. Beierle, J. Dörre, U. Pletat, C. Rollinger, P.H. Schmitt, and R.Studer. *The Knowledge Representation Language* L$_{LILOG}$. LILOG-Report 41, IBM Germany, Scientific Center, July 1988.

[Bollinger et al., 1988] T. Bollinger, U. Hedtstück, and C.-R. Rollinger. *Reasoning in Text Understanding: Knowledge Processing in the LILOG Prototype.* LILOG-Report 49, IBM Deutschland, Scientific Center, October 1988.

[Brachman and Schmolze, 1985] R.J. Brachman and J.G. Schmolze. An Overview of the KL-ONE Knowledge Representation System. *Cognitive Science*, 9(2):171–216, April 1985.

[Cohn, 1987] A. G. Cohn. A More Expressive Formulation of Many Sorted Logic. *Journal of Automated Reasoning*, 3:113–200, 1987.

[Glubrecht et al., 1983] J.-M. Glubrecht, A. Oberschelp, and G. Todt. *Klassenlogik*. BI, Mannheim, Wien, Zürich, 1983.

[Goguen and Meseguer, 1986] J.A. Goguen and J. Meseguer. Remarks on Remarks on Many-Sorted equational Logic. In *Bulletin of the EATCS*, 1986.

[Stickel, 1985] M.E. Stickel. Automated Deduction by Theory Resolution. *Journal of Automated Reasoning*, 1:333–355, 1985.

[Walther, 1987] C. Walther. *A Many-Sorted Calculus Based on Resolution and Paramodulation. Research Notes in Artificial Intelligence*, Pitman, London, and Morgan Kaufmann, Los Altos, Calif., 1987.

Types, Modules and Databases in the Logic Programming Language PROTOS-L

Christoph Beierle

IBM Germany

Scientific Center

Institute for Knowledge Based Systems

P.O. Box 80 08 80

D-7000 Stuttgart 80, West Germany

e-mail: BEIERLE at DS0LILOG.BITNET

Abstract

Whereas in many programming languages types play a central role, logic programming languages often do not have a typing concept at all. After a survey on many-sorted, order-sorted and polymorphic approaches to types in logic programming the basic components of the typed logic programming language PROTOS-L are discussed. It has a polymorphic order-sorted type concept and a module concept that allows the integration of external data bases. PROTOS-L is implemented on the PAM, an abstract machine that extends the Warren Abstract Machine in particular by polymorphic order-sorted unification and by a data base component.

1 Introduction

Looking at the history of programming language development the logic programming paradigm has opened a new dimension in this field. Based on the Horn clause subset of first-order predicate logic it naturally provides a simple declarative meaning. Moreover, there is a clear separation between declarative and operational semantics where the latter is also defined in a - compared to other programming language like Pascal or Ada - very simple and natural way, namely SLD-resolution [Lloyd, 1984]. Thus, it is argued that logic programs are easier both to write and to understand and that therefore reasoning about them and modifying them is also easier than, say, in an imperative programming language.

These observations render a logic programming language a candidate for software development. The most prominent representative of the logic programming family, Prolog, realizes the logic programming paradigm, but most Prolog implementations compromise the idea of "pure" logic programming by trading correctness and completeness for efficiency (e.g. by the depth-first search strategy and by leaving out the "occur check" in the unification, c.f. [Lloyd, 1984]) and by introducing "impure" features like the cut operation. With these additions Prolog has already been used for quite large applications, especially in the area of Artificial Intelligence.

On the other hand, comparing Prolog to other programming languages it seems doubtful whether in the long run Prolog or any other logic programming language will really be a serious alternative to a programming language that supports major software engineering principles as long as these are lacking in logic programming. For instance, there is virtually no typing mechanism in Prolog nor any modularization, data abstraction or any other structuring mechanism. However, these mechanisms have generally been accepted as being indispensable for large software applications.

Consequently, there have been numerous suggestions and attempts to overcome these deficiencies of Prolog. The purpose of this paper is to look at some particular aspects of some of them and to discuss some basic components of the typed logic programming language PROTOS-L.

In Sections 2 and 3, a survey of many-sorted, order-sorted and polymorphic type concepts for logic programming are given. In Section 4, the major aspects of PROTOS-L are presented. PROTOS-L has a polymorphic order-sorted type concept derived from TEL ([Smolka, 1988b]) and a module concept that allows the integration of external data bases. PROTOS-L is implemented on the PAM, an abstract machine that extends the Warren Abstract Machine in particular by polymorphic order-sorted unification and by a data base component (Section 5). In Section 6, some extensions to the current work are suggested.

Acknowledgements: I would like to thank Harald Ganzinger, Michael Hanus and Gert Smolka for many detailed and helpful comments and suggestions and for pointing out some shortcomings in an earlier version of this paper; my thanks go also to all my colleagues in the PROTOS project at IBM in Stuttgart for numerous discussions. PROTOS is an international EUREKA project (EU56) in which the work reported has been carried out.

2 Many-sorted and order-sorted approaches

2.1 A many-sorted approach

In unsorted predicate logic unary predicates define a subset of the universe for any interpretation. For instance, a Prolog program containing the clauses

```
car(opel).
car(ford).
car(mercedes).
```

where car does not occur in the head of any other clause will have the three constants opel, ford, and mercedes in the set assigned to car in its standard interpretation (i.e. its least Herbrand model).

Within Prolog it is not possible to reason on this more abstract level of "cars"; rather the level of the individuals of this set must be used. However, within the framework of algebraic abstract data type specifications ([Goguen *et al.*, 1978], [Ehrig and Mahr, 1985]) the declaration

```
sort        car.
operations  opel:    → car.
            ford:    → car.
            mercedes:  → car.
```

within a specification yields essentially the same effect. Provided there are no other operations with target sort car and no equations are imposed on the constants, the set assigned to car in the standard interpretation of the specification (i.e. its initial model) contains the three constants opel, ford, mercedes.

A first extension of Prolog to sorts could thus be to allow sort declarations together with the operations that yield objects of the respective sorts. Following terminology used in abstract data type specifications we call these operations *constructors* since they construct the elements of the given sort. Thus, the fragment of a Prolog program given above would correspond to

```
sort         car.
constructors opel, ford, mercedes.
```

in a sorted version. In the following we will use the more compact notation

```
sort  car := { opel, ford, mercedes }.
```

Similarly,

> sort boat := { ferry, steamer, sailing boat }.

introduces the sort boat with its elements ferry, steamer, sailing_boat.

Given such sort declarations we can attach to every predicate p a declaration stating how many arguments p takes and of which sorts the arguments of p must be. For instance, the declaration

> predicate go_from_to_with: city x city x vehicle.

would require the first two arguments of go_from_to_with to be of sort city and the third argument to be of sort vehicle.

Requiring such declarations for all predicates in a logic program and insisting on well-typed terms and literals already has some important consequences that lead us away from the usual Prolog situation:

- The distinction between declarations and clauses defining a predicate together with a notion of well-typedness requires a *type checker*.

- Since type checking takes place with respect to a set of declarations the unit for type checking cannot be a single clause or declaration but should be a program part containing both a set of declarations and a set of clauses (e.g. a module).

- Since only well-typed programs should be executed there is a distinction between the type checking phase and the execution phase of a program. In analogy to classical typed programming languages we will refer to these phases as *compile time* and *run time*, respectively. If all type checking can be done at compile-time we have *static* type checking whereas type checking done at run time is called *dynamic*.

In Prolog there is essentially no type-checking, and every clause added to a program file can be given to the interpreter immediately. However, the third distinction listed above can be exploited in terms of efficiency. In the situation discussed so far type checking is completely a compile-time activity and no type information whatsoever has to be present at run time. Crucial for this observation is the fact that up to now we have only considered sorts without the possibility of having subsort relations between them. Again using abstract data type terminology this case is referred to as the *many-sorted* case. The possible gains in run-time efficiency are based upon the fact that on the one hand no overhead due to type information occurs at run time, and that on the other hand due to well-typedness any argument of a predicate is guaranteed to be of the specified sort. Therefore, the underlying unification does not have to deal with the whole universe of terms but only with the terms which are of the given argument sort. An example for a logic programming language based on the many-sorted approach is Turbo-Prolog.

Generally one can say that a many-sorted approach does only require a static type-checking phase - which can be realized by a preprocessor - and that every implementation for the unsorted case also works fine for the many-sorted case. In particular, an abstract machine like the Warren Abstract Machine (WAM) [Warren, 1977], [Warren, 1983] does not have to be modified for the many-sorted case; however, it could be modified for efficiency reasons as indicated above.

2.2 An order-sorted approach

An extension to the many-sorted approach is to allow subsort relationships between sorts. For instance the sorts car and boat could both be seen as subsorts of a common supersort vehicle. Likewise, there could also be a sort amphibious_vehicle as a common subsort to both car and boat, yielding the following sort hierarchy:

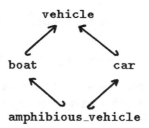

Order-sorted approaches allowing such sort hierarchies have been suggested in different areas like abstract data type specifications, automatic theorem proving, or logic programming. An order-sorted logic was already given in [Oberschelp, 1962]. Order-sorted algebra has its origin in [Goguen, 1978] and was developed further by Goguen et al. (see e.g. [Goguen and Meseguer, 1987], [Smolka et al., 1987]) and is the basis for OBJ ([Futatsugi et al., 1985]) and Eqlog ([Goguen and Meseguer, 1986]). [Gogolla, 1986] studies and extends order-sorted algebra in particular with respect to error handling approaches. [Walther, 1987], [Walther, 1988] investigates order-sorted unification and gives correctness and completeness results for resolution and paramodulation with order-sorted unification, and [Huber and Varsek, 1987] uses this approach for an extended Prolog with order-sorted resolution.

Extending the approach from the previous section we add to every sort decla-ration an enumeration of its (direct) subsorts separated by ++, if there are any. In the upper part of Figure 1 we have listed some sort declarations extending the examples from the previous section and leading to the sort hierarchy depicted in Figure 2.

Additionally, Figure 1 contains the predicate go_direct which takes two argu-ments of sort city and a third argument of sort vehicle. go_direct(a, b, v) means that there is a direct way of getting from city a to city b using only vehicle v. The predicate go_to_from_with takes the same arguments as go_direct. The intended meaning of go_to_from_with(a, b, v) is that there is a way of going from city a to city b using only vehicle v, but possibly going via some other cities. Correspondingly, there are two clauses defining the predicate go_to_from_with as given in Figure 1. This simple travelling world example is an extended and modified version of an example given in [Müller, 1988].

The unsorted version of the order-sorted travelling world program is given in Figure 3. Note that a sort restriction on a variable like X:airplane corresponds to a subgoal airplane(X) in the unsorted version.

Now consider the query

$$?- \text{go\_from\_to\_with}(\text{stuttgart}, \text{london}, V) \tag{1}$$

asking for a vehicle V which can be used to go from stuttgart to london.

Let us first look at the unsorted version. Assuming the usual Prolog evaluation strategy (i.e. left-to-right and depth-first search) a Prolog implementation yields

    V = bo747

as the first answer. Repeatedly enforcing backtracking yields the solutions

    V = dc10
    V = airbus
    V = amphi1
    V = amphi2
    V = amphi3

which are all the individual solutions to the given query.

In order to be able to compare these solutions to the sorted version we also assume the left-to-right and depth-first search strategy. But instead of ordinary term unification we now use order-sorted unification [Walther, 1988]. The original

```
sort   vehicle   := airplane ++ boat ++ car.

sort   airplane := { bo747, dc10, airbus }.

sort   car       := amphibious_vehicle
                     ++ { opel, ford, mercedes }.

sort   boat      := amphibious_vehicle
                     ++ { ferry, steamer, sailing_boat }.

sort   amphibious_vehicle := { amphi1, amphi2, amphi3 }.

sort   city      := { stuttgart, frankfurt,
                      london, calais, dover }.

predicate  go_direct:  city x city x vehicle.
     go_direct(stuttgart, frankfurt, X:airplane).
     go_direct(frankfurt, london, X:airplane).
     go_direct(stuttgart, calais, X:car).
     go_direct(dover, london, X:car).
     go_direct(calais, dover, X:boat).

predicate  go_from_to_with:  city x city x vehicle.
     go_from_to_with(From, To, With) :-
         go_direct(From, To, With).
     go_from_to_with(From, To, With) :-
         go_direct(From, Over, With) &
         go_from_to_with(Over, To, With).
```

Figure 1: An order-sorted logic program

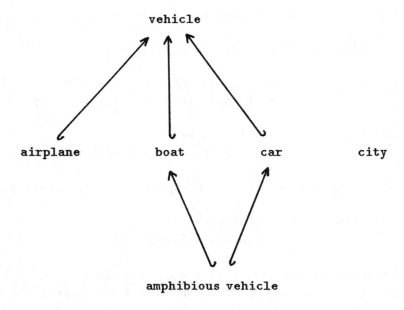

Figure 2: The sort hierarchy (c.f. Figure 1)

```
vehicle(X) :- airplane(X).
vehicle(X) :- boat(X).
vehicle(X) :- car(X).

airplane(bo747).
airplane(dc10).
airplane(airbus).

car(opel).
car(ford).
car(mercedes).
car(X) :- amphibious_vehicle(X).

boat(ferry).
boat(steamer).
boat(sailing_boat).
boat(X) :- amphibious_vehicle(X).

amphibious_vehicle(amphi1).
amphibious_vehicle(amphi2).
amphibious_vehicle(amphi3).

city(stuttgart).
city(frankfurt).
city(london).
city(calais).
city(dover).

go_direct(stuttgart, frankfurt, X) :- airplane(X).
go_direct(frankfurt, london, X) :- airplane(X).
go_direct(stuttgart, calais, X) :- car(X).
go_direct(dover, london, X) :- car(X).
go_direct(calais, dover, X) :- boat(X).

go_from_to_with(From, To, With) :-
    go_direct(From, To, With).
go_from_to_with(From, To, With) :-
    go_direct(From, Over, With) &
    go_from_to_with(Over, To, With).
```

Figure 3: The unsorted version (c.f. Figure 1)

sort restriction of the variable V in query (1) is vehicle due to the declaration of go_to_from_with. The first solution to (1) leaves the variable V uninstantiated but sharpens its sort restriction to airplane:

    V = X:airplane

Obviously, this solution represents a more abstract answer than enumerating all elements of the sort airplane as is done in the unsorted case. Enforcing backtracking will now first reset the sort of V to vehicle. Resolving with the clause

    go_direct(stuttgart, calais, X₁:car).

restricts the sort of V to car. Further resolving the remaining subgaol with

    go_direct(calais, dover, X₂:boat).

leads to the unification of the variable V with sort restriction car and the variable X₂ with sort restriction boat. Order-sorted unification yields a variable whose sort restriction is the greatest common subsort of the two sorts in the given sort hierarchy, i.e. amphibious_vehicle in this case. After one more resolution step the second solution

    V = X:amphibious_vehicle

to query (1) is given.

Comparing the unsorted and the order-sorted case we can observe that the sorted version

- provides a higher level of abstraction and thus
- provides more compact solutions,
- avoids backtracking and
- is "more complete" than Prolog.

To illustrate the last point let us assume that the sort airplane contains an additional constructor numbered_airplane

    sort airplane := { bo747, dc10, airbus,
                        numbered_airplane: nat }.

where the sort nat denotes the natural numbers as given by

```
sort   nat := { zero,
               succ:   nat }.
```

Now the set of terms of sort `airplane` also contains `numbered_airplane(zero)`, `numbered_airplane(succ(zero))`, `numbered_airplane(succ(succ(zero)))`, etc.

In the sorted version the restriction of a variable `V` to sort `airplane` still works fine, and backtracking over this sort restriction just yields the previous sort restriction for that variable (like `vehicle` in the example discussed above). In the Prolog situation this sort restriction corresponds to a subgoal `airplane(V)` which of course can easily be satisfied. However, upon backtracking we would now enter a nonterminating sequence of backtracking steps since there are infinitely many solutions to that subgoal.

Whereas the observations made for the many-sorted case above about type checking at compile-time, type-checkable units etc. essentially still hold in the order-sorted case this is not true in the absence of type information at run time.

The rules for ordinary term unification are given in Figure 4 in the style of equation solving ([Martelli and Montanari, 1982]) with the rules for elimination (E), decomposition (D), variable binding (B), and orientation (O). The idea is to start with a set of equations $E = \{t_1 \doteq t_1', \ldots, t_n \doteq t_n'\}$ representing the pairs of terms to be unified and to transform E into a set of equations E' that is in *solved form* by using the four given rules.

There are two conditions indicating that there exists no solution:

C1   If there is an equation $f(t_1, \ldots, t_n) \doteq g(t_1', \ldots, t_m')$ in E with $f \neq g$ then there exists no solution.

C2   If there is an equation $x \doteq t$ in E such that $x \neq t$ and x occurs in t then there exists no solution.

E' is in solved form if it is of the form $E' = \{z_i \doteq t_i \mid i \in \{1, \ldots, n\}\}$ where z_i are variables that do not occur elsewhere in E'. In this case the substitution represented by $\{z_1/t_1, \ldots, z_n/t_n\}$ is the most general unifier of the unification problem given by the original set of equations E.

Moving to the order-sorted case the rules for order-sorted unification are given in Figure 5. Note that the rules for elimination, decomposition and orientation look exactly as in the unsorted case. However, now every variable is of a particular fixed sort, and the difference from the unsorted case comes in the rules (B1) and (B2) when binding a variable.

$$(E) \qquad \frac{E \ \& \ x \dot{=} x}{E} \qquad\qquad \text{if } x \text{ is a variable}$$

$$(D) \qquad \frac{E \ \& \ f(t_1, \ldots, t_n) \dot{=} f(t'_1, \ldots, t'_n)}{E \ \& \ t_1 \dot{=} t'_1 \ \& \ \cdots \ \& \ t_n \dot{=} t'_n}$$

$$(B) \qquad \frac{E \ \& \ x \dot{=} t}{\sigma(E) \ \& \ x \dot{=} t} \qquad\qquad \text{if } x \text{ is a variable, } t \text{ is a variable or a non-variable term, and } x \text{ occurs in } E \text{ but not in } t, \text{ and where } \sigma = \{x/t\}$$

$$(O) \qquad \frac{E \ \& \ t \dot{=} x}{E \ \& \ x \dot{=} t} \qquad\qquad \text{if } x \text{ is a variable and } t \text{ is not a variable}$$

Figure 4: The rules for ordinary term unification

A precondition for these rules to work correctly is that the sort structure satisfies the following conditions:

- There are only finitely many sorts, and the subsort relationship is a partial order such that two different sorts have at most one common maximal subsort.

Additionally, there are the following requirements:

- There is no overloading of function symbols, i.e. for every function symbol f with arity n there is exactly one arity declaration $f : s_1 \ldots s_n \to s$.

- Sorts are not empty, i.e. for every sort there is a ground term of that sort.

- The unification problems considered involve only well-sorted terms, i.e. the argument sort of t_i in $f(t_1, \ldots, t_n)$ must be of a subsort of s_i for $f : s_1 \ldots s_n \to s$.

Note that the first condition could be weakened by allowing certain cases of overloading, see for instance the recent paper [Waldmann, 1989].

Now there are two additional situations compared to the unsorted case such that there is no solution:

(E)
$$\frac{E \ \& \ x \doteq x}{E}$$
if x is a variable

(D)
$$\frac{E \ \& \ f(t_1,\ldots,t_n) \doteq f(t'_1,\ldots,t'_n)}{E \ \& \ t_1 \doteq t'_1 \ \& \ \cdots \ \& \ t_n \doteq t'_n}$$

(B1)
$$\frac{E \ \& \ x \doteq t}{\sigma(E) \ \& \ x \doteq t}$$
if x is a variable of sort s, t is a variable or a non-variable term of sort s', $s' \leq s$, and x occurs in E but not in t and where $\sigma = \{x/t\}$

(B2)
$$\frac{E \ \& \ x \doteq y}{\sigma(E) \ \& \ x \doteq z \ \& \ y \doteq z}$$
if x is a variable of sort s, y is a variable of sort s', $x \neq y$, $s' \not\leq s$, z is a new variable of sort s'' where s'' is the greatest common subsort of s and s', and where $\sigma = \{x/z, y/z\}$

(O)
$$\frac{E \ \& \ t \doteq x}{E \ \& \ x \doteq t}$$
if x is a variable and t is not a variable

Figure 5: The rules for order-sorted unification

C3   If there is an equation $x \doteq t$ in E such that the sort of t is strictly greater than the sort of x then there exists no solution.

C4   If there is an equation $x \doteq y$ in E such that the sorts of x and y do not have a common subsort then there exists no solution.

If neither of the conditions **C1** - **C4** is satisfied and E' is in solved form $E' = \{z_i \doteq t_i \mid i \in \{1, \ldots, n\}\}$ then

$$\{z_i/t_i \mid z_i \doteq t_i \in E' \text{ and } z_i \text{ occurs in the original set of equations } E\}$$

represents a most general unifier of the original unification problem E.

Consider the rules for order-sorted unification as given in Figure 5. Comparing the rules to the unsorted version (Figure 4) reveals two places where sort information is involved, namely (B1) and (B2). In (B1), a subsort test $s' \leq s$ is required involving the sort s of the variable x and the sort s' of t. Thus, if t is a variable then s' is its sort, and if t is the non-variable term $f(t_1, \ldots, t_n)$ then s' is the target sort of the top-level function symbol $f : s_1 \ldots s_n \rightarrow s'$ of t.

To give an example, binding the term opel (of sort car) to variable Y:amphibious_vehicle would not be possible under rule (B1), but binding the term amphi1 of sort amphibious_vehicle to Y:car would be possible.

However, it is interesting to note that under the requirements given above - in particular no overloading and only well-sorted unification problems - when trying to apply (B1), a variable of sort city will never be tried to be bound to a term of sort car. Vice versa, a variable of sort vehicle will never be tried to be bound to a term of sort city. Therefore, whenever the variable X involved in (B1) is either of sort vehicle or of sort city - i.e. the sort of X is maximal with respect to the subsort relationship - then the subsort test will always be successful.

A similar observation is true for the other binding rule (B2). Under the given requirements, if either one of the involved variables x and y in the equation $x \doteq y$ is of a maximal sort (vehicle or city in our example) then the other one is either of the same sort or of a subsort thereof. If the sort s of x is maximal, rule (B2) cannot be applied but (B1) has to be used instead. Otherwise, if in (B2) the sort s' of y is maximal and s is non-maximal we have $s < s'$ and the sort of the resulting variable z (i.e. s) is determined completely by the non-maximal sort. In fact we could simplify (B2) in this case by not introducing a new variable but by orienting the equation to $y \doteq x$ and just doing the binding as in (B1).

We conclude that in both places where sort information is involved the sort information for variables of maximal sorts is redundant. Thus, leaving out the redundant sort tests for variables of maximal sorts as indicated above both (B1) and (B2) reduce to special cases of the binding rule (B) in the unsorted case.

Thus, we could do the following optimization for a given well-sorted unification problem $E = \{t_1 \doteq t_1', \ldots, t_n \doteq t_n'\}$

1. Transform E into E' by replacing every variable x of a maximal sort by an 'unsorted' variable x_u.

2. Transform E' into E'' using the rules in Figure 5 with the modified version of (B1) and (B2), - i.e. applying effectively rule (B) - when binding an 'unsorted' variable.

3. Replace every 'unsorted' variable x_u from E' by the original sorted variable x.

Considering the degenerated order-sorted case where all sorts are maximal all sort information therefore becomes redundant. This means that in step 2 above we do just ordinary unsorted unification. Of course, this comes at no surprise since that case is just the many-sorted case where - for well-sorted unification problems - the unsorted unification rules work fine. In [Smolka, 1989] this optimization is suggested and elaborated in detail for the more general polymorphic order-sorted case which we will discuss in the following section.

3   Polymorphic type concepts

A serious drawback of both a many-sorted approach and an order-sorted approach as discussed so far is the fact that it is not possible to achieve the effect of parameterizing a structured sort over some other sort. For instance, in the many-sorted setting lists must be defined explicitly for every sort. To give an example, the sort list_nat of lists over natural numbers would be given by

```
sort  list_nat := { []_nat,
                    • _nat:  nat x list_nat }.
```

If we additionally want to use lists over the sort car we have to define a new sort and new constructors:

```
sort  list_car := { []_car,
                    • _car:  car x list_car }.
```

Moreover, for every predicate operating on lists we have to have an extra version for every list variant, e.g.

```
predicate   append_nat:   list_nat x list_nat x list_nat.
predicate   append_car:   list_car x list_car x list_car.
...
```

where every predicate version must be defined explicitly.

One method to overcome this difficulty is to introduce *parameterization*. This concept has been studied extensively in abstract data type theory (c.f. [Ehrig and Mahr, 1985]) and has been applied to logic programming in Eqlog ([Goguen and Meseguer, 1986]). For instance, in Eqlog one can introduce a parameterized module

module LIST(X::ELEM)

which refers to a formal parameter specification ELEM. The functions and predicates operating on lists are defined by referring to the names introduced in ELEM. Instances of a parameterized module are obtained by replacing the formal parameter X::ELEM with an actual parameter, e.g. NAT or CITY, yielding the resulting instance like LIST(NAT) or LIST(CITY). Thus, there needs to be only one definition for every list predicate like append, namely in the parameterized LIST module.

An advantage of such a parameterized concept is that apart from formal sort, function and predicate names the parameter specification may also contain requirements for these; for instance, a parameterized specification for sets would require an equality relation on the elements. However, every instance of such a parameterized module must be generated explicitly, and for all names stemming from the parameterized specification different new names must be used in every instance.

A related way of avoiding multiple definitions for the same structure is the concept of polymorphism ([Milner, 1978], [Damas and Milner, 1982]). Compared to the approach of parameterized specifications as outlined above there are also parameter variables in the polymorphic case. However, they range over all sorts and it is not possible to restrict them by a parameter specification. Thus, there are no function or predicate symbols in the formal parameters nor any constraints. On the other hand, in the polymorphic case it is not necessary to generate any instance explicitly and the same names from a polymorphic definition are used in every instance.

In the following, we will denote such sort variables by α and β. A polymorphic sort declaration for lists would then be

$$\text{sort} \quad \text{list}(\alpha) := \{ \ [], \qquad\qquad\qquad\qquad\qquad (2)$$
$$[\_|\_]: \quad \alpha \ \text{x} \ \text{list}(\alpha) \ \}.$$

Similarly, the polymorphic sort definition

> sort pair(α,β) := { mkpair: α x β }.

defines ordered pairs over two sorts.

Having such polymorphic sort functions at hand, all instances of them are automatically available, e.g.

```
list(city)
list(airplanes)
list(boat)
list(list(city))
pair(boat,city)
pair(car,list(city))
...
```

In [Mycroft and O'Keefe, 1984] a polymorphic type system for Prolog is proposed that allows parametric polymorphism (for a discussion of different polymorphism concepts see [Cardelli and Wegner, 1985]). In this approach the user has to give sort declarations for the function and predicate symbols similar to the many-sorted setting (see Section 2.2) but with the additional polymorphic sort declarations. [Mycroft and O'Keefe, 1984] defines a notion of well-typedness such that - again as in the the many-sorted case - static type checking is sufficient and no run-time type information is needed. Using (2) and the syntax of Section 2 the polymorphic definition of the append predicate would be

$$\begin{aligned}&\textbf{predicate append: list}(\alpha) \text{ x list}(\alpha) \text{ x list}(\alpha).\\&\quad \text{append([], L, L).} \qquad\qquad\qquad\qquad\qquad\qquad (3)\\&\quad \text{append([H|T], L, [H|TL]) :- append(T, L, TL).}\end{aligned}$$

There have been several other approaches to a polymorphic type system for logic programming. [Dietrich and Hagl, 1988] extends the approach of [Mycroft and O'Keefe, 1984] to an order-sorted setting. In order to ensure that static type checking is sufficient dataflow information within the program clauses is required in certain cases. Such dataflow could be provided by a mode system or in some cases it could be provided by global analysis, but the necessity for such information restricts the generality of this approach.

Another extension of [Mycroft and O'Keefe, 1984] also dealing with subsorts is reported in [Dayantis, 1988]. However, in this case dynamic type checking is needed. The dynamic type checking is not achieved by using a special unification algorithm (as in Section 2.2) but by a reduction to the ordinary Prolog case: A preprocessor inserts system-defined literals into the body of a clause in order to ensure sort-correct instantiations, reporting a runtime error otherwise. For instance, given the predicate declarations

```
predicate  p:  car.
    ...
predicate  q:  vehicle.
    ...
```

the clause

```
p(X) :- q(X).
```

in the definition of p will be translated to

```
p(X) :- instantiated(X) & q(X).
```

Therefore, computing with uninstantiated sort-restricted variables as demonstrated in the travel world example in Section 2.2 is not possible and the advantages of order-sortedness are thus severely restricted in this framework.

The three approaches to a polymorphic type system discussed so far are all operational approaches: They do not provide a semantic notion of a type but only a syntactic one. Their aim is to guarantee by static type checking that no type error can occur at run time, and the operational semantics is the same as the semantics of the untyped version (except for the third approach as indicated by the example given above). On the other hand there are approaches to a polymorphic type system for logic programming that provide also a semantic notion of a type, namely [Hanus, 1988] and [Smolka, 1989]. Moreover, in contrast to the syntactic approaches discussed above there is a need to take into account the type information at run time, requiring a special unification algorithm as in the order-sorted case.

The approach of [Hanus, 1988] extends the polymorphic concept of [Mycroft and O'Keefe, 1984]. By removing the restrictions of [Mycroft and O'Keefe, 1984] on the use of type expressions [Hanus, 1988] also allows for ad-hoc polymorphism ([Cardelli and Wegner, 1985]). Besides a model-theoretic semantics, sound and complete deduction relations involving a special polymorphic unification are given. For instance, consider the polymorphic append predicate given in (3). Then adding the specialized clause (here and later on we will sometimes use the standard Prolog notation for lists)

$$\text{append([opel, ford], [mercedes], [opel, ford, mercedes]).} \qquad (4)$$

is possible in this approach but not allowed in [Mycroft and O'Keefe, 1984] since in that framework the clause (4) is not well-typed because the types of the arguments in the head of (4) - list(car) - are not of the most general type - list(α) - declared for the predicate append in (3).

Whereas [Hanus, 1988] uses ad-hoc polymorphism and exploits it e.g. for higher-order programming techniques but does not consider subsorts, the approach of [Smolka, 1989] combines parametric polymorphism with an order-sorted approach. In this framework a sound and complete deduction relation for the corresponding model-theoretic semantics is defined that uses polymorphically order-sorted unification. To give an example, let us extend the travel world program in Figure 1 by the standard definition of polymorphic lists as given in (3). As in the polymorphic cases above we now have not only a finite set of sorts but infinitely many sorts denoted by sort terms which show up in the unification procedure: Besides a sort constant like car or vehicle the sort restriction of a variable can be a sort term like list(car) or list(list(vehicle)), etc. To illustrate polymorphically order-sorted unification we list the unification results for different values of the terms t1 and t2:

t1	t2	result:
X:list(car)	[ford, opel]	[ford, opel]
X:list(car)	[ford, airbus, opel]	*fail*
X:list(car)	[ford, Y:vehicle, opel]	[ford, Y:car, opel]
X:list(car)	Y:list(boat)	Z:list(amphibious_vehicle)

The partial order on the (monomorphic) sort constants induces a partial order on sort terms as present in the example above. For instance, list(amphibious_vehicle) is a subsort of list(car) since amphibious_vehicle is a subsort of car.

In addition to this induced sort relationships the approach of [Smolka, 1989] allows explicit definitions of subsort relationships between polymorphic sorts like

```
sort   lp(α,β).
subsorts    list(α), pair(α,β).
```

Thus, in the general case the computation of the greatest lower bound of two sort restrictions requires taking into account such explicit subsort relationships between polymorphic sorts which can be handled by sort rewriting systems. For instance, when trying to unify `X:lp(car,airplane)` and `Y:list(boat)` the sort restriction of `X` would have to be rewritten first as `list(car)`. There are several conditions for such sort rewriting systems that have to be satisfied ([Smolka, 1989]) and their use in a logic programming language could slow down the unification procedure. On the other hand, the important optimization of neglecting any maximal sort information is blocked in many cases: For instance, if $\text{list}(\alpha)$ does not have any supersort then `list(vehicle)`, `list(city)`, `list(list(city))`, etc. will also be maximal so that such sort information can be neglected as well. However, if $\text{list}(\alpha)$ is a subsort of e.g. $\text{lp}(\alpha,\beta)$ then no list instance at all will be maximal. In the following, we will now present the rules for polymorphically order-sorted unification for the case where polymorphic sorts do not have any subsorts; this is the case that has been incorporated into PROTOS-L (see Section 4) whereas the general case together with correctness and completeness proofs is given in [Smolka, 1989].

First, let us recall the order-sorted unification rules discussed in the previous section. There, we considered every variable to be of a fixed sort. However, one could also use unsorted variables and introduce the sort restrictions on the variables by a special predicate ([Smolka, 1988a]). This yields certain technical advantages; for instance, when unifying two variables `X:car` and `Y:boat` one can produce a new sort restriction for the variables (i.e. `X:amphibious_vehicle`) instead of introducing a new variable as done by the rules in Figure 5. We have already used this approach informally in our discussion above.

Thus, in addition to the set E of equations we will now also have a set $P = \{t_1{:}\tau_1, \ldots, t_n{:}\tau_n\}$ of sort restrictions with value terms t_i and sort terms τ_i. We call P a *prefix* if all t_i are variables that are pairwise distinct. As in the order-sorted case we consider only well-sorted unification problems P & E which now means that P is a prefix having a sort restriction for every variable in E and that for every equation $t = t'$ in E there exists a sort term τ such that both t and t' belong to τ under the sort restrictions given in P.

For the optimized version of the unification rules such a well-sorted unification problem P & E will first be transformed such that maximal sort information will be neglected: every maximal sort is replaced recursively by the special symbol $\top$. For a sort term τ the *approximation* of τ is defined by

$$
\begin{array}{lll}
{\downarrow}\tau & = \top & \text{if } \tau \text{ is a sort variable} \\
{\downarrow}\tau & = \top & \text{if } \tau \text{ is a sort constant that is maximal} \\
& & \text{in the partial order on the sorts} \\
{\downarrow}\xi(\tau_1,\ldots,\tau_n) & = \top & \text{if } {\downarrow}\tau_1 = \ldots = {\downarrow}\tau_n = \top \\
& \bot & \text{if } \xi(\tau_1,\ldots,\tau_n) \text{ cannot be instantiated} \\
& & \text{(see below)} \\
& \xi({\downarrow}\tau_1,\ldots,{\downarrow}\tau_n) & \text{otherwise}
\end{array}
$$

The approximation $\downarrow P$ of a prefix P is obtained by replacing all sort terms in P by their approximations.

Analogously to the computation of greatest common subsorts in the order-sorted case one needs now the computation of the infimum of two sort terms. For instance, the infimum of list(car) and list(boat) yields list(amphibious_vehicle). However, whereas there is no common subsort of e.g. airplane and car the infimum of list(airplane) and list(car) should be well-defined, namely the set consisting of exactly the empty list.

Thus, another special symbol $\perp$ is introduced above that denotes the 'empty' sort which becomes a subsort of every other sort. Then the infimum of list(airplane) and list(car) ist list($\perp$) since there is a ground term of sort list($\perp$), namely the empty list []. In other words, list($\perp$) can be instantiated or is *inhabited*. However, given the definition of standard pairs as above the infimum of e.g. pair(airplane,city) and pair(car,city) is not inhabited since pair($\perp$,city) cannot be instantiated. Thus, the infimum of the two sort terms is $\perp$.

In general the infimum $inf(\tau, \tau')$ of two sort term approximations τ and τ' is given by:

$$
\begin{aligned}
inf(\top, \tau) &= \tau \\
inf(\tau, \top) &= \tau \\
inf(\tau, \tau') &= \tau''
\end{aligned}
$$

if τ and τ' are sort constants with maximal common subsort τ''

$$
inf(\xi(\tau_1, \ldots, \tau_n), \xi(\tau_1', \ldots, \tau_n')) = \xi(inf(\tau_1, \tau_1'), \ldots, inf(\tau_n, \tau_n'))
$$

if this term can be instantiated

$$
inf(\tau, \tau') = \perp
$$

otherwise

Using this infimum operation the rules for polymorphic order-sorted unification are given in Figure 6, again under the assumption that there is no overloading of function symbols and that only well-sorted unification problems are considered. Let P & E be a well-sorted unification problem and let these rules transform the approximation P' & E into the form P'' & E''. Then P'' & E'' is in solved form and presents a solution if E'' is in solved form (in the sense of Section 2.2), P'' is a prefix that does not contain $x : \perp$, the variables in P'' do not occur on the left hand sides of an equation in E'', and every right hand side of an equation in E'' is well-sorted under the prefix P''.

Analogously to the unsorted case (compare the three steps 1 - 3 at the end of Section 2.2) one can now solve a well-sorted unification problem by:

1. Transform P into the approximation P'.

(E)
$$\frac{P \;\&\; E \;\&\; x \doteq x}{P \;\&\; E}$$

(D)
$$\frac{P \;\&\; E \;\&\; f(t_1,\ldots,t_n) \doteq f(t_1',\ldots,t_n')}{P \;\&\; E \;\&\; t_1 \doteq t_1' \;\&\; \cdots \;\&\; t_n \doteq t_n'}$$

(B)
$$\frac{P \;\&\; x:\tau \;\&\; E \;\&\; x \doteq t}{P' \;\&\; \sigma(E) \;\&\; x \doteq t}$$
if x occurs in E but not in t, and where $\sigma = \{x/t\}$ and $P \;\&\; t : \tau$ reduces to the prefix P' using the rules (ES) ... (DS)

(O)
$$\frac{P \;\&\; E \;\&\; t \doteq x}{P \;\&\; E \;\&\; x \doteq t}$$
if t is not a variable

(ES)
$$\frac{P \;\&\; f(t_1,\ldots,t_n):s}{P}$$
if $f: s_1 \ldots s_n \to s'$ and $s' \le s$.

(ES')
$$\frac{P \;\&\; f(t_1,\ldots,t_n):\top}{P}$$

(MS)
$$\frac{P \;\&\; x:\tau \;\&\; x:\tau'}{P \;\&\; x:inf(\tau,\tau')}$$

(DS)
$$\frac{E \;\&\; f(t_1,\ldots,t_n):\xi(\tau_1,\ldots,\tau_m)}{E \;\&\; t_1: \downarrow\theta(\tau_1') \;\&\; \cdots \;\&\; t_n: \downarrow\theta(\tau_n')}$$
if $f : \tau_1' \ldots \tau_n' \to \xi(\alpha_1,\ldots,\alpha_m)$ and where $\theta = \{\alpha_1/\tau_1,\ldots,\alpha_m/\tau_m\}$

Figure 6: The rules for polymorphic order-sorted unification

2. Transform P' & E into P'' & E'' using the rules (E) - (O) and (ES) - (DS) from Figure 6.

3. Transform P'' into P''' by replacing the approximations in P'' by the actual sort information from the original prefix P and the substitution defined by E''.

The third step uses the *retract* operation that transforms an approximation τ and a sort term τ' that is an upper bound of τ into the retraction $\tau\uparrow\tau'$ as follows:

$$
\begin{aligned}
\top\uparrow\tau &= \tau \\
\tau\uparrow\tau' &= \tau \\
& \qquad \text{if } \tau \text{ and } \tau' \text{ are sort constants and} \\
& \qquad \tau \text{ is a subsort of } \tau' \\
\xi(\tau_1,\ldots,\tau_n)\uparrow\xi(\tau_1',\ldots,\tau_n') &= \xi(\tau_1\uparrow\tau_1',\ldots,\tau_n\uparrow\tau_n') \\
\tau\uparrow\tau' & \qquad \text{undefined otherwise}
\end{aligned}
$$

The upper bounds for the sort term approximations in P'' are obtained by applying the substitution $\sigma_{E''}$ defined by E'' to the original prefix P and transforming $\sigma_{E''}(P)$ again into a prefix, say $P_{E''}$. If now $x : \tau$ is in P'' and $x : \tau'$ is in $P_{E''}$ then $x : \tau\uparrow\tau'$ will be in P'''.

4 The logic programming language PROTOS-L

The logic programming language PROTOS-L is currently being developed within the Eureka Project PROTOS (EU56) ([Appelrath, 1987], [Beierle and Böttcher, 1989]). The general objectives are:

1. Provide a logic programming language with a type concept.

2. Provide a module concept that allows separate type checking and compilation of modules.

3. Integrate access to existing external databases in a type-safe way and support the storage of large programs on external databases.

4. Provide an implementation model realizing the type concept, the separate handling of modules as well as the database aspects.

In the following three subsections we will briefly characterize types, modules and databases in PROTOS-L, while the implemenation model of PROTOS-L is treated in Section 5.

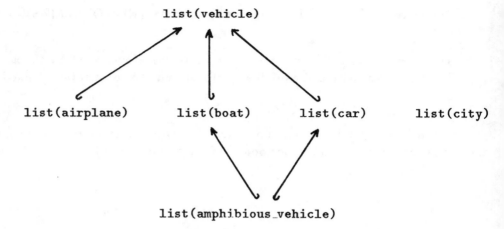

Figure 7: Induced subsort relationships (c.f. Figure 2)

4.1 Types

Due to the greater flexibility and naturalness of an order-sorted approach over a many-sorted approach PROTOS-L allows a partial order on the sort constants, and also provides parametric polymorphism. In the version of the language currently being implemented the type concept is inherited from TEL ([Smolka, 1988b]) but with a slightly simplified version of polymorphism: The TEL case may be argued to be too permissive in allowing the user to define explicitly subsort relationships between polymorphic sorts since this complicates the unification procedure considerably and prohibits the maximal-sort optimization in many cases (c.f. Section 3).

Therefore, in PROTOS-L there are no explicit subsort relationships between polymorphic sorts. Thus, the subsort relationship between sort terms is completely determined by the partial order on the sort constants and the monotonicity of the sort functions. For instance, given the sort hierarchy of Figure 2 and the polymorphic definition of standard lists some of the induced subsort relationships between sort terms are given in Figure 7.

An example of a simple PROTOS-L program without polymorphism (and without the modularization aspects) is the program given in Figure 1. Note that it is not necessary to give a sort restriction for every variable. Instead, the appropriate sort restrictions are derived from the declarations of the respective predicates. A simple polymorphic definition in PROTOS-L is the definition of the **append** predicate given in the previous section.

The underlying polymorphic order-sorted unification for the type concept of PROTOS-L has been given at the end of Section 3. Here we restricts ourselves to summarizing the major conditions the collection of sort and predicate definitions in a PROTOS-L program must satisfy, most of which have already been mentioned:

First of all, all occuring sort, function and predicate symbols must be introduced in a (unique) declaration. In particular, no overloading of symbols is allowed. The sorts may not be empty, i.e. for every sort there is a ground term of that sort. The partial order on the monomorphic sorts must have greatest lower bounds for sorts having common subsorts - the usual requirement for the uniqueness of order-sorted most general unifiers, c.f. [Walther, 1988]. A polymorphic sort declaration may not contain an explicit subsort declaration, and for a polymorphic sort declaration

$$\text{sort} \quad \text{ps}(\alpha_1, \ldots, \alpha_n) \; := \; \ldots$$

the α_i must be pairwise distinct variables - prohibiting for instance a declaration like sort pair(α, α) ....

Sort restrictions for variables may be declared in the form of X:car (c.f. Figure 1). For every clause C a prefix P_C is derived from both the sort restrictions given exlicitly in the clause and the declarations of the involved function and predicate symbols. All program clauses (and queries) must be well-typed w.r.t. the derived prefix. If C belongs to a non-polymorphic predicate both the head and the body of C must be well-typed in the usual order-sorted sense (w.r.t. P_C). If C belongs to a polymorphic predicate the body of C may contain literals that belong to a proper instance of the most general type, but the head of C must be of the most general type (c.f. Section 3).

4.2 Modules

In order to support separate type checking and compilation each PROTOS-L program consists of a set of modules. Each module in turn consists of an *interface* and a *body*. The purpose of the interface of a module is to define the set of imported names and the names that are introduced newly in this module and that are also exported. The user of a module sees only its interface, and not the body. Of course, when comparing this situation to the work done in the area of specification languages it is unsatisfactory insofar as the interface does not contain also a formal specification of the exported predicates. The incorporation of such a feature is the subject of further work.

In the module body the newly introduced names are defined, e.g. the clauses for a predicate p are given (for the so-called database bodies see the next subsection).

Naturally, such an implementation of a predicate must be compatible with its declaration in the interface.

A compilation unit is either an interface or a body. In order to compile such a compilation unit the interfaces of all imported modules must have been compiled, but not their bodies. The compilation of a module body of course additionally requires that its own interface has been compiled.

To give a simple example, suppose we want to have a module PRICES offering a relation price that gives the buying price and the retail price for several products. The interface of PRICES could be given by

```
interface PRICES.
    predicate  price:  string x nat x nat.
endinterface.
```

which uses the built-in sorts string and nat. A corresponding module body defining the retail price to be twice the buying price for all products is given by

```
module_body PRICES.
        price(Product, Buying_price, Retail_price) :-
            times(Buying_price, 2, Retail_price).
endmodule_body.
```

Note that not only an interface but also a module body may import modules, so-called local imports. This allows for nested module hierarchies, e.g. as sketched in Figure 8. In the interfaces of module M1 and M2 only M3 is imported. M3 in turn realizes a module hierarchy in its body. If now M4 or M5 are changed or m3 additionally imports locally in its body a new module M6, only the body of M3 has to be checked and recompiled, but not M1 and M2.

4.3 Databases

There are two different database aspects in PROTOS-L that must be distinguished.

One is to provide the integration of already existing external databases in order to gain access to large amounts of knowledge that otherwise might be unusable for e.g. a knowledge based sytem. However, the idea is to keep the database access as transparent as possible. In order to achieve this goal there is the concept of a *database body*. This means that besides the usual module bodies PROTOS-L also has database modules that provide an implementation of the predicates in its interface in terms of relations in an external database.

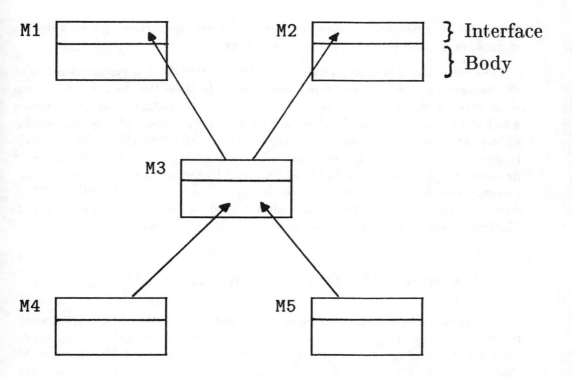

Figure 8: A module hierarchy

For instance, consider the interface definition for PRICES given above that introduces the predicate price and exports this predicate to all modules using PRICES. However, a user of PRICES does not know whether price is implemented by program clauses or by an external database relation. The latter is the case if the corresponding module body looks like

```
database_body PRICES.
    using db3.
    price is_implemented_by r4.
enddatabase_body.
```

which says that the predicate price is being implemented by the relation r4 in database db3.

In order to guarantee that such an access to an already existing external database is type-safe the argument types of a predicate implemented by a database relation are restricted by the argument types supported by the underlying database system. In the current implementation of PROTOS-L these are integer and string. However, the PROTOS-L system is also capable of correctly handling free variables in database bodies that have sort restrictions steming from subsorts of `integer` such as `pos_integer` or `neg_integer`.

The other use of external databases within the PROTOS-L system that has to be distinguished from the first one given above is to allow the storage of programs in an external database. As opposed to the access of existing external databases which is documented in the database body the external storage of programs is completely transparent to the user. At compile time the code of a PROTOS-L module is divided automatically into clusters that are stored in the database and accessed dynamically at runtime. Special clustering and buffer management strategies are designed in order to minimize the necessary database accesses at runtime. Thus, data retrieval on the one hand and code retrieval on the other are provided (c.f. [Böttcher, 1988], [Böttcher and Beierle, 1989], [Garidis, 1988]).

5 An abstract machine implementation

In this section we sketch an implementation model that satisfies the requirements listed at the beginning of Section 4. The heart of this implementation model for PROTOS-L is an abstract machine (called Protos Abstract Machine - PAM) that extends the Warren Abstract Maschine (WAM) in particular by polymorphic order-sorted unification and by a data base component.

5.1 The Warren Abstract Machine

For a detailed description of the WAM we refer to [Warren, 1977], [Warren, 1983] or [Gabriel *et al.*, 1985]. Here, we only give a brief overview on the machine model and describe the extensions and modifications that led to the PAM.

The machine model of the WAM consists essentially of the following:

1. The *code area* containing the machine code of the program.

2. Three stacks:

   - The *global stack* containing all structures that are generated during program execution. Structures are represented by the top-level function symbol followed by the arguments of the structure.

   - The *local stack* containing information on the execution structure of the program and backtracking information. In particular, the local stack contains *choice points* for alternative clauses for a goal and *environments* containing the status information needed for evaluating the rest of a clause.

- The *trail stack* containing the addresses of the variables that have been bound during program execution and that have to be reset upon backtracking.

3. A set of *registers* defining the current machine state, e.g. a next-instruction pointer, pointers to the top of the three stacks, a pointer to the last choicepoint, etc. and a special set of argument registers.

In addition to that an actual implementation will have further components like a *symbol table* containing the arity and the print names for every function and predicate symbol occuring in the program, etc.

The instruction set of the WAM can be divided into 5 classes:

1. The *get*-instructions are used for the arguments in the head of a clause in order to unify them with incoming arguments.

2. The *put*-instructions are used for the arguments in the body of a clause in order to build up the arguments of the subgoals.

3. The *unify*-instructions are used for the deeper-nested arguments within a structure.

4. The *environment* and *choice*-instructions are used for the management of procedure calls, choicepoints, and environments.

5. The *switch*-instructions are used to select only a subset of all alternative clauses from the set of all clauses of a predicate, depending on the value of the given arguments.

Additionally, there are some lower level instructions that are called by various of the instructions above, e.g. the low level instruction *fail* is called in order to initiate backtracking.

5.2 The PROTOS Abstract Machine

When implementing a first order-sorted version of PROTOS-L we experimented with an interpreting approach. The interpreter given in [Schütze, 1988] realizes order-sorted unification where each sort is being implemented by a bit-vector representing its subsorts, c.f. [Aït-Kaci *et al.*, 1989]. It is not necessary that any two sorts have at most one common maximal subsort since the partial order on the sorts given by the user is transformed automatically so that this condition is satisfied.

A first abstract machine implementation of an order-sorted version of PROTOS-L (without polymorphism) is described in [Müller, 1988]; this approach is similar to the one suggested in [Huber and Varsek, 1987]; c.f. also [Bürckert, 1985]. Here (and in the following) greatest lower bounds in the sort hierarchy are required. The subsort relationships between the sorts are compiled into a square matrix such that both a subsort test and the computation of the greatest lower bound of two sorts can be done in constant time. However, the matrix implementation is not easily extendible to separate compilation of modules, and it requires space that is quadratic in the number of sorts.

The current version of the PAM is described in [Semle, 1989]. It has the following distinguishing features:

- Polymorphic order-sorted unification.

- Support for the separate compilation of modules.

- Access to external data bases.

In the following, we will sketch the differences to the WAM.

The first difference is that the sort structure of a program has to be stored in the machine. This is mainly done in a new *sort table* containing the following information:

- The sort constants and their subsort relationships.

- The arity of the polymorphic sorts.

- Information on how sort terms can be instantiated.

For example, the instantiation information for the sort term $\text{list}(\alpha)$ states that α may be the 'empty' sort $\bot$ since [] is a term of sort $\text{list}(\bot)$. For the sort term $\text{pair}(\alpha,\beta)$ it is required that both α and β are instantiated by a sort other than $\bot$ since otherwise there would be no term of that pair sort.

Additionally, the symbol table must contain the arity of the function symbols e.g. in order to determine the target sort of the top-level function symbol of a term.

The second major difference is the representation of variables. In the WAM variables are represented by pointers. Binding a variable just requires pointing to the value (e.g. a structure in the global stack) and a free variable is a pointer to itself.

In the non-polymorphic version of the PAM the self-reference of a free variable was replaced by the sort constant representing the current sort restriction of that variable. The trail stack recording the variables that have been bound was extended in order to keep also the previous sort restriction of the variables (which have to be restored upon backtracking).

In the polymorphic PAM the variables are divided into three classes:

- The *free* variables which are treated as the free variables in the ordinary Prolog case. These are the variables for which no sort information is relevant at runtime, i.e. the approximation of their sort restriction is $\top$.

- The *mono* variables whose sort restriction is a (monomorphic) sort constant. These are represented as in the non-polymorphic PAM.

- The *poly* variables whose sort restriction is a non-constant sort term. The sort restriction of these variables is a pointer to the respective sort term. The sort terms themselves are represented just as usual terms, i.e. in the global stack where the polymorphic sort symbol is followed by its arguments.

The trail stack must record the previous sort restrictions of the variables which in the case of *poly* variables may now be a pointer to the global stack.

In order to determine the changes for the instruction set consider the unification rules given in Figure 6. The rules (E) - (O) are effectively as in the unsorted case (Figure 4) so that they do not bring any changes except that the binding rule (B) refers to the rules (ES) - (DS). The rules (ES) - (DS) manipulate the sort restrictions. For the elimination of a monomorphic sort restriction (ES) the target sort of the function symbol (which is recorded in the symbol table) must be in the subsort relationship (documented in the sort table) to the required sort. For the other elimination rule (ES') no action is required at all.

The merging of two sort restrictions (MS) requires the computation of the infimum of the two sort terms. The information required is the subsort relationship for the monomorphic sorts and the instantiability of polymorphic sorts which are both contained in the sort table.

The decomposition of a sort restriction (DS) requires that the arguments of a polymorphic sort term - like pair(car,airplane) - are propagated to the arguments of a value term - like mkpair(X:vehicle,Y:vehicle) -, yielding mkpair(X:car,Y:airplane) in the given example.

The instructions that have to be modified because of these observations are the *get-*, *put-*, and *unify*-instructions which are the instructions used for the unification of terms. They are generated by the compiler depending on the occurences of a variable or a term in a clause.

For example, suppose that the i-th argument of the head of a clause is a (temporary) variable represented by X_n. All one has to do in the unsorted case is to set X_n to the value of the i-th incoming argument A_i since at this point X_n is guaranteed to be free. This is achieved by the WAM instruction

 get_x_variable X_n, A_i

However, in the PAM case the sort restriction of X_n has to be taken into account. Following the distinctions made above X_n is either a *free*, a *mono* or a *poly* variable. Accordingly, the WAM instruction get_x_variable is replaced by three PAM instructions:

1. get_x_free X_n, A_i
2. get_x_mono X_n, A_i, s
3. get_x_poly X_n, A_i, st

The first instruction corresponds exactly to the WAM case since no sort-related action has to be done. The second instruction is generated for a variable with the sort constant s as its restriction. The execution of this instruction creates a new valuecell on the global stack with sort restriction s which is then unified with A_i in order to ensure that A_i is of sort s (possibly by further restricting the sort of A_i if it is a variable). Analogously, the third instruction generates a new valuecell with the polymorphic sort restriction st where st is a pointer to a sort term on the global stack. The unification with A_i then ensures that A_i satisfies the sort restriction.

Thus, the three PAM instructions (1) - (3) represent three special cases of the general unification procedure where at compile time it is guaranteed that one of the two terms to be unified is a free variable. On the other hand, for the second and any subsequent occurrence of X_n in the head of a clause the full unification is required. In the WAM case this is achieved by the instruction

 get_x_value X_n, A_i

and the only difference from the PAM case is that polymorphic order-sorted unification is used instead of ordinary term unification.

Analogous modifications are made for all other *get-*, *put-*, and *unify*-instructions: If a WAM instruction is used for the first occurence of a variable (i.e. get_y_variable, put_x_variable, put_y_variable, unify_x_variable, unify_y_variable) it is replaced by three PAM instructions for a *free*, a *mono* and a *poly* variable, respectively. For these as well as for all remaining instructions the extended unification is used.

However, except for the extended trail information the entire execution control including the management of choicepoints and enviroments can be treated as in the WAM case. For the *switch*-instructions an additional optimization is possible. For the exclusion of alternative clauses it is now also possible to take into account the sort restrictions for the variables. The sort-related modifications require also instructions for the retraction of the actual sort of a variable from its original sort and its approximation, as well as some additional lower level instructions, e.g. for the computation of the infimum of two sort terms.

The module support for the separate compilation of modules is given essentially by the structure of the sort table that allows adding new sorts and subsort relations incrementally and by an indirect addressing capability. Thus, the module concept can be realized mainly by the compiler and is not visible at the machine level. The address space of the PAM is a two-dimensional one where each address consists of a cluster identifier and an offset so that clusters can be loaded into main memory dynamically at runtime.

The PAM is implemented on an IBM RT/PC 6150. Although a detailed evaluation has still to be done our first results show the following observations for our first implementation ([Semle, 1989]):

- If a program uses only maximal sorts and no polymorphism we obtain the same efficiency as in the unsorted case. The same is true for polymorphic programs using only maximal sorts.

- The use of non-maximal polymorphic sorts may require time and space (global stack) depending on the depth of the involved structures.

- The time required by use of non-maximal sorts without polymorphism does not depend on the size of the involved terms. No additional global stack is required.

- In all cases no additional local stack is required.

- The number of executed abstract machine instructions is not changed by the use of sorts.

Thus, the maximal-sort optimization may already yield the same efficiency as in the unsorted case. However, the sort-related actions can still be reduced considerably which in this first implementation is necessary especially for the combination of polymorphic sorts with non-maximal sorts as arguments. For instance, the propagation of a sort restriction - e.g. list(car) - to a value term - e.g. a list with n elements - can be optimized if that term is already of that sort. In the given example this would save the propagation of the sort restriction car to the n elements of the list. These and other optimizations are currently being worked out and integrated into the PAM.

6   Extensions

The design of the current version of PROTOS-L and the PAM was influenced by various constraints that led to the exclusion of many desirable and interesting features. Some extension are given in the following.

A logic programming language should also include functions. There has already been a lot of work in the area of integrating logic and functional programming (see e.g. [DeGroot and Lindstrom, 1986]) but efficient implementations have still to be worked out. Also, the module concept should be elaborated. Drawing from work done in specifications languages e.g. the interface of a module should not only contain the names and the arity of the operations but also a formal specification of their behaviour. Another extension would refer to the type system although it is already quite powerful. One especially interesting direction for the development of knowledge based systems would be the inclusion of attributes or feature structures known from object oriented programming and knowledge representation. The data base access should also be extended in order to allow more powerful queries involving e.g. views, recursion etc. The PAM needs some further optimizations as already indicated in the previous section, and moreover, some of the extensions mentioned above will require a modified abstract machine concept.

References

[Aït-Kaci *et al.*, 1989] H. Aït-Kaci, R. Boyer, P. Lincoln, and R. Nasr. Efficient lattice operations. *ACM Transactions on Programming Languages and Systems*, 11(1):115–146, January 1989.

[Appelrath, 1987] H.-J. Appelrath. Das EUREKA-Projekt PROTOS. In W. Brauer and W. Wahlster, editors, *Proceedings GI-Kongreß Wissensbasierte Systeme*, pages 1–11, Springer-Verlag, Informatik-Fachberichte 155, 1987.

[Beierle and Böttcher, 1989] C. Beierle and S. Böttcher. PROTOS-L: Towards a knowledge base programming language. In *Proceedings GI-Kongreß Wissensbasierte Systeme*, Springer-Verlag, 1989. (to appear).

[Böttcher, 1988] S. Böttcher. The architecture of the PROTOS-L system. In H.-J. Appelrath, A. B. Cremers, and H. Schiltknecht, editors, *Proc. First PROTOS Workshop*, Morcote, Switzerland, 1988.

[Böttcher and Beierle, 1989] S. Böttcher and C. Beierle. Data base support for the PROTOS-L system. In *Proceedings EUROMICRO-89*, North Holland, 1989. (to appear).

[Bürckert, 1985] H.-J. Bürckert. *Extending the WARREN Abstract Machine to Many-Sorted Prolog*. SEKI-Memo-85-07, FB Informatik, Universität Kaiserslautern, 1985.

[Cardelli and Wegner, 1985] L. Cardelli and P. Wegner. On understanding types, data abstraction and polymorphism. *ACM Computing Surveys*, 17(4):471–522, December 1985.

[Cohn, 1987] A. G. Cohn. A more expressive formulation of many sorted logic. *Journal of Automated Reasoning*, 3:113–200, 1987.

[Damas and Milner, 1982] L. Damas and R. Milner. Principal type-schemes for functional programs. In *Proceedings of the 9th ACM Symposium on Principles of Programming Languages*, pages 207–212, 1982.

[Dayantis, 1988] G. Dayantis. Types, modules and abstraction in logic programming. In J. Grabowski, P. Lescanne, and W. Wechler, editors, *Algebraic and Logic Programming*, pages 127–136, Akademie-Verlag, Berlin, 1988.

[DeGroot and Lindstrom, 1986] Douglas DeGroot and Gary Lindstrom, editors. *Functional and Logic Programming*. Prentice Hall, 1986.

[Dietrich and Hagl, 1988] R. Dietrich and F. Hagl. A polymorphic type system with subtypes for Prolog. In *Proceedings of the 2nd European Symposium on Programming*, pages 79–93, Lectures Notes in Computer Science, Volume 300, Springer-Verlag, Berlin, Heidelberg, New York, 1988.

[Ehrig and Mahr, 1985] Harmut Ehrig and Bernd Mahr. *Fundamentals of Algebraic Specification 1 - Equations and Initial Semantics*. EATCS Monographs on Theoretical Computer Science, Volume 6, Springer-Verlag, Berlin, Heidelberg, New York, 1985.

[Futatsugi et al., 1985] K. Futatsugi, J. Goguen, J.-P. Jouannaud, and J. Meseguer. Principles of OBJ2. In B. Reid, editor, *Proceedings of 12th ACM Conference on Principles of Programming Languages*, pages 52–66, ACM, 1985.

[Gabriel et al., 1985] J. Gabriel, T. Lindholm, E. L. Lusk, and R. A. Overbeek. *A Tutorial on the Warren Abstract Machine*. Technical Report, Mathematics and Computer Science Department, Argonne National Laboratory, Argonne, Illinois, 1985.

[Garidis, 1988] C. Garidis. *Logisches Clustering von PROTOS-L Prozeduren*. PROTOS Document B.4, IBM Germany, Stuttgart, 1988.

[Gogolla, 1986] M. Gogolla. *Über partiell geordnete Sortenmengen und deren Anwendung zur Fehlerbehandlung in abstrakten Datentypen*. PhD thesis, TU Braunschweig, 1986.

[Goguen, 1978] J. Goguen. *Order Sorted Algebra*. Semantics and Theory of Computation Report No. 14, University of California, Los Angeles, 1978.

[Goguen and Meseguer, 1986] Joseph A. Goguen and Jose Meseguer. Eqlog: equality, types, and generic modules for logic programming. In Douglas De-Groot and Gary Lindstrom, editors, *Functional and Logic Programming*, pages 295–363, Prentice Hall, 1986.

[Goguen and Meseguer, 1987] J. A. Goguen and J. Meseguer. *Order-Sorted Algebra I: Partial and Overloaded Operators, Errors and Inheritance*. Technical Report, Computer Science Lab., SRI International, Menlo Park, 1987.

[Goguen et al., 1978] Joseph A. Goguen, James W. Thatcher, and Eric G. Wagner. An initial algebra approach to the specification, correctness and implementation of abstract data types. In R. Yeh, editor, *Current Trends in Programming Methodology IV: Data and Structuring*, pages 80–144, Prentice Hall, 1978.

[Hanus, 1988] M. Hanus. *Horn Clause Specifications with Polymorphic Types*. PhD thesis, FB Informatik, Universität Dortmund, 1988.

[Hanus, 1989] M. Hanus. Horn clause programs with polymorphic types. In *Proceedings TAPSOFT'89*, Springer-Verlag, 1989.

[Huber and Varsek, 1987] M. Huber and I. Varsek. Extended Prolog for order-sorted resolution. In *Proceedings of the 4th IEEE Symposium on Logic Programming*, pages 34–45, San Francisco, 1987.

[Lloyd, 1984] J. W. Lloyd. *Foundations of Logic Programming*. Symbolic Computation, Springer-Verlag, Berlin, Heidelberg, New York, 1984.

[Martelli and Montanari, 1982] A. Martelli and U. Montanari. An efficient unification algorithm. *ACM Transactions on Programming Languages and Systems*, 4(2):258–282, 1982.

[Milner, 1978] R. Milner. A Theory of Type Polymorphism in Programming. *Journal of Computer and System Science*, 17:348–375, 1978.

[Müller, 1988] B. Müller. *Entwurf und Implementierung einer abstrakten Maschine für ordnungsortierte Logikprogramme*. Studienarbeit Nr. 711, Universität Stuttgart und IBM Deutschland GmbH, Stuttgart, October 1988.

[Mycroft and O'Keefe, 1984] A. Mycroft and R. A. O'Keefe. A polymorphic type system for Prolog. *Artificial Intelligence*, 23:295–307, 1984.

[Oberschelp, 1962] A. Oberschelp. Untersuchungen zur mehrsortigen Quantorenlogik. *Mathematische Annalen*, 145:297–333, 1962.

[Schmidt-Schauß, 1985] M. Schmidt-Schauß. A many-sorted calculus with polymorphic functions based on resolution and paramodulation. In *Proceedings of the 9th International Conference on Artificial Intelligence*, pages 1162–1168, Kaufmann, 1985.

[Schütze, 1988] H. Schütze. *Erweiterung eines Prolog-Interpreters um Typbehandlung*. Studienarbeit Nr. 710, Universität Stuttgart und IBM Deutschland GmbH, Stuttgart, August 1988.

[Semle, 1989] H. Semle. *Erweiterung einer abstrakten Maschine für ordnungssortiertes Prolog um die Behandlung polymorpher Sorten*. Diplomarbeit Nr. 583, Universität Stuttgart und IBM Deutschland GmbH, Stuttgart, April 1989.

[Smolka, 1988a] G. Smolka. Logic programming with polymorphically ordersorted types. In J. Grabowski, P. Lescanne, and W. Wechler, editors, *Algebraic and Logic Programming*, Akademie-Verlag, Berlin, 1988.

[Smolka, 1988b] G. Smolka. *TEL (Version 0.9), Report and User Manual*. SEKI-Report SR 87-17, FB Informatik, Universität Kaiserslautern, 1988.

[Smolka, 1989] G. Smolka. *Logic Programming over Polymorphically Order-Sorted Types*. PhD thesis, FB Informatik, Univ. Kaiserslautern, 1989.

[Smolka et al., 1987] G. Smolka, W. Nutt, J. A. Goguen, and J. Meseguer. *Order-Sorted Equational Computation*. SEKI-Report SR 87-14, FB Informatik, Universität Kaiserslautern, 1987. (to appear in: H. Ait-Kaci, M. Nivat (eds): Resolution of Equations in Algebraic Structures, Academic Press).

[Waldmann, 1989] U. Waldmann. *Unification in Order-Sorted Signatures*. Forschungsbericht Nr. 298, FB Informatik, Universität Dortmund, 1989.

[Walther, 1987] C. Walther. *A Many-Sorted Calculus Based on Resolution and Paramodulation*. Research Notes in Artificial Intelligence, Pitman, London, and Morgan Kaufmann, Los Altos, Calif., 1987.

[Walther, 1988] C. Walther. Many-sorted unification. *Journal of the ACM*, 35(1):1–17, January 1988.

[Warren, 1977] D. Warren. *Compiling Predicate Logic Programs*. D.A.I. Research Report, University of Edinburgh, 1977.

[Warren, 1983] D. Warren. *An Abstract PROLOG Instruction Set.* Technical
    Report 309, SRI, 1983.

II. On Sorts and Types in Knowledge Representation including Qualitative Reasoning

Representation and Reasoning
with Attributive Descriptions

Bernhard Nebel and Gert Smolka*

IWBS, IBM Deutschland[†]

Abstract

This paper surveys terminological representation languages and feature-based unification grammars pointing out the similarities and differences between these two families of attributive description formalisms. Emphasis is given to the logical foundations of these formalisms.

1 Introduction

Research in knowledge representation and linguistics has led to the development of two families of formalisms which can jointly be characterized as *attributive description formalisms*. The members of the first family are known as *terminological representation languages* and are offsprings of Brachman's KL-ONE [9], which grew out of research in semantic networks and frame systems. The second family whose members are known as *unification grammars* originated with Kaplan and Bresnan's Lexical-Functional Grammar [17] and Kay's Functional Unification Grammar [19,21,20].

This paper surveys terminological representation languages and unification grammars in an attempt to clarify the similarities and differences between the two approaches. Both approaches

1. rely on attributes as the primary notational primitive for representing knowledge

2. are best formalized as first-order logics with Tarski-style models

3. employ compositional set descriptions.

The two approaches differ significantly, however, both in the representational constructs and the reasoning operations they provide. Unification grammars employ functional attributes called *features* while terminological representation languages rely on more general relational attributes called *roles*. The semantically minor-looking difference between features and roles results in very different computational properties. Moreover, terminological representation systems infer set inclusion and set membership relations on user-defined symbols, while parsers based on unification grammars solve constraints in a domain consisting of so-called feature graphs.

---

*Funded by the EUREKA Project Protos (EU 56).

[†]Address for correspondence: IBM Deutschland, IWBS, Postfach 80 08 80, 7000 Stuttgart 80, West Germany; e-mail: nebel@ds0lilog.bitnet, smolka@ds0lilog.bitnet.

Surprisingly, the similarities of terminological representation languages and unification grammars have not been recognized for quite a while. The main reason for this ignorance is probably that the communities of researchers working in the respective fields are almost non-overlapping although recently ideas from feature constraint languages used with unification grammars found their way into terminological representation systems. Nevertheless, so far, a paper putting both approaches into perspective by giving a comprehensible survey of the commonalities and differences is missing—a situation we hope to remedy with this paper.

We emphasize the logical foundations of the two approaches since it is here that the similarities and differences show up most clearly. Moreover, only the development of the logical foundations of terminological representation languages and unification grammars provided the base for the recent surge of important results on and extensions of these formalisms.

The structure of the paper is straightforward. Section 2 discusses terminological representational languages and Section 3 discusses feature-based unification grammars.

Acknowledgement. Our presentation of unification grammars profited from discussions with Jochen Dörre and Bill Rounds.

2   Terminological Representation and Reasoning

One important task in modeling an application domain in artificial intelligence systems is to fix the vocabulary intended to describe the domain—the *terminology*—and to define interrelationships between the atomic parts of the terminology. Representation systems supporting this task are KL-ONE [9] and its descendants [8,51,33,29,24]. The main idea is to introduce each concept by a *terminological axiom* that associates the new concept with a *concept description* specifying the intended meaning in terms of other concepts.

If we intend to talk, for instance, about persons, we may introduce the concepts Person, Adult, Man, and Woman by relating them to each other as follows:

$$
\begin{aligned}
\text{Adult} &\sqsubseteq \text{Person} \\
\text{Woman} &\sqsubseteq \text{Adult} \\
\text{Man} &\doteq \text{Adult} \sqcap \neg\text{Woman}.
\end{aligned}
$$

In other words, Adult is introduced as something specializing Person, Woman as specializing Adult, and men as adults who aren't women.

2.1   The Representational Inventory

Formal *terminologies*, such as the one above, are composed out of *terminological axioms* (*TA*) which relate a concept (the left hand side) to a concept description (the right hand side) using the *specialization operator* "$\sqsubseteq$" and the *equivalence operator* "$\doteq$":

$$TA \rightarrow A \sqsubseteq C \mid A \doteq C$$

with the additional restriction that no concept may occur more than once as a left hand side in a terminology.

For the moment, let us assume that the right hand sides of terminological axioms—the *concept descriptions* (denoted by C and D)—are composed out of *concepts* (denoted by A and B) and the following *description-forming operators*:

$$C, D \rightarrow A \mid C \sqcap D \mid C \sqcup D \mid \neg C.$$

In order to specify the meaning of terminologies formally, we define an **interpretation** $\mathcal{I}$ as a pair $\langle \mathbf{D}^{\mathcal{I}}, \cdot^{\mathcal{I}} \rangle$ with $\mathbf{D}^{\mathcal{I}}$ an arbitrary set—the *domain*—and $\cdot^{\mathcal{I}}$ a function from concepts to subsets of $\mathbf{D}^{\mathcal{I}}$—the *interpretation function*. Based on that, the interpretation of concept descriptions is defined inductively:

$$
\begin{aligned}
(C \sqcap D)^{\mathcal{I}} &= C^{\mathcal{I}} \cap D^{\mathcal{I}} \\
(C \sqcup D)^{\mathcal{I}} &= C^{\mathcal{I}} \cup D^{\mathcal{I}} \\
(\neg C)^{\mathcal{I}} &= \mathbf{D}^{\mathcal{I}} \setminus C^{\mathcal{I}}.
\end{aligned}
$$

This means that a concept is interpreted as standing for a *set* of objects—its *extension*—and a concept descriptions is interpreted as standing for the set resulting from straightforward applications of set operations corresponding to the description-forming operations. The concepts $\top$ and $\bot$ will be used as abbreviations for $A \sqcup \neg A$ and $A \sqcap \neg A$, respectively, where A is any concept. Thus, $\top$ is interpreted as the set of everything and $\bot$ is interpreted as the empty set.

An interpretation $\mathcal{I}$ **satisfies a terminological axiom** σ, written $\models_{\mathcal{I}} \sigma$, iff the sets denoted by the right hand and left hand side relate to each other as suggested by the symbols:

$$
\begin{aligned}
&\models_{\mathcal{I}} A \doteq D \quad \text{iff} \quad A^{\mathcal{I}} = D^{\mathcal{I}} \\
&\models_{\mathcal{I}} A \sqsubseteq D \quad \text{iff} \quad A^{\mathcal{I}} \subseteq D^{\mathcal{I}}.
\end{aligned}
$$

Furthermore, an interpretation $\mathcal{I}$ is a **model of a terminology** T, written $\models_{\mathcal{I}} T$, iff all terminological axioms in T are satisfied by $\mathcal{I}$.

Having defined the formal meaning of terminologies, we can now say which *terminological formulas* are *entailed* by a terminology. Here, we will permit arbitrary formulas $C \doteq D$ and $C \sqsubseteq D$. Such a formula τ is **entailed** by a terminology T, written $T \models \tau$, iff τ is satisfied by all models of T. Applying this definition to the introductory example, it is easy to see that the following formulas are entailed:

$$
\begin{aligned}
\text{Man} &\sqsubseteq \text{Person} \\
\text{Man} \sqcap \text{Woman} &\doteq \bot \\
\text{Man} \sqcup \text{Woman} &\doteq \text{Adult}.
\end{aligned}
$$

Based on the entailment relation between terminologies and formulas, it is possible to define a relation on the set of concept descriptions, namely, the **subsumption relation** $\preceq_T$ defined as

$$C \preceq_T D \quad \text{iff} \quad T \models C \sqsubseteq D.$$

This relation is obviously a preorder (that is, transitive and reflexive) on the set $Cd(T)$ of concept descriptions composed from symbols appearing in T, and a partial order on the quotient of $Cd(T)$ with respect to the equivalence relation $\approx_T$ defined by

$$C \approx_T D \quad \text{iff} \quad T \models C \doteq D.$$

Although the language defined so far gives a good first impression of the general idea behind terminological representation formalisms, one essential ingredient is missing. All such languages contain constructs to describe concepts by specifying *attributes*—a property which led to the title of this paper.

Reconsidering the introductory example, we note that this terminology contains a certain asymmetry; the meaning of Man is derived from the meaning of Woman but not *vice versa*. In order to remove this asymmetry it is tempting to define these concepts by referring to the sex attribute of a person. Before we can do this, our description language has to be extended, however.

First of all, we need *features* (denoted by f, g and h), which we will interpret as unary partial functions $f^{\mathcal{I}}: \mathbf{D}(f^{\mathcal{I}}) \to \mathbf{D}^{\mathcal{I}}$ with $\mathbf{D}(f^{\mathcal{I}}) \subseteq \mathbf{D}^{\mathcal{I}}$. Second, we need *constants* (denoted by a, b, and c), such as male and female, which are interpreted as elements of $\mathbf{D}^{\mathcal{I}}$ under the *unique name assumption*, that is, different constants are assumed to denote different objects:

$$\text{if } a^{\mathcal{I}} = b^{\mathcal{I}} \text{ then } a = b.$$

Employing these new syntactic categories, the *description-forming language* is extended to

$$C, D \to \dots \mid c \mid f : C$$

with the interpretation

$$c^{\mathcal{I}} = \{c^{\mathcal{I}}\}$$
$$(f : C)^{\mathcal{I}} = \{d \in \mathbf{D}(f^{\mathcal{I}}) \mid f^{\mathcal{I}}(d) \in C^{\mathcal{I}}\}.$$

Note that our notation is overloaded since a constant c is interpreted as the singleton $\{c^{\mathcal{I}}\}$ if it appears as a concept description.

With this machinery, the introductory example can be rephrased as

$$
\begin{aligned}
\text{Person} &\sqsubseteq \text{sex} : (\text{male} \sqcup \text{female}) \\
\text{Adult} &\sqsubseteq \text{Person} \\
\text{Woman} &\doteq \text{Adult} \sqcap \text{sex} : \text{female} \\
\text{Man} &\doteq \text{Adult} \sqcap \text{sex} : \text{male}.
\end{aligned}
$$

Although describing concepts by placing restrictions on features results already in a powerful description language, it is possible to generalize this a bit further. Instead of permitting only single-valued attributes, we may conceive multi-valued attributes. For instance, when we want to talk about the children of a person, we cannot represent them as values of a feature. Accounting for this, we introduce multi-valued attributes, called *roles* and denoted by r, that are interpreted as total functions $r^{\mathcal{I}}: \mathbf{D}^{\mathcal{I}} \to 2^{\mathbf{D}^{\mathcal{I}}}$ from elements of $\mathbf{D}^{\mathcal{I}}$ to subsets of $\mathbf{D}^{\mathcal{I}}$. The set $r^{\mathcal{I}}(d)$ will be called the *role-filler set* of d for role r.

Using roles for forming concept description, at least something like "all objects of a role-filler set are of a certain type" and "there exist objects of a certain type" seem to make sense. Thus, let us extend our syntax by *role restrictions*

$$C, D \to \dots \mid \forall r : C \mid \exists r : C$$

and assume the interpretations

$$(\forall r\colon C)^{\mathcal{I}} \;=\; \{d \in \mathbf{D}^{\mathcal{I}} \mid r^{\mathcal{I}}(d) \subseteq C^{\mathcal{I}}\}$$
$$(\exists r\colon C)^{\mathcal{I}} \;=\; \{d \in \mathbf{D}^{\mathcal{I}} \mid r^{\mathcal{I}}(d) \cap C^{\mathcal{I}} \neq \emptyset\}.$$

With this extension of our concept description language[1] we can, for instance, describe the important persons in a family:

Person	$\sqsubseteq$	sex: (male $\sqcup$ female)
Parent	$\doteq$	Person $\sqcap$ $\exists$child: $\top$ $\sqcap$ $\forall$child: Person
Mother	$\doteq$	Parent $\sqcap$ sex: female
Grandparent	$\doteq$	Person $\sqcap$ $\exists$child: Parent $\sqcap$ $\forall$child: Person
Grandmother	$\doteq$	Grandparent $\sqcap$ sex: female.

Although this terminology looks quite natural, one might ask why we did not mention the role child when introducing the Person concept, that is,

$$\text{Person} \quad \sqsubseteq \quad \text{sex: (male} \sqcup \text{female)} \sqcap \forall \text{child: Person.}$$

Actually, this would have been possible. However, it would have resulted in a **terminological cycle**, a direct (or indirect) occurrence of the concept introduced on the left hand side in the concept description on the right hand side. Such cycles, which are usually not supported in terminological representation systems, are problematical for at least two reasons:

1. the intuitive meaning of a concept containing a terminological cycle is not fully clear

2. it is not straightforward to design subsumption algorithms (see Section 2.2) which deal correctly with terminological cycles.

Concerning the first problem, the semantics of concepts containing cycles, we will note here only that the formal semantics provided so far is adequate to deal with such structures in a satisfying way (see [27, Chapter 5]). A solution for the second problem will be sketched in Section 2.3.

While so far we have talked about how to represent knowledge about concepts by employing terminological axioms, the question comes up: what can you do with your represented knowledge— what are the *computational services* provided by a terminological representation system?

2.2 Computational Services

One service terminological representation systems such as KL-ONE provide, called *classification*, is the computation of a *concept taxonomy*, such as the one in Figure 1, which represents the subsumption relation between concepts for the terminology in the previous subsection. Such a concept taxonomy represents the subsumption relation in a quite dense format. Defining $\lhd_T$ as a **base** of $\preceq_T$, that is,

---

[1]Note that the story we told so far is oversimplified. Terminological formalisms used in existing representation systems such as KL-ONE [9], KANDOR [33], BACK [29], CLASSIC [6] use different formalisms. In particular, *features* are used only in CLASSIC. For the purpose of giving the general idea, the current presentation should suffice, though.

a *minimal relation* such that the reflexive, transitive closure of $\unlhd_T$ is identical with $\preceq_T$, the relation computed by the classification process can be described as the *base of the subsumption relation on the quotient of the set of concepts with respect to* $\approx_T$. Since the base of a finite partial order is always unique, for every terminology there is a unique concept taxonomy.

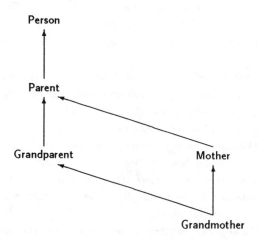

Figure 1: A concept taxonomy

Evidently, subsumption and classification are intertwined. In order to compute the concept taxonomy, subsumption between concepts must be determined. Once the concept taxonomy has been computed, subsumption between concepts can be read off from the taxonomy,[2] that is, classification can be regarded as a kind of assert-time inference technique.

As can be seen from the example in Figure 1, classification is a non-trivial service. It has to take into account all concept descriptions used in terminological axioms and has to compare them in order to determine the subsumption relationship. For instance, the relationship between Grandparent and Parent, which is not explicitly stated in the terminology, has to be derived by comparing the restrictions on the child role. Although in this example the subsumption relation seems to be quite obvious, the task of determining subsumption can become arbitrarily difficult as we will see in Section 2.3.

Ignoring this unpleasant situation for the moment, we note that classification is a versatile service for the knowledge acquisition task. Classification points out all implicit relationships between concepts which might have been missed when introducing a concept. As shown in [1,10], classification can be used to drive the knowledge acquisition process by employing a number of reasonable heuristics such as that different concepts should not denote the same set and that no concept should be **incoherent**, that is, equivalent to $\perp$. Note that such incoherent concepts are quite useless because they denote the empty set, but they do not "infect" the knowledge base in the sense a contradictory proposition in logic does. Terminologies always have at least one model, namely, the trivial one

---

[2] Actually, in some representation systems not only the concept taxonomy but also the transitive closure is stored.

interpreting every concept, feature, and role as the empty set.

Proposition 2.1 *Every terminology has a model.*

Knowledge acquisition is not the only application where classification can be put to use. In general, any problem requiring *classification-based reasoning* [15] can exploit this service. This kind of reasoning proceeds along the following line. Given some concept description, identify the concepts which most accurately characterize the given description and use information associated with the identified concepts to do something meaningful, that is, the concepts are used as a kind of *conceptual coat rack* [52].

Making this idea less abstract, let us assume that we want to identify a plan in order to solve a problem. Now, we may define a hierarchy of problem concepts associating with each such problem concept a plan for solving the problem. Thus, given a particular problem description, classification can determine the most specialized set of problem concepts for which plans are known in order to solve the given problem [30]. Such an organization of problem-solving knowledge is not only very elegant and natural, but also makes maintenance of such a knowledge base easier and supports explanation facilities. Other examples of where this kind of representation and reasoning can be profitably exploited are computer configuration [31], natural language generation [49], presentation planning [4], and information retrieval [50,35,5].

However, in most of the applications cited above, one does not start with a description of, say, a particular problem, but one has a collection of *objects* and *relationships* between them. Given such a *world description*,[3] one wants to know the set of concepts most accurately describing those objects. In order to capture this formally, let us again extend our formalism. This time, however, we do not add new description-forming expressions or terminological axioms, but something, which will be called *world axioms* (*WA*) in order to describe objects by naming the concepts they shall be an *instance* of and to describe relationships between two objects by specifying features or roles:

$$WA \rightarrow C(c) \mid f(c,d) \mid r(c,d).$$

Using the interpretation of constants, concepts, roles, and features given above, we will say that **a world axiom is satisfied by an interpretation**, written $\models_{\mathcal{I}} \omega$, under the following conditions:

$$\models_{\mathcal{I}} C(c) \quad \text{iff} \quad c^{\mathcal{I}} \in C^{\mathcal{I}}$$
$$\models_{\mathcal{I}} f(c,d) \quad \text{iff} \quad f^{\mathcal{I}}(c^{\mathcal{I}}) = d^{\mathcal{I}}$$
$$\models_{\mathcal{I}} r(c,d) \quad \text{iff} \quad d^{\mathcal{I}} \in r^{\mathcal{I}}(c^{\mathcal{I}}).$$

Similar to the definition of a model of a terminology, we can say what we mean by a **model of a world description** W or by a **model of a world description** W **combined with a terminology** T, namely, an interpretation which satisfies all world axioms in W or all terminological and world axioms in $T \cup W$, respectively. Furthermore, we will say that c is an **instance** of C iff $T \cup W \models C(c)$.

Using this extension of our formalism, we can describe a world specifying the objects and relationships of interest by a world description W employing a terminology T. For instance, we may use

---

[3]In the KL-ONE terminology, such a world description is usually called ABox—the *assertional box*—contrasting it with the *terminological box*, the TBox.

the family terminology in order to describe a particular family constellation:

$$Parent(harry)$$
$$(Parent \sqcap sex: female)(mary)$$
$$child(mary, tom)$$
$$child(mary, harry).$$

From this constellation it follows that mary is a Grandmother and that tom is a Person.

In general, representation systems supporting the reasoning with terminological and world axioms provide a computational service called **realization** which computes for each constant c the set of most specialized concepts $MSC(c)$ the constant is an instance of. Formally, $MSC(c)$ is a minimal set of concepts[4] such that

$$if \ A \in MSC(c) \quad then \quad T \cup W \models A(c) \tag{1}$$
$$if \ T \cup W \models B(c) \quad then \quad there \ exists \ A : A \in MSC(c) \ and \ T \cup W \models A \sqsubseteq B. \tag{2}$$

In our case, the second condition (2) can be simplified because world descriptions form an (almost) *conservative extension* of terminologies, that is, entailment of terminological formulas depends (in all interesting cases) only on the terminology and not on the world description.

Theorem 2.2 *If a world description W and a terminology T are jointly satisfiable, and if no constant that is used in a terminological axiom occurs in a world axiom, then*

$$T \cup W \models \tau \ iff \ T \models \tau$$

for all terminological formulas τ.

Proof: The "if" direction is obvious. For the "only if" direction let us assume that there are two descriptions C and D with $T \cup W \models C \sqsubseteq D$ but $T \not\models C \sqsubseteq D$. Since we can extend any model of $T \cup W$ by a disjoint union of a model of T (only agreeing in the interpretation of constants used in terminological axioms), there must be a model of $T \cup W$ which does not satisfy $C \sqsubseteq D$. Thus, our assumption must be wrong. ∎

This means that we can formulate condition (2) equivalently as

$$if \ T \cup W \models B(c) \quad then \quad there \ exists \ A : A \in MSC(c) \ and \ A \preceq_T B,$$

provided the assumptions of the theorem hold. This means in particular that after $MSC(c)$ has been computed, instance relationships for c can be determined by looking up subsumption in the concept taxonomy.[5]

In a presentation planning application (see, for instance, [4]) the information associated with the concepts in MSC might be used to decide how to represent a given object. In a database or

---

[4] Actually, of equivalence classes of concepts in order to guarantee uniqueness.

[5] Note that this property heavily depends on the fact that the world axioms have very limited expressiveness. If arbitrary first-order formulas are permitted, as in KRYPTON, the computation of instance relationships becomes much more complicated.

information retrieval application, *MSC* can be used to *index* the data objects by the concepts in *MSC* [35,5]. *Query processing* can then be implemented as *classification* of a *query concept*, *retrieval* of all objects indexed by the immediate superconcepts of the query concept in the concept taxonomy, and *filtering* by testing each retrieved object against the query concept.

There are a number of points we have omitted in the presentation of the representational inventory provided by terminological representation systems. In particular, most terminological formalisms allow for a richer repertoire of description-forming operators. First, instead of a simple existential quantification, numerical quantification is usually permitted, for instance, "at least 2 children." Second, roles are not necessarily treated as primitive entities, but they can be defined similarly to the way concepts can be defined. For instance, defining an unspecific subrole of a role or defining a role by restricting the range of another role is a common operation. An example for the latter is the definition of a role son, which can be done by restricting the range of child to sex: male. Third, so-called *role-value-maps*—equality constraints on role-filler sets—are often used, which correspond to *agreements* to be discussed in Section 3.3.

2.3 Algorithmic Considerations and Complexity

Although we have talked about computational services, we haven't given algorithms which do the actual computations. However, instead of specifying inference algorithms (see, e.g. [23,32,29,27]), we will investigate the computational properties of the problems. In particular, it will be shown that terminological reasoning is inherently intractable.

As we have seen in the previous section, subsumption determination is the central operation in a terminological knowledge representation system. This point is reinforced by the fact that all other interesting properties and relations, such as *equivalence of two concepts* ($C \approx_T D$), *incoherency of a concept* ($C \approx_T \perp$), and *disjointness of two concepts* ($(C \sqcap D) \approx_T \perp$) can be reduced to subsumption in linear time.

Proposition 2.3 *Given a terminology T and two concept descriptions C, D:*

1. $C \approx_T \perp$ *iff* $C \preceq_T \perp$
2. $C \approx_T D$ *iff* $C \preceq_T D$ *and* $D \preceq_T C$

Similarly, subsumption can be reduced to incoherency and equivalence.

Proposition 2.4 *Given a terminology T and two concept descriptions C, D:*

1. $C \preceq_T D$ *iff* $(\neg C \sqcap D) \approx_T \perp$.
2. $C \preceq_T D$ *iff* $C \approx_T (C \sqcap D)$.

In other words, when looking for efficient inference algorithms for terminological reasoning systems, we have to find efficient subsumption, equivalence, or incoherency detection algorithms. In order to simplify matters, we will show how subsumption in arbitrary terminologies can be reduced to subsumption in the *empty terminology*, denoted by $\emptyset$. First, we will show that the specialization operator "$\sqsubseteq$" is not essential for the expressiveness of terminological formalisms. Terminologies without this operator will be called **equational** terminologies.

Lemma 2.5 *Any terminology T can be transformed in linear time into a equational terminology T'' such that for all $C, D \in Cd(T)$:*

$$C \preceq_T D \quad \text{iff} \quad C \preceq_{T'} D.$$

Proof: Rewrite each axiom of the form $A \sqsubseteq D$ to $A \doteq \overline{A} \sqcap D$, where $\overline{A}$ is a fresh concept, called *primitive component* of A. Obviously, for any model $\mathcal{I}$ of T there exists a model $\mathcal{I}'$ of T'' (setting $\overline{A}^{\mathcal{I}} = A^{\mathcal{I}}$) such that

$$C^{\mathcal{I}} = C^{\mathcal{I}'} \quad \text{for all } C \in Cd(T)$$

and *vice versa*. Since the interpretation of all concept descriptions is identical, the subsumption relation on $Cd(T)$ is identical. ∎

As the second step, a function Exp from concept descriptions and equational terminologies to concept descriptions is defined. This function repeatedly replaces all concepts A appearing in a given concept description C by the right hand side of the introduction of A until no further replacements are possible. Thus, $Exp(C, T)$ contains only concepts that do not appear as the left hand side of a terminological axiom in T. This function clearly terminates if T does not contain terminological cycles. Moreover, for every model $\mathcal{I}$ of T

$$C^{\mathcal{I}} = (Exp(C, T))^{\mathcal{I}} \tag{3}$$

since the replaced subexpressions and the replacing expressions in C are identically interpreted in T. From this observation it is almost immediate that subsumption in terminologies can be reduced to subsumption in the empty terminology $\emptyset$.

Theorem 2.6 *Given a equational terminology T and two concept descriptions C, D:*

$$C \preceq_T D \quad \text{iff} \quad Exp(C, T) \preceq_\emptyset Exp(D, T).$$

Proof: For the "if" direction note that any interpretation is a model of the empty terminology $\emptyset$. Thus, subsumption in $\emptyset$ implies subsumption in any particular terminology. Furthermore, because of (3) the "if" direction holds. The converse direction follows from (3) and the fact that subsumption between concept descriptions of the form $Exp(C, T)$ depend only on interpretations of concepts unconstrained by T. ∎

This means when developing inference algorithms, we have to consider only the description-forming part of the formalism. Concentrating on the description-forming language introduced in Section 2.1, which will be called $\mathcal{ALC}$ following [43],[6] it is easy to see that subsumption is decidable for this language.

Lemma 2.7 *Let C and D be $\mathcal{ALC}$ concept descriptions, then it is decidable whether $C \preceq_\emptyset D$.*

---

[6] Actually, in [43] the description-forming language does not contain features. These do not add to the principal complexity of the subsumption problem, however.

Proof Sketch: In order to decide $C \preceq_{\emptyset} D$, it suffices to check whether $(\neg C \sqcap D)^{\mathcal{I}} = \emptyset$ for all $\mathcal{I}$. Note that if there is a non-empty interpretation of $(\neg C \sqcap D)$, that is, there is an $x \in (\neg C \sqcap D)^{\mathcal{I}}$, then there must be an interpretation of $(\neg C \sqcap D)$ such that there are only n elements y with $f^{\mathcal{I}}(x) = y$ or $y \in r^{\mathcal{I}}(x)$, where n is the number of feature and role restrictions in the top-level expression of $(\neg C \sqcap D)$. Furthermore, this holds also for the (finite number of) concept descriptions embedded in feature and role restrictions. This means, if $(\neg C \sqcap D)$ has a non-empty interpretation at all, then there is also a finite non-empty interpretation bounded in size by $(\neg C \sqcap D)$. Thus, it suffices to check only a finite number of finite interpretations in order to decide whether $(\neg C \sqcap D)$ has a necessarily empty interpretation, and, by that, to decide subsumption. ■

Based on this, it is easy to see that subsumption determination for the terminological formalism based on $\mathcal{ALC}$ is decidable—provided there are no terminological cycles in the terminology

Theorem 2.8 *Let T be a cycle-free $\mathcal{ALC}$ terminology and let C, D be concept descriptions. Then it is decidable whether $C \preceq_T D$.*

Proof: Immediate by Lemma 2.5, Theorem 2.6, and Lemma 2.7. ■

Interestingly, decidability is preserved even if terminological cycles are introduced. Although in this case subsumption in a terminology cannot be reduced to subsumption in the empty terminology because *Exp* doesn't terminate, it is possible to define a similar expansion function which expands the concept descriptions only to a finite depth. The main argument for decidability is that the descriptive power of $\mathcal{ALC}$ does not allow to distinguish between infinite interpretations and finite interpretations containing "assertional cycles."[7]

Now there may be the question, what kind of additional description-forming operator would result in the undecidability of subsumption. As has been shown in some recent papers [41,42,34], if some very natural looking extensions are added to our language, for instance, equality constraints on role-filler sets or role-negation and role-composition, then subsumption becomes undecidable, even if we consider only the empty terminology.

Despite the positive result concerning decidability, $\mathcal{ALC}$ has an unsatisfying property. Subsumption in the empty terminology is PSPACE-complete as shown in [43]. Since a terminological representation system is supposed to provide its computational services in reasonable time, this is an unacceptable state of affairs. Most terminological representation systems support therefore only less expressive description-forming languages [33] or limit the inference capabilities in a way such that only "interesting" inferences are drawn [26], where "interesting" may be defined by an alternative, weaker set-theoretic semantics [32] or by enumerating (perhaps only implicitly) the possible inference rules.

Following the first suggestion of limiting the expressiveness of the description-forming language, let us analyze some subsets of $\mathcal{ALC}$. First, it should be obvious that any subset containing $\neg$, $\sqcup$, and $\sqcap$ has still an intractable subsumption problem, because the satisfiability problem of propositional logic can be reduced to the problem of determining coherency of a concept description.

---

[7]See [27, Chapter 5] for a decidability proof for a related language.

Proposition 2.9 *Given a description-forming language that includes ¬, ⊔, and ⊓, it is co-NP-hard to decide whether $C \preceq_\emptyset D$.*

For this and other reasons, disjunction is usually banned from terminological formalisms and negation is only permitted on *primitive components* (see proof of Lemma 2.5). However, even with these severe restrictions we do not necessarily achieve tractable subsumption. For the description-forming language containing only ∀, ∃, ⊓, and ¬ on primitive components, it is unknown whether a polynomial subsumption algorithm exists. Only in the case when the second argument of the ∃ operator is always ⊤, subsumption is known to be polynomial [23].[8] Although this language, which has been called $\mathcal{FL}^-$ in [7,23], can be slightly extended without "falling off the computational cliff," the expressiveness is severely restricted if we confine ourselves to description-forming languages that are tractable with respect to subsumption.

Moreover, even if we adopt the point of view that subsumption determination in the empty terminology should be tractable, does that really help? When reducing subsumption in a terminology to subsumption in the empty terminology in Theorem 2.6, nothing was said about time and space bounds of the function *Exp*. As a matter of fact, the application of *Exp* can lead to expressions that are not polynomially bounded in the size of its arguments, as the following example demonstrates:

$$C_1 \doteq \forall r: C_0 \sqcap \forall r': C_0$$
$$C_2 \doteq \forall r: C_1 \sqcap \forall r': C_1$$
$$\vdots$$
$$C_n \doteq \forall r: C_{n-1} \sqcap \forall r': C_{n-1}.$$

Here, the size of $Exp(C_n, T)$ is obviously proportional to 2^n. Of course, better algorithms are conceivable. But as it turns out, the subsumption problem in terminologies cannot be reduced generally to the subsumption problem in the empty terminology in linear time, as the following theorem shows [28]:

Theorem 2.10 *Given a terminological formalism containing only the operators $\doteq$, ⊓, and ∀, the problem of deciding whether $C \preceq_T D$ in cycle-free terminologies is co-NP-complete.*

Proof Sketch: The problem of *inequivalence of nondeterministic finite automatons* that generate finite languages, which is known to be NP-complete [11, p. 265], can be reduced to inequivalence of concept descriptions in a cycle-free terminology by a straightforward mapping of states to concepts, input symbols to roles, and transitions to ∀ role restrictions.[9] Concepts which correspond to final states of the automatons are labeled with a primitive component. Then two automatons are inequivalent if, and only if, the two concepts corresponding to the respective initial states are inequivalent. For this reason, equivalence determination in terminologies, which is a special case of subsumption, is co-NP-hard. Since nonsubsumption can be easily shown to be in NP, the subsumption problem itself is co-NP-complete. ■

---

[8] Again, the addition of features would not add to the principal complexity.
[9] One could use features instead without invalidating the argument.

From a theoretical point of view this result means that "the goal of forging a powerful system out of tractable parts" [23, p. 89] cannot be achieved in the area of terminological reasoning.[10] Furthermore, it means that almost all terminological reasoning systems described in the literature that have been conjectured or proven to be tractable with respect to subsumption over the description forming language (in the empty terminology) can be blown up with a carefully thought out example. However, nobody seems to have noticed this fact, and, indeed, terminologies occurring in applications appear to be well-behaved. In this respect and with regard to the structure, our problem is similar to the type inference problem in ML, which seems to be solvable in linear time in all practical applications encountered so far, but is PSPACE-hard in general [16]. The conclusion one can draw from this strange situation is that although the theory of computational complexity can shade some light on the structure of a problem, one should not be scared by intractability in the first place. It may well be the case that it is possible to find algorithms that are well-behaved in all "normal cases."

3   Unification Grammars

In the last decade a new type of grammar formalisms, now commonly referred to as unification grammars, has evolved from research in linguistics, computational linguistics and artificial intelligence.[11] In contrast to augmented transition networks, one of their precursors, unification grammar formalisms provide for the declarative or logical specification of linguistic knowledge. Nevertheless, unification grammars are aimed towards operational use in parsing and generating natural language.

Like any grammar formalism, unification grammars define sets of sentences. But in addition, a unification grammar assigns to every grammatical sentence one or several so-called feature graphs representing syntactic and semantic information. For instance, the syntactic structure of the sentence

John sings a song

can be represented by the feature graph in Figure 2. This graph states that the sentence consists of a subject (John), a predicate (sings) and an object (a song). It also states that the agent of the singing is given by the subject of the sentence and what is sung is given by the object of the sentence. Moreover, the graph states that the tense of the sentence is present.

The linguistic knowledge represented by a unification grammar can be used for parsing and generating sentences in natural language. Suppose "$U(S,G)$" is the relation between strings of symbols and feature graphs specified by some grammar U. Then a string of symbols S is a *sentence of U* if and only if there exists at least one feature graph G such that $U(S,G)$ holds. A parser computes for a given string of symbols S the set $\{G \mid U(S,G)\}$ of all feature graphs related to S, and a generator computes for a given feature graph G the set $\{S \mid U(S,G)\}$ of all sentences related to G.

Unification grammars are written in some formalism providing for the declarative specification of grammar relations "$U(S,G)$". Ideally, a unification grammar formalism should be general enough to

---

[10]Under the reasonable assumption that we have $\doteq$, $\sqcap$, and $\forall$ restrictions (respectively, feature restrictions) with the standard semantics—which is probably something nobody wants to give up on.

[11]For the sake of a short name, we use the term unification grammars to stand for feature-based unification grammars, which excludes term-based unification grammars such as Definite Clause Grammars [37].

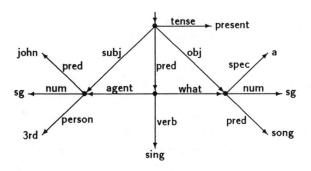

Figure 2: A feature graph

adopt easily to different linguistic theories. Since unification grammars are supposed to be declarative specifications, seeing a unification grammar formalism as a logic for specifying grammar relations makes good sense.

Obviously, there is a strong analogy between logic programs and unification grammars. A logic program is a specification in Horn clause logic that can be employed operationally in various ways. In fact, Definite Clause Grammars [37], which are specified as Horn clauses (which are one possible form of definite clauses), are an early form of unification grammars lacking the notion of features and feature graphs, which evolved independently in (computational) linguistics. So what we need to capture unification grammars as logical specifications is an integration of features and feature graphs with definite clauses. This can be nicely accomplished with the new model of logic programming called Constraint Logic Programming (CLP) [13,12]. In contrast to Kowalski's [22] classical Horn clause model, CLP is parameterized with respect to a general notion of constraint language. Thus, in order to capture unification grammars as logical specifications, one needs a suitable constraint language for describing feature graphs. Feature constraint languages are in fact the essence of unification grammar formalisms.

The thing feature-based unification grammars have in common with the terminological axioms discussed in the previous section is that both formalisms rely mainly on attributes for representing knowledge. However, the descriptions used and the reasoning services supported are quite different. In KL-ONE-like systems two kinds of attributes—features and roles—are used and the primary reasoning services are determining inclusion between concepts and membership of constants in concepts. Descriptions in unification grammars employ only features and the primary reasoning operation is constructing the most general feature graphs satisfying a description.

The outline of this section is as follows. We start with a formal definition of feature graphs and then devise a suitable constraint language for partially describing feature graphs. Then we show how unification grammars can be written as definite clauses over this constraint language, discuss possible extensions of the basic model, and give pointers to the literature. Finally, we prove the most important properties of our constraint language and give a constraint solving algorithm.

Our view of unification grammars as definite clauses over feature constraint languages is by no means standard and, to our knowledge, has not appeared explicitly in the literature. Our presentation is often informal and relies on previous work on feature logic [48] and definite relations over constraint languages [12].

3.1 A Simple Feature Logic

We assume that an infinite set of **variables** (denoted by x, y, z), a set of **features** (denoted by f, g, h), and a set of **constants** (denoted by a, b, c) are given.

A feature graph is a finite, rooted, connected and directed graph whose edges are labeled with feature symbols such that the labels of the edges departing from a node are pairwise distinct. Moreover, every inner node of a feature graph is a variable and every terminal node is either a constant or a variable.

Formally, an f-**edge from** x **to** s is a triple xfs such that x is a variable, f is a feature, and s is either a variable or a constant. A **feature graph** is either a pair $(a, \emptyset)$, where a is a constant and $\emptyset$ is the empty set, or a pair (x_0, E), where x_0 is a variable (the **root**) and E is a finite, possibly empty set of edges such that

1. the graph is determinate, that is, if $xfs \in E$ and $xft \in E$, then $s = t$

2. the graph is connected, that is, if $xfs \in E$, then E contains edges leading from the root x_0 to the node x.

To obtain the right notion of feature graph, we will identify all feature graphs that are equal up to consistent variable renaming.

A **feature algebra** is a pair $(\mathbf{D}^{\mathcal{I}}, \cdot^{\mathcal{I}})$ consisting of a nonempty set $\mathbf{D}^{\mathcal{I}}$ (the **domain** of $\mathcal{I}$) and an **interpretation function** $\cdot^{\mathcal{I}}$ assigning to ever constant a an element $a^{\mathcal{I}} \in \mathbf{D}^{\mathcal{I}}$ and to every feature f a unary partial function $f^{\mathcal{I}}$ from $\mathbf{D}^{\mathcal{I}}$ to $\mathbf{D}^{\mathcal{I}}$ such that the following conditions are satisfied:

1. if $a \neq b$, then $a^{\mathcal{I}} \neq b^{\mathcal{I}}$ (*unique name assumption*)

2. for no feature f and no constant a, the partial function $f^{\mathcal{I}}$ is defined on $a^{\mathcal{I}}$ (*constants are primitive*).

Note that feature algebras are interpretations in the sense of the previous section, where there are no roles and concepts, and no feature is defined on a constant. Feature algebras can also be seen as Tarski interpretations of the predicate calculus, if we view features equivalently as binary predicates that must be interpreted as functional relations.

There are three reasons for introducing feature algebras. First, we will see that the set of all feature graphs can be regarded naturally as a feature algebra. Second, the constraint language we are going to develop can be interpreted over every feature algebra and we will see that the algebra of all feature graphs takes a prominent position with respect to this constraint language. And third, this approach makes explicit the close connection between Predicate Logic and the particular feature logic we are going to develop.

The **feature graph algebra** $\mathcal{F}$ is obtained as follows:

1. $\mathbf{D}^{\mathcal{F}}$ is the set of all feature graphs

2. $a^{\mathcal{F}}$ is the feature graph $(a, \emptyset)$

3. $f^{\mathcal{F}}$ is defined on a feature graph if and only if there is an f-edge departing from its root

4. if $G = (x, E)$ and $xfs \in E$, then $f^{\mathcal{F}}(G)$ is the largest feature graph (s, E') such that $E' \subseteq E$.

One verifies easily that $\mathcal{F}$ is a feature algebra.

Next we define the constraint language. From features, constants and variables we obtain **terms** as in Predicate Logic. To ease our notation, we omit parentheses and write fs for $f(s)$. A **feature equation** is a pair $s \doteq t$ consisting of two terms. Feature equations are the only primitive constraints we need.

Let $\mathcal{I}$ be a feature algebra. As in Predicate Logic, an $\mathcal{I}$-**assignment** is a mapping from the set of all variables to the domain of $\mathcal{I}$. The interpretation of terms in $\mathcal{I}$ under an $\mathcal{I}$-assignment α are defined as one would expect:

$$
\begin{aligned}
\mathcal{I}[\![x]\!]_\alpha &= \alpha(x) \\
\mathcal{I}[\![a]\!]_\alpha &= a^{\mathcal{I}} \\
\mathcal{I}[\![fs]\!]_\alpha &= f^{\mathcal{I}}(\mathcal{I}[\![s]\!]_\alpha).
\end{aligned}
$$

Note that the interpretation of a term fs is not always defined since features are interpreted as partial functions. In particular, no term having a subterm fa has an interpretation. A **solution** of an equation $s \doteq t$ in $\mathcal{I}$ is an $\mathcal{I}$-assignment α such that $\mathcal{I}[\![s]\!]_\alpha$ and $\mathcal{I}[\![t]\!]_\alpha$ are defined and $\mathcal{I}[\![s]\!]_\alpha = \mathcal{I}[\![t]\!]_\alpha$.

A **feature clause** is a finite, possibly empty set of feature equations representing their conjunction. Consequently, a **solution** of a feature clause C in a feature algebra $\mathcal{I}$ is an $\mathcal{I}$-assignment that solves every feature equation in C. If $\mathcal{I}$ is a feature algebra, C is a feature clause and α is a solution of C in $\mathcal{I}$, then we call $\alpha(x)$ a **solution of x in C and $\mathcal{I}$**.

One verifies easily that the feature graph in Figure 2 is a solution of x in the feature graph algebra $\mathcal{F}$ and the feature clause C consisting of the equations

pred subj x $\doteq$ john	agent pred x $\doteq$ subj x	spec obj x $\doteq$ a
num subj x $\doteq$ sg	verb pred x $\doteq$ sing	num obj x $\doteq$ sg
person subj x $\doteq$ 3rd	what pred x $\doteq$ obj x	pred obj x $\doteq$ song

$$\text{tense } x \doteq \text{present.}$$

This clause admits infinitely many other solutions for x that are obtained by adding further edges to the inner nodes of the feature graph in Figure 2. However, the graph in Figure 2 is the most general solution of x in C in that it realizes exactly what is required by the clause and nothing else.

To capture the idea of being more general formally, we define a partial order on feature graphs usually called subsumption.[12] A **morphism** is a function that maps every variable to a variable or a constant, and that maps every constant to itself. If (s, E) and (s', E') are feature graphs, we

---

[12]Note that the subsumption relation on feature graphs defined here is different from the subsumption relation on concept descriptions defined in Section 2.

write $(s, E) \leq (s', E')$ and say that (s', E') is **more specific** than (s, E) or conversely that (s, E) is **more general** than (s', E') if and only if there exists a morphism γ such that $\gamma(s) = s'$ and $\gamma(E) \subseteq E'$, where $\gamma(E) = \{\gamma(x)f\gamma(t) \mid xft \in E\}$. Since we identify feature graphs that are equal up to consistent variable renaming, this definition yields in fact a partial order on the set of all feature graphs.

Proposition 3.1 *For every feature graph G and every variable x one can compute in linear time a feature clause C such that G is the most general solution of x in C and $\mathcal{F}$.*

Proof: If $G = (a, \emptyset)$, then $\{x \doteq a\}$ is a clause as claimed. If $G = (x, E)$, then $\{fy \doteq s \mid yfs \in E\}$ is a clause as claimed. Otherwise, if some variable other than x is the root node of G, x can be made the root node by consistent variable renaming. ∎

The most important properties of our constraint language are stated by the following theorem, which we will prove in Subsection 3.4.

Theorem 3.2 *Let C be a feature clause, x be a variable, and $\mathcal{F}$ be the feature graph algebra. Then the following conditions are equivalent:*

1. *C has a solution in some feature algebra*
2. *C has a solution in $\mathcal{F}$*
3. *x has a most general solution in C and $\mathcal{F}$.*

Furthermore, there is a quadratic time algorithm that, given a clause C and a variable x, either returns fail if C has no solution or returns the most general feature graph solution of x in C.

3.2 Unification Grammars as Definite Clauses over Feature Logic

We will now outline a simple unification grammar formalism in which a grammar is given as a set of definite clauses over the feature logic just presented. Such definite clauses can be seen as context-free syntax rules constrained with feature equations.

Figure 3 shows how ordinary context-free grammars translate into Horn clauses. This translation, which is well-known from Definite Clause Grammars [37], assumes that a string of symbols $w_1, \ldots, w_n$ is represented as the term $w_1 \cdot \cdots \cdot w_n.\text{nil}$, where the dot is a binary function symbol written in right-associative infix notation. A string of symbols S is a sentence of the grammar if and only if the statement $\mathsf{s}(S)$ follows logically from the Horn clause translation of the grammar.

Given the Horn clause translation of a context-free grammar, we can introduce for every phrase predicate an additional argument that ranges over feature graphs. Furthermore, in the body of every clause constraints for these additional feature graph arguments can be given. For instance, consider the clauses

$$np(X, Y, NP) \leftarrow d(X, Z, D) \wedge n(Z, Y, N) \wedge NP \doteq D \wedge D \doteq N$$
$$d(a.X, X, D) \leftarrow spec(D) \doteq a \wedge num(D) \doteq sg$$
$$n(song.X, X, N) \leftarrow pred(N) \doteq song \wedge num(N) \doteq sg,$$

which may be written in the following more intelligible syntax:

$$S \longrightarrow NP\ VP \qquad\qquad s(X) \leftarrow np(X, Y) \wedge vp(Y, nil)$$
$$VP \longrightarrow V\ NP \qquad\qquad vp(X, Y) \leftarrow v(X, Z) \wedge np(Z, Y)$$
$$NP \longrightarrow D\ N \qquad\qquad np(X, Y) \leftarrow d(X, Z) \wedge n(Z, Y)$$
$$NP \longrightarrow john \qquad\qquad np(john.X, X) \leftarrow$$
$$V \longrightarrow sings \qquad\qquad v(sings.X, X) \leftarrow$$
$$D \longrightarrow a \qquad\qquad d(a.X, X) \leftarrow$$
$$N \longrightarrow song. \qquad\qquad n(song.X, X) \leftarrow\ .$$

Figure 3: A context-free grammar and its translation into Horn clauses

$$
\begin{array}{lll}
NP \longrightarrow D\ N & D \longrightarrow a & N \longrightarrow song \\
\quad NP \doteq D \doteq N & \quad spec\ D \doteq a\ \wedge & \quad pred\ N \doteq song\ \wedge \\
& \quad num\ D \doteq sg & \quad num\ N \doteq sg.
\end{array}
$$

These clauses already illustrate one possible way unification grammars can enforce agreement between the numerus of a noun and its determiner. Since the determiner and the noun are forced by the first clause to carry the same feature graphs (which must also be the feature graphs of the entire noun phrase), it suffices if the rules for the determiner and the noun constrain the feature num independently. In case they constrain the feature num with conflicting values, there exists no feature graph satisfying both constraints.

Saying that a unification grammar consists of definite clauses over a feature constraint language is a little bit oversimplified. Our rules actually employ a three-sorted constraint language providing the sort of all feature graphs, the sort of all words, and the sort of all strings of words. It is straightforward to accommodate this technically. It is also possible to have a single-sorted approach if one codes strings of words as feature graphs.

Figure 4 shows a unification grammar that covers our example sentence "John sings a song". The grammar is given in a sugared syntax that can be translated automatically into definite clauses.[13]

Since our unification grammars are definite clauses over feature logic, they enjoy a logical semantics provided by the constraint logic programming model [12]. Let x be a fixed variable and let, for every feature graph G, $C[x, G]$ denote a feature clause such that G is the unique most general solution of x in $C[x, G]$ and $\mathcal{F}$. Then a grammar U with the sentence predicate s_U defines a relation "$U(S, G)$" between strings of symbols and feature graphs as follows:

$$
\begin{aligned}
U(S, G) \iff & \ U \models C[x, G] \rightarrow s_U(S, x)\ \wedge \\
& \ \forall G': \ U \models C[x, G'] \rightarrow s_U(S, x)\ \Rightarrow\ G \leq G',
\end{aligned}
$$

where $U \models C[x, G] \rightarrow s_U(S, x)$ means that the implication $C[x, G] \rightarrow s_U(S, x)$ follows logically from U and $G \leq G'$ means that G' is more specific than G.

---

[13] An equation, say, agent pred $V \doteq$ subj V in our term-oriented syntax would be written in PATR's [47] syntax as < V pred agent > = < V subj >.

$$S \longrightarrow NP\ VP$$
$$S \doteq VP \wedge subj\ S \doteq NP$$

$$VP \longrightarrow V\ NP$$
$$VP \doteq V \wedge obj\ VP \doteq NP$$

$$NP \longrightarrow D\ N$$
$$NP \doteq D \doteq N$$

$$NP \longrightarrow john$$
$$pred\ NP \doteq john \wedge$$
$$num\ NP \doteq sg \wedge$$
$$person\ NP \doteq 3rd$$

$$V \longrightarrow sings$$
$$tense\ V \doteq present \wedge$$
$$verb\ pred\ V \doteq sing \wedge$$
$$agent\ pred\ V \doteq subj\ V \wedge$$
$$what\ pred\ V \doteq obj\ V \wedge$$
$$num\ subj\ V \doteq sg \wedge$$
$$person\ subj\ V \doteq 3rd$$

$$D \longrightarrow a$$
$$spec\ D \doteq a \wedge$$
$$num\ D \doteq sg$$

$$N \longrightarrow song$$
$$pred\ N \doteq song \wedge$$
$$num\ N \doteq sg$$

Figure 4: A unification grammar

It is possible to verify that the unification grammar in Figure 4 relates the sentence "John sings a song" to the feature graph in Figure 2 and to no other feature graph.

The word problem of a grammar U is to decide for a given string S of symbols whether there exists a feature graph G such that $U(S,G)$ holds. One can show that our formalism allows for grammars having an undecidable word problem by adapting proofs given by Johnson [14] and Rounds and Manaster-Ramer [40] for slightly different formalisms.

Every grammar U in our formalism comes with a context-free grammar $CF[U]$ such that U is obtained from $CF[U]$ by adding feature equations to the rules of U. A grammar U satisfies the *off-line parsability constraint* [17] if the number of different derivations of a string of symbols in $CF[U]$ is bounded by a computable function of the length of that string. Using the operational semantics of definite clauses [12], it is easy to see that for a grammar U satisfying the off-line parsability constraint the word problem is decidable, and that, for every string of symbols S, the set $\{G \mid U(S,G)\}$ is finite and can be computed from S. The off-line parsability constraint is, for instance, satisfied if the right-hand side of every rule of $CF[U]$ contains either at least one terminal or at least two nonterminals.

3.3 Background and Extensions

The simple unification grammar formalism sketched here bears much resemblance with the PATR formalism developed at SRI International by Stuart Shieber and his colleagues [47,44]. It is also closely related to Kaplan and Bresnan's Lexical-Functional Grammar formalism (LFG) [17]. LFG and PATR were conceived and developed at a time when the constraint logic programming model was not available and have been described quite differently by their inventors. In fact, even the most recent attempts at formalizing unification grammar formalisms [14,46] still don't make use of the constraint logic programming model.

Shieber's [45] introduction to unification-based approaches to grammar is an excellent survey of existing formalisms and provides the linguistic motivations our presentation is lacking. Other state of the art guides into this fascinating area of research are [38] and [36]. Johnson's thesis [14] gives a formal account of LFG and investigates a feature constraint language with disjunctions and negations. Furthermore, Shieber's [46] thesis gives a rigorous formalization of the PATR formalism.

Why are unification grammars called "unification" grammars? In this context unification is understood as the operation that, given two feature graphs G_1 and G_2, decides whether there exists a feature graph that is more specific than both G_1 and G_2 and, if so, returns the most general such feature graph (which then in fact uniquely exists). Feature graph unification can be seen as an operation combining the information given by two feature graphs provided it doesn't conflict. Due to the close relation between clauses and feature graphs, feature graph unification can be employed as the central operation of a parser emulating a unification grammar, an implementation technique that is elaborated carefully in [46]. Nevertheless, the name "unification grammar", which can be traced back to Martin Kay's [19,21,20] Functional Unification Grammar, is rather misleading since it is derived from an operation that may or may not be employed in implementations of these grammar formalisms. Incidentally, in Kaplan and Bresnan's [17] clear and insightful presentation of LFG feature graph unification is not even mentioned.

Kay's Functional Unification Grammar (FUG) [19,21,20] is more general than our formalism since it is not based on context-free rules but uses more flexible mechanisms for establishing word order. Rounds and Manaster-Ramer [40] present a logical formalization of FUG.

The operational semantics of the constraint logic programming model [13,12] given by goal reduction results in a top-down parsing strategy when applied to our unification grammar formalism. The role of term unification in ordinary Horn clause programming is taken by a constraint solver that simplifies sets of feature equations and thereby checks their solvability. Such a constraint solver generalizes feature graph unification. Existing unification grammar systems often employ chart parsing techniques realizing a bottom up strategy.

The feature constraint language presented here restricts constraints to conjunctions of equations. One obvious extension is to admit other logical connectives such as disjunction, negation or implication. For instance, a lexical rule for the verb "sing" may come with an implicational constraint:

V $\longrightarrow$ sing
    (person subj V $\doteq$ 3rd $\rightarrow$ num subj V $\doteq$ pl) $\wedge$ ... .

Deciding for such general constraints whether they have a solution is an NP-complete problem [14,48]. Furthermore, such a general constraint can have more than one most general feature graph solution, but at most finitely many. Feature constraint languages with all propositional connectives have been investigated by Johnson [14] and Smolka [48].

The integration of Prolog-like logic programming with feature constraint languages seems to be a very promising line of research. Concrete language proposals based on this idea are LOGIN [3] and CIL [25]. The theoretical foundations for this kind of languages have been established in [48,13,12].

Kasper and Rounds [18,39] were the first to develop a constraint logic for feature graphs. Their logic accounts for concept descriptions interpreted in the feature graph algebra $\mathcal{F}$. Their concept descriptions, which we call **feature terms**, are given by the abstract syntax rule

$$S, T \longrightarrow a \mid p \downarrow q \mid f : S \mid S \sqcap T \mid S \sqcup T.$$

$$
\begin{bmatrix}
\text{tense: present} \\
\text{subj:} \begin{bmatrix} \text{pred: john} \\ \text{num: sg} \\ \text{person: 3rd} \end{bmatrix} \\
\text{pred: verb: sing} \\
\text{agent pred} \downarrow \text{subj} \\
\text{what pred} \downarrow \text{obj} \\
\text{obj:} \begin{bmatrix} \text{pred: song} \\ \text{num: sg} \\ \text{spec: a} \end{bmatrix}
\end{bmatrix}
$$

Figure 5: A feature term in matrix notation

The new construct $p \downarrow q$, which we call **agreement**, consists of two strings p and q of features (called **paths**). Given a feature algebra $\mathcal{I}$, the interpretation $p^{\mathcal{I}}$ of a string of features $p = f_n \cdots f_1$ is the partial function obtained as the composition of the partial functions $f_n^{\mathcal{I}}, \ldots, f_1^{\mathcal{I}}$, where $f_1^{\mathcal{I}}$ is applied first. The empty string denotes the identity function of $\mathbf{D}^{\mathcal{I}}$. Now the interpretation of an agreement $p \downarrow q$ in $\mathcal{I}$ is the greatest subset of $\mathbf{D}^{\mathcal{I}}$ on which $p^{\mathcal{I}}$ and $q^{\mathcal{I}}$ agree, that is,

$$(p \downarrow q)^{\mathcal{I}} = \{d \in \mathbf{D}(p^{\mathcal{I}}) \cap \mathbf{D}(q^{\mathcal{I}}) \mid p^{\mathcal{I}}(d) = q^{\mathcal{I}}(d)\},$$

where $\mathbf{D}(p^{\mathcal{I}})$ and $\mathbf{D}(q^{\mathcal{I}})$ are the domains of the partial functions $p^{\mathcal{I}}$ and $q^{\mathcal{I}}$, respectively. Agreements are needed for expressing *coreferences* in feature graphs, that is, the fact that two paths lead to the same node. Interpreted in the feature graph algebra $\mathcal{F}$, every feature term denotes a set of feature graphs.

Figure 5 gives an example of a feature term written in matrix notation. The feature terms given as the rows of a matrix are connected by intersection. The feature graph in Figure 2 is the most general element of the set of feature graphs denoted by the feature term in Figure 5 in the feature graph algebra $\mathcal{F}$.

Agreements generalize to roles and are known as role value maps in KL-ONE [9]. A recent paper of Schmidt-Schauß [42] shows that role value maps result in an undecidable subsumption relation on concept descriptions. This is in sharp contrast to agreements for feature terms, which don't cause a blow-up of the computational complexity.

Feature terms can be used for constraining variables if the logic provides for memberships $x : S$ consisting of a variable and a feature term, where an $\mathcal{I}$-assignment α is a solution of $x : S$ in a feature algebra $\mathcal{I}$ if and only if $\alpha(x) \in S^{\mathcal{I}}$. One can show that memberships not employing unions have the same expressivity as conjunctions of feature equations. Smolka [48] investigates various feature term languages and shows how they reduce to equational constraint languages.

Unification grammars usually have most of their structural information in their lexical rules. Since lexica for realistic subsets of natural language are large, techniques are needed for expressing lexical generalizations so as to allow lexical entries to be written in a compact notation. One such technique is the use of so-called templates in the PATR formalism, which turn out to be noncyclic equational terminological axioms based on feature terms. Figure 6 gives an example.

$$\text{present3rdsg} \doteq \left[\begin{array}{l} \text{tense: present} \\ \text{subj:} \left[\begin{array}{l} \text{num: sg} \\ \text{person: 3rd} \end{array} \right] \end{array} \right] \qquad \text{V} \longrightarrow \text{sings}$$

$$\text{V:} \left[\begin{array}{l} \text{pred: verb: sing} \\ \text{transitive} \\ \text{present3rdsg} \end{array} \right]$$

$$\text{transitive} \doteq \left[\begin{array}{l} \text{agent pred} \downarrow \text{subj} \\ \text{what pred} \downarrow \text{obj} \end{array} \right]$$

Figure 6: A lexical entry using templates

3.4 Solving Feature Clauses

We now prove Theorem 3.2 by exhibiting a quadratic-time algorithm for solving feature clauses. The algorithm is given abstractly as a collection of simplification rules providing for the solution-preserving transformation of feature clauses to a solved form.[14]

We use $\mathcal{I}[C]$ to denote the set of all solutions of a feature clause C in a feature algebra $\mathcal{I}$. Two feature clauses C and D are called **equivalent** if $\mathcal{I}[C] = \mathcal{I}[D]$ for every feature algebra $\mathcal{I}$. Let V be a set of variables. Then two feature clauses C and D are called V-**equivalent** if for every feature algebra $\mathcal{I}$ the following two conditions are satisfied:

1. if $\alpha \in \mathcal{I}[C]$, then there exists $\beta \in \mathcal{I}[D]$ such that α and β agree on V

2. if $\alpha \in \mathcal{I}[D]$, then there exists $\beta \in \mathcal{I}[C]$ such that α and β agree on V.

Proposition 3.3 *Let C and D be V-equivalent feature clauses, $x \in V$ and $\mathcal{I}$ be a feature algebra. Then $d \in \mathbf{D}^{\mathcal{I}}$ is a solution of x in C and $\mathcal{I}$ if and only if d is a solution of x in D and $\mathcal{I}$.*

Our solution algorithm for feature clauses consists of a linear-time unfolding phase followed by a quadratic-time normalization phase.

A feature clause is **unfolded** if each of its equations has either the form $s \doteq t$ or $fs \doteq t$, where s and t range over variables and constants. In other words, a feature clause is unfolded if each of its equations contains either no feature or exactly one feature that occurs on its left-hand side.

Let C be a feature clause. Then we use

1. $\mathcal{V}C$ to denote the set of all variables occurring in C

2. $[x/s]C$ to denote the clause that is obtained from C by replacing every occurrence of the variable x with the term s

3. $s \doteq t \,\&\, C$ to denote the feature clause $\{s \doteq t\} \cup C$ provided $s \doteq t \notin C$.

Unfolding is done by iteratively replacing a term fs in an unfolded position with a fresh variable x and adding the constraint $fs \doteq x$. For instance, the clause $\{fgx \doteq hy\}$ can be unfolded to

$$\{fz \doteq u, \quad gx \doteq z, \quad hy \doteq u\}$$

by introducing the new variables u and z. Unfolding steps are justified by the following proposition:

---

[14]Theorem 3.2 and the algorithm are taken from [48].

Proposition 3.4 *Let D be a feature clause, $fs \doteq x \in D$, $C = [x/fs](D - \{fs \doteq x\})$, and let x not occur in fs. Then C and D are $\mathcal{VC}$-equivalent.*

Proposition 3.5 *For every feature clause C one can compute in linear time a $\mathcal{VC}$-equivalent unfolded feature clause D.*

Next we present the solved form for feature clauses to which the normalization phase of our solution algorithm attempts to transform unfolded feature clauses.

A feature clause C is **solved** if it satisfies the following conditions:

1. every equation in C has one of the following forms: $x \doteq y$, $x \doteq a$, $fx \doteq y$ or $fx \doteq a$

2. if $fx \doteq s$ and $fx \doteq t$ are in C, then $s = t$

3. if $x \doteq s$ is in C, then x occurs only once in C.

Let C be a solved feature clause. Then $x \to_C y \iff \exists\, fx \doteq y \in C$ defines a binary relation $\to_C$ on the variables occurring in C. We use $\to_C^*$ to denote the reflexive and transitive closure of $\to_C$ on the set of all variables. If x is a variable, then

$$\mathrm{FG}[x, C] := \begin{cases} (a, \emptyset) & \text{if } x \doteq a \in C \\ \mathrm{FG}[y, C] & \text{if } x \doteq y \in C \\ (x, \{yfs \mid fy \doteq s \in C \,\wedge\, x \to_C^* y\}) & \text{otherwise} \end{cases}$$

defines a feature graph.

Theorem 3.6 *If C is a solved feature clause and x is a variable, then $\mathrm{FG}[x, C]$ is the most general solution of x in C and $\mathcal{F}$.*

The proof of this theorem is straightforward.

Next we present the normalization phase of the algorithm, which is given by the following solution-preserving simplification rules for unfolded feature clauses:

1. $s \doteq s \,\&\, C \;\to\; C$

2. $x \doteq s \,\&\, C \;\to\; x \doteq s \,\&\, [x/s]C \qquad$ if $x \in \mathcal{VC}$ and $x \neq s$

3. $s \doteq x \,\&\, C \;\to\; x \doteq s \,\&\, C \qquad$ if s is not a variable

4. $fx \doteq s \,\&\, fx \doteq t \,\&\, C \;\to\; fx \doteq s \,\&\, s \doteq t \,\&\, C.$

A clause is called **normal** if it is unfolded and no normalization rule applies to it.

Proposition 3.7 *Let C be an unfolded feature clause. Then:*

1. *if D is obtained from C by a normalization rule, then D is an unfolded feature clause that is equivalent to C*

2. *there is no infinite chain of normalization steps issuing from C.*

Proof: The verification of the first claim is straightforward. To show the second claim, suppose there is an infinite sequence $C_1, C_2, \cdots$ of unfolded feature clauses such that, for every $i \geq 1$, C_{i+1} is obtained from C_i by a normalization rule. First note that every variable occurring in some C_i must also occur in C_1, that is, normalization steps don't introduce new variables. A variable x is called isolated in a clause C if C contains an equation $x \doteq s$ and x occurs exactly once in C. Now observe that no normalization rule decreases the number of isolated variables, and that the second normalization rule increases this number. Hence we can assume without loss of generality that the infinite sequence doesn't employ the second normalization rule. However, it is easy to see that the remaining normalization rules cannot support an infinite sequence. ∎

Proposition 3.8 *For every feature clause one can compute in quadratic time a $\mathcal{VC}$-equivalent normal feature clause.*

Proof: Let C be a clause. We have seen that we can compute in linear time an unfolded clause D that is $\mathcal{VC}$-equivalent to C. By the previous proposition we know that we can compute a normal feature clause E that is equivalent to D using the normalization rules. The normalization of D to E can be done in quadratic time by employing the normalization rules together with an efficient union-find method [2] for maintaining equivalence classes of variables and constants. ∎

A **clash** is an equation that has either the form $fa = s$ or the form $a \doteq b$, where a and b are different constants. A feature clause is **clash-free** if it contains no clash.

Proposition 3.9 *If a feature clause has a solution in some feature algebra, then it is clash-free. Furthermore, a feature clause is solved if and only if it is normal and clash-free.*

Now Theorem 3.2 follows easily from Propositions 3.8 and 3.9 and Theorem 3.6.

References

[1] G. Abrett and M. H. Burstein. The KREME knowledge editing environment. *International Journal of Man-Machine Studies*, 27(2):103–126, 1987.

[2] A. V. Aho, J. E. Hopcroft, and J. D. Ullman. *The Design and Analysis of Computer Algorithms.* Addison-Wesley, Reading, Mass., 1974.

[3] H. Aït-Kaci and R. Nasr. LOGIN: a logic programming language with built-in inheritance. *The Journal of Logic Programming*, 3:185–215, 1986.

[4] Y. Arens, L. Miller, S. C. Shapiro, and N. K. Sondheimer. Automatic construction of user-interface displays. In *Proceedings of the 7th National Conference of the American Association for Artificial Intelligence*, pages 808–813, Saint Paul, Minn., Aug. 1988.

[5] H. W. Beck, S. K. Gala, and S. B. Navathe. Classification as a query processing technique in the CANDIDE semantic data model. In *Proceedings of the International Data Engineering Conference, IEEE*, pages 572–581, Los Angeles, Cal., Feb. 1989.

[6] R. J. Brachman, A. Borgida, D. L. McGuinness, and L. A. Resnick. The CLASSIC knowledge representation system, or, KL-ONE: the next generation. In *Preprints of the Workshop on Formal Aspects of Semantic Networks*, Two Harbors, Cal., Feb. 1989.

[7] R. J. Brachman and H. J. Levesque. The tractability of subsumption in frame-based description languages. In *Proceedings of the 4th National Conference of the American Association for Artificial Intelligence*, pages 34–37, Austin, Tex., Aug. 1984.

[8] R. J. Brachman, V. Pigman Gilbert, and H. J. Levesque. An essential hybrid reasoning system: knowledge and symbol level accounts in KRYPTON. In *Proceedings of the 9th International Joint Conference on Artificial Intelligence*, pages 532–539, Los Angeles, Cal., Aug. 1985.

[9] R. J. Brachman and J. G. Schmolze. An overview of the KL-ONE knowledge representation system. *Cognitive Science*, 9(2):171–216, Apr. 1985.

[10] T. W. Finin and D. Silverman. Interactive classification as a knowledge acquisition tool. In L. Kerschberg, editor, *Expert Database Systems—Proceedings From the 1st International Workshop*, pages 79–90, Benjamin/Cummings, Menlo Park, Cal., 1986.

[11] M. R. Garey and D. S. Johnson. *Computers and Intractability—A Guide to the Theory of NP-Completeness*. Freeman, San Francisco, Cal., 1979.

[12] M. Höhfeld and G. Smolka. *Definite Relations over Constraint Languages*. LILOG Report 53, IBM Deutschland, Stuttgart, West Germany, Oct. 1988.

[13] J. Jaffar and J. Lassez. Constraint logic programming. In *Proceedings of the 14th ACM Symposium on Principles of Programming Languages*, pages 111–119, ACM, Munich, West Germany, Jan. 1987.

[14] M. Johnson. *Attribute-Value Logic and the Theory of Grammar. CSLI Lecture Notes 16*, Center for the Study of Language and Information, Stanford University, 1987.

[15] T. S. Kaczmarek, R. Bates, and G. Robins. Recent developments in NIKL. In *Proceedings of the 5th National Conference of the American Association for Artificial Intelligence*, pages 978–987, Philadelphia, Pa., Aug. 1986.

[16] P. C. Kanellakis and J. C. Mitchell. Polymorphic unification and ML typing. In *Proceedings of the 16th ACM Symposium on Principles of Programming Languages*, pages 5–15, Jan. 1989.

[17] R. Kaplan and J. Bresnan. Lexical-Functional Grammar: a formal system for grammatical representation. In J. Bresnan, editor, *The Mental Representation of Grammatical Relations*, pages 173–381, MIT Press, Cambridge, Mass., 1982.

[18] R. T. Kasper and W. C. Rounds. A logical semantics for feature structures. In *Proceedings of the 24th Annual Meeting of the ACL, Columbia University*, pages 257–265, New York, N.Y., 1986.

[19] M. Kay. Functional grammar. In *Proceedings of the Fifth Annual Meeting of the Berkeley Linguistics Society*, Berkeley Linguistics Society, Berkeley, Cal., 1979.

[20] M. Kay. Parsing in functional unification grammars. In D. Dowty and L. Karttunen, editors, *Natural Language Parsing*, Cambridge University Press, Cambridge, England, 1985.

[21] M. Kay. *Unification Grammar*. Technical Report, Xerox PARC, Palo Alto, Cal., 1983.

[22] R. A. Kowalski. Algorithm = Logic + Control. *Communications of the ACM*, 22(7):424–436, 1979.

[23] H. J. Levesque and R. J. Brachman. Expressiveness and tractability in knowledge representation and reasoning. *Computational Intelligence*, 3:78–93, 1987.

[24] R. MacGregor and R. Bates. *The Loom Knowledge Representation Language*. Technical Report ISI/RS-87-188, University of Southern California, Information Science Institute, Marina del Rey, Cal., 1987.

[25] K. Mukai. *Anadic Tuples in Prolog*. Technical Report TR-239, ICOT, Tokyo, Japan, 1987.

[26] B. Nebel. Computational complexity of terminological reasoning in BACK. *Artificial Intelligence*, 34(3):371–383, Apr. 1988.

[27] B. Nebel. *Reasoning and Revision in Hybrid Representation Systems*. PhD thesis, Universität des Saarlandes, Saarbrücken, West Germany, June 1989.

[28] B. Nebel. *Terminological Reasoning is Inherently Intractable*. IWBS Report, IWBS, IBM Deutschland, Stuttgart, West Germany, Aug. 1989. In preparation.

[29] B. Nebel and K. von Luck. Hybrid reasoning in BACK. In Z. W. Ras and L. Saitta, editors, *Methodologies for Intelligent Systems*, pages 260–269, North-Holland, Amsterdam, Holland, 1988.

[30] R. Neches, W. R. Swartout, and J. D. Moore. Explainable (and maintainable) expert systems. In *Proceedings of the 9th International Joint Conference on Artificial Intelligence*, pages 382–389, Los Angeles, Cal., Aug. 1985.

[31] B. Owsnicki-Klewe. Configuration as a consistency maintenance task. In W. Hoeppner, editor, *GWAI-88. 12th German Workshop on Artificial Intelligence*, pages 77–87, Springer-Verlag, Berlin, West Germany, 1988.

[32] P. F. Patel-Schneider. A four-valued semantics for terminological logics. *Artificial Intelligence*, 38(3):319–351, Apr. 1989.

[33] P. F. Patel-Schneider. Small can be beautiful in knowledge representation. In *Proceedings of the IEEE Workshop on Principles of Knowledge-Based Systems*, pages 11–16, Denver, Colo., 1984.

[34] P. F. Patel-Schneider. Undecidability of subsumption in NIKL. *Artificial Intelligence*, 39(2):263–272, June 1989.

[35] P. F. Patel-Schneider, R. J. Brachman, and H. J. Levesque. ARGON: knowledge representation meets information retrieval. In *Proceedings of the 1st Conference on Artificial Intelligence Applications*, pages 280–286, Denver, Col., 1984.

[36] F. Pereira. Grammars and logics of partial information. In *Proceedings of the 4th International Conference on Logic Programming*, pages 989–1013, MIT Press, Cambridge, Mass., 1987.

[37] F. Pereira and D. Warren. Definite clause grammars for language analysis—a survey of the formalism and a comparison with augmented transition networks. *Artificial Intelligence*, 13:231–278, 1980.

[38] C. Pollard and I. Sag. *An Information-Based Syntax and Semantics. CSLI Lecture Notes 13*, Center for the Study of Language and Information, Stanford University, 1987.

[39] W. Rounds and R. Kasper. A complete logical calculus for record structures representing linguistic information. In *Proceedings of the 1st IEEE Symposium on Logic in Computer Science*, pages 38–43, Boston, Mass., 1986.

[40] W. C. Rounds and A. Manaster-Ramer. A logical version of functional grammar. In *Proceedings of the 25th Annual Meeting of the ACL, Stanford University*, pages 89–96, Stanford, Cal., 1987.

[41] K. Schild. *Undecidability of U*. KIT Report 67, Department of Computer Science, Technische Universität Berlin, Berlin, West Germany, Oct. 1988.

[42] M. Schmidt-Schauß. Subsumption in KL-ONE is undecidable. In R. J. Brachman, H. J. Levesque, and R. Reiter, editors, *Proceedings of the 1st International Conference on Principles of Knowledge Representation and Reasoning*, pages 421–431, Toronto, Ont., May 1989.

[43] M. Schmidt-Schauß and G. Smolka. *Attributive Concept Descriptions with Unions and Complements*. SEKI Report SR-88-21, Department of Computer Science, Universität Kaiserslautern, Kaiserslautern, West Germany, Dec. 1988.

[44] S. M. Shieber. The design of a computer language for linguistic information. In *Proceedings of the 10th International Conference on Computational Linguistics*, pages 362–366, Stanford, Cal., 1984.

[45] S. M. Shieber. *An Introduction to Unification-Based Approaches to Grammar. CSLI Lecture Notes 4*, Center for the Study of Language and Information, Stanford University, 1986.

[46] S. M. Shieber. *Parsing and Type Inference for Natural and Computer Languages*. Technical Note 460, SRI International, Artificial Intelligence Center, Menlo Park, Cal., March 1989.

[47] S. M. Shieber, H. Uszkoreit, F. Pereira, J. Robinson, and M. Tyson. The formalism and implementation of PATR-II. In J. Bresnan, editor, *Research on Interactive Acquisition and Use of Knowledge*, SRI International, Artificial Intelligence Center, Menlo Park, Cal., 1983.

[48] G. Smolka. *A Feature Logic with Subsorts*. LILOG Report 33, IBM Deutschland, Stuttgart, May 1988.

[49] N. K. Sondheimer and B. Nebel. A logical-form and knowledge-base design for natural language generation. In *Proceedings of the 5th National Conference of the American Association for Artificial Intelligence*, pages 612–618, Philadelphia, Pa., Aug. 1986.

[50] F. N. Tou, M. D. Williams, R. E. Fikes, A. Henderson, and T. Malone. RABBIT: an intelligent database assistant. In *Proceedings of the 2nd National Conference of the American Association for Artificial Intelligence*, pages 314–318, Pittsburgh, Pa., Aug. 1982.

[51] M. B. Vilain. The restricted language architecture of a hybrid representation system. In *Proceedings of the 9th International Joint Conference on Artificial Intelligence*, pages 547–551, Los Angeles, Cal., Aug. 1985.

[52] W. A. Woods. What's important about knowledge representation. *IEEE Computer*, 16(10):22–29, Oct. 1983.

Knowledge Representation in LILOG

Udo Pletat, Kai von Luck
IBM Germany
Scientific Center
Institute for Knowledge Based Systems
Project LILOG
P. O. Box 80 08 80
D-7000 Stuttgart 80

Abstract

This paper introduces the knowledge representation language L_{LILOG}. The language is being developed in the framework of the LILOG project and serves for modelling the semantic backgrond knowledge of the LILOG natural language understanding system. Moreover, it is also used as the target language for representing information extracted from German texts in a logical form. The aspects of L_{LILOG} discussed here focus on the sort concept of L_{LILOG} and its means for structuring knowledge bases. The sort concept of L_{LILOG} integrates ideas from the KL-ONE family of languages and other feature term languages having their origin in the area of computational linguistics into the framework of an order-sorted predicate logic. The structuring concept introduced for L_{LILOG} is a simple form of separating logical theories into modules.

1   Introduction

The knowledge representation language L_{LILOG} to be presented in this paper is being developed in the framework of the LILOG project on natural language understanding (see [Herzog et al. 86]). The language is used for two purposes within the project: on the one hand we employ it for modelling the semantic background knowledge required for the process of understanding German texts. On the other hand it serves as the target language for representing information extracted from German texts in a logical form. In this context, the knowledge representation language for LILOG has to satisfy several requirements:

- it has to be expressive enough to capture a wide range of natural language phenomena,

- it has to serve as a communication medium between the linguistic and the logic part of the project,

- it has to offer structuring principles for knowledge to be defined in L_{LILOG} , because applying the language in the natural language understanding context involves developing and handling large knowledge bases,

- it has to be defined in terms of a formal syntax and semantics.

The first version of L_{LILOG} (see [Beierle et al. 88a]) has been a first step towards satisfying these requirements. In the development of L_{LILOG} I, the major efforts have been spent on integrating order-sorted predicate logic and (parts of) the feature graph description language STUF (see [Bouma et al. 88] and [Beierle et al. 88b]). STUF is used as the representation language for the linguistic knowledge of the LILOG system: the grammar and the lexicon. The semantics of STUF characterizes it as a set definition language and therefore the integration of order-sorted logic and STUF was achieved by allowing STUF expressions to occur as sorts in L_{LILOG} I. While L_{LILOG} I did not fully integrate STUF as a sort description language into an order-sorted logic, L_{LILOG} II is more elaborate in this respect. The language is again an order-sorted logic, but in contrast to ordinary order-sorted logic, where sorts can only be described by sort names, L_{LILOG} II offers complex sort descriptions which have been adapted from the KL-ONE family of languages (see e.g. [Brachman, Schmolze 85], [Luck, Owsnicki-Klewe 89]), and Feature Logic ([Smolka 88]) where disjunction and negation have been integrated into a language like STUF. The complex sort description language offered by L_{LILOG} II allows us to form sort expressions representing the intersection or the union of two sorts; we may form the complement of a sort expression or define a sort by enumerating its elements. Using the concept of features we may form expressions like

> *person* ⊓ **with-feature** *age* **in** [*13 .. 19*]

describing the sort of all *persons* of *age* between 13 and 19. A definition like this only makes sense if the *age* feature is applicable to persons. This would be the case if the following additional information were part of the sort hierarchy of a knowledge base:

> **sort** *physical-object*
>   **features** *age : integer*
>
> **sort** *person* < *physical-object*

The above pieces of knowledge introduce a sort *physical-object* with the feature *age*; the sort *person* is declared to be a subsort of *physical-object*.

In contrast to the approaches of the KL-ONE family of languages, but also of STUF and Feature Logic, features to be used in L_{LILOG} II may have more specific source sorts than the implicitly given top sort ⊤ of the lattice of sort expressions. This is a reasonable generalization when embedding features from set-description languages into a logic framework, since features are (distinguished) one-place functions attached to the sorts.

The complex sort expressions describe sets with a certain structure while sort names stand for arbitrary sets about which no structural information is available. In order to impose such a structure on a set represented by a sort name L_{LILOG} offers the concept of sort constraints. These constraints allow us to require a sort name to be interpreted by the same set as a sort expression, or to state that a sort is a subsort of a sort expression. E. g., we can define a sort containing all physical objects whose age is between 13 any 19 (years) which are also persons by

sort *teenager = person* ⊓ **with-feature** *age* **in** [*13 .. 19*],

i.e., the sort *teenager* stands for the same set of objects as the sort expression

person ⊓ **with-feature** *age* **in** [*13 .. 19*].

The above sketches the basic ideas of the enriched sort description language of L_{LILOG} II. The logic part is almost a standard order-sorted logic built on top of this extended sort definition language. The most important difference from order-sorted logic are the sort literals. Sort literals are literals whose predicate symbol is a sort expression and state that a term t is of a certain sort although the sort declaration of the topmost symbol of t makes t belong to another sort. The sort literals are mainly motivated by the natural language understanding context in which L_{LILOG} is currently applied. Thus we want to illustrate the use of sort literals by having a look at what may happen when analysing a natural language text.

If we hear that *John* works at the university we might be a bit narrow-minded and consider him to be a *professor*; this will cause the declaration of a reference object

reference object *John : professor*

Further information about *John* extracted from a text may cause us to change our opinion about *John* being a *professor* to *John* being a *student*. This could be expressed by an axiom like

axiom *X : person . student(X)* ← *further-info(X)*

where *student* is a sort which is the reason why we call *student(X)* a sort literal. If we now add a further piece of information about *John* to our knowledge by stating

axiom *further-info(John)*

we obtain both *student(John)* and *professor(John)* giving us *professor* ⊓ *student(John)*. So *John* is both a *professor* and a *student* and thus an element of the intersection of the two sorts. Since sort literals may be used arbitrarily, they are an elegant means for expressing conditional sort membership, a situation which occurs frequently in the framework of processing natural language.

Sort literals have also been introduced in the order-sorted logics of Oberschelp ([Oberschelp 62]) and Cohn ([Cohn 87]). [Beierle et al. 89] discusses inference rules for an order-sorted logic with sort literals which will form the backbone for processing L_{LILOG} II.

Besides a sophisticated sort concept, L_{LILOG} offers a simple concept for modularizing knowledge bases. The natural language processing framework in which L_{LILOG} will be used requires large background knowledge bases for understanding a natural language text. The amount of background knowledge needed suggests organizing this knowledge within modules. The approach of

[Wachsmuth 88] suggests structuring knowledge bases into so-called knowledge packets, which are the modules of a knowledge base. This is in the tradition of [Sussman, McDermott 72] and [Hendrix 79] where the concept of context has been discussed as a means for separating knowledge bases into clusters of knowledge entities. The knowledge packets of a knowledge base are related by an accessibility relation expressing from which modules a module imports knowledge. This accessibility relation is given in terms of a partial order on the knowledge packets of a knowledge base. The four knowledge packets kp_1, kp_2 kp_3, and kp_4 of a knowledge base could be arranged as follows:

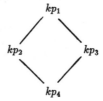

This accessibility relation expresses that kp_4 can import knowledge from kp_2 and kp_3 which both import knowledge from kp_1. Importing knowledge has a dynamic aspect and depends on the focused knowledge packet. The focused knowledge packet selects a part of the knowledge base: those knowledge packets which are greater - according to the partial order representing the accessibility relation - than the focused knowledge packet. From a dynamic point of view this is interpreted in such a way that focusing on a knowledge packet means importing all the knowledge of the selected part of the knowledge base into the focused knowledge packet.

The rest of the paper is organized as follows. Section 2 introduces the sort language of L_{LILOG} and section 3 is dedicated to imposing constraints on the partial order over the sort expressions. Then we proceed to signatures in section 4 and define terms and knowledge bases in sections 5 and 6, respectively. In section 7 we present the structuring mechanism for L_{LILOG} knowledge bases: the knowledge packets. In the final chapter we draw some conclusions and refer to subjects of future research in the framework of L_{LILOG} .

2 Sort Signatures and Sort Expressions

A sort signature contains those symbols over which sort expressions serving for the description of sets can be formed. Classical order-sorted logic ([Oberschelp 62], [Walther 87], [Goguen, Meseguer 87]) considers only very restricted sort signatures consisting just of a collection of sort names which are used for describing sets. The language L_{LILOG} has adopted concepts from elaborate set description languages that originated from the KL-ONE family of languages ([Brachman, Schmolze 85]) and languages for defining feature structures which have their origin in the field of computational linguistics ([Kasper, Rounds 86], [Shieber 86], [Bouma et al. 88], [Smolka 88]). According to all these influences on L_{LILOG} , a sort signature contains, in addition to sort names, atoms, features, and roles. In contrast to approaches of the KL-ONE-style and also the feature term languages, we adopt the ideas of sorted logic and assume that features, roles, and the atoms of a sort signature are tagged with sort information. The rationale for this is to attach features, roles and atoms explicitly to sorts.

Definition 2.1 *A sort signature* Σ^{SORT} *is a quadruple* $\Sigma^{SORT} = \langle S, F, R, A \rangle$ *where*

- *S is a set of sort names such that* $\{\bot, bool, int, \top\} \subseteq S$

- $F = \langle F_{s \to ese} \rangle_{s \in S, ese \in ESE(\Sigma^{SORT})}$ *is a family of sets of* <u>features</u>
- $R = \langle R_{s,ese} \rangle_{s \in S, ese \in ESE(\Sigma^{SORT})}$ *is a family of sets of* <u>roles</u>
- $A = \langle A_s \rangle_{s \in S}$ *is a family of* <u>atoms</u>

Sort expressions to be defined in the next step represent complex sorts constructed over the alphabet given by a sort signature.

Definition 2.2 *Let* $\Sigma^{SORT} = \langle S, F, R, A \rangle$ *be a sort signature.*

- *The set of* <u>sort expressions</u> $SE(\Sigma^{SORT})$ *over* Σ^{SORT} *contains*
    - S
    - $se \sqcap se'$
    - $se \sqcup se'$
    - $\neg se$
    - $\{a_1, ..., a_k\}$
    - **with-feature** f **in** ese
    - **with-role** r **in** ese
    - **some** r
    - **agree** $fp\ fp'$
    - **disagree** $fp\ fp'$

  where $se, se' \in SE(\Sigma^{SORT})$, $ese \in ESE(\Sigma^{SORT})$, $r \in R$, $a_i \in A$ *and* $fp, fp' \in FP(\Sigma^{SORT})$.
- *The set of* <u>extended sort expressions</u> $ESE(\Sigma^{SORT})$ *is*
    $$ESE(\Sigma^{SORT}) = SE(\Sigma^{SORT}) \cup \{se^*, se^+ \mid se \in SE(\Sigma^{SORT})\}$$
  and contains, besides the sort expressions introduced above, expressions se^* *and* se^+ *which describe sorts whose objects are sets of at least one/two elements of* se, *respectively.*
- *The family of* <u>feature paths</u> $FP(\Sigma^{SORT})$ *over* Σ^{SORT} *is defined as*
    $$FP(\Sigma^{SORT}) = \langle FP(\Sigma^{SORT})_{s \to ese} \rangle_{s \in S, ese \in ESE(\Sigma^{SORT})}$$
  where
    - $F_{s \to ese} \subseteq FP(\Sigma^{SORT})_{s \to ese}$
    - $f \in F_{s \to s'}$ *and* $p \in FP(\Sigma^{SORT})_{s' \to ese}$ *implies* $f; p \in FP(\Sigma^{SORT})_{s \to ese}$

  Note that this definition of feature paths allows set valued features only at the end of a feature path, i.e. at the leaves if we consider the feature path to be part of a feature graph.

Sort expressions are complex set descriptions and integrate concepts which may be found in other set description languages as well. The sort expressions se^* and se^+ stand for the sort of subsets of at least one (at least two) element(s) of the sort represented by se, respectively. These constructs have been adopted from the Plural Logic developed in [Link 83] and will be used for representing plural objects to be handled in the natural language understanding framework where L_{LILOG} is applied.

Before we define the formal semantics of these syntactic constructs, let us give an idea of what can be formulated with them.

The basic idea is to describe sets by means of sort expressions. Compared to other set description formalisms like the KL-ONE family of languages, STUF, or Feature Logic, features and roles can be attached to arbitrary sorts, not only the top sort $\top$ of the lattice of sort expressions. This is a straightforward generalization when embedding these concepts into a sorted logic, since features and roles are (distinguished) one-place functions and two-place relations, respectively.

Using an ad hoc concrete syntax for providing our examples of how to use L_{LILOG} we can define the sort of vehicles as

sort *vehicle*

    features *wheels* : [*2 .. 16*]
                 doors : [*0 .. 4*]
                 type : { *bike, sedan, cabrio, truck* }

where *bike, sedan, cabrio,* and *truck* would be atoms of another sort, maybe *vehicle-type*; i.e. we would have another sort declaration

sort *vehicle-type*

    atoms *bike, cabrio, sedan, truck*

The *integers* are built-in and thus don't have to be declared. We only want to mention that intervals like [0 .. 4] are short forms of sort expressions formed by enumerating atoms of the sort *int*: { 0, 1, 2, 3, 4 } in this case, where the numbers are defined to be atoms of the built-in sort *int*. We now may describe special sets by forming expressions like

with-feature *wheels* **in** [*4 .. 16*]

which might describe all kinds of cars, or

with-feature *type* **in** { *cabrio* }

describing all vehicles having a convertible roof.

Because features are attached to sorts, an expression like

with-feature *wheels* **in** [*4 .. 16*]

always describes a set which is a subset of the source sort of the feature occurring in the expression, *wheels* in this case, i.e. the sort *vehicle* subsumes the sort expression **with-feature** wheels **in** [4 .. 16]. This is again a generalization of the situation in KL-ONE, STUF, and Feature Logic where, for example, features are implicitly given as $f : \top \longrightarrow \top$ and the expression **with-feature** wheels **in** [4 .. 16] is also subsumed by the source sort of the feature, namely by $\top$.

Finally we want to comment on the extended sort expressions se^* and se^+. They have been introduced to treat plural phenomena according to the Plural Logic of [Link 83]. The standard example of a piano being carried by some people could be modeled by defining a sort *piano* which foresees some *porters* of a piano by introducing a feature taking its values in the sort *person** when light pianos which can be carried by just one person are considered. Alternatively we might think of heavy pianos which can only be carried by at least two people: this could be expressed by chosing *person*$^+$ as the target sort of the feature *porters*.
Thus the corresponding sort declarations could be

sort *light-piano*

    features *porters* : *person**

or

> **sort** *heavy-piano*
>> **features** *porters : person*[+]

For the syntactic concepts of sort signatures and sort expressions we now provide the semantic notion of universes which are the structures in which we interpret them.

In order to give a meaning to the sort expressions se^* and se^+, we introduce the following operation on sets and their powersets:

Let M be any set.

$$\oplus : (2^M \cup M) \times (2^M \cup M) \longrightarrow 2^M$$

$$
\begin{aligned}
\oplus(m, m') &= m \cup m' &&\text{, if } m, m' \in 2^M \\
\oplus(m, m') &= m \cup \{m'\} &&\text{, if } m \in 2^M, m' \in M \\
\oplus(m, m') &= \{m\} \cup m' &&\text{, if } m' \in 2^M, m \in M \\
\oplus(m, m') &= \{m\} \cup \{m'\} &&\text{, if } m, m' \in M
\end{aligned}
$$

We adopt the following convention for finite sums formed by using $\oplus$:

$$\oplus_{i=1}^{n} m_i = m_1, \text{ for } n = 1.$$

The model-theoretic concept of a universe consists of a collection of sets such that for each sort expression over a sort signature there is a corresponding set interpreting it. Moreover, the sort expressions impose a special structure on the sets in the universe. Starting with the interpretation of the sort expressions, we will then say how to interpret the other components of a sort signature. Features and roles are interpreted by one-place functions and two-place relations, respectively, and atoms are elements of the sets interpreting the sort they are attached to.

Definition 2.3 *Let $\Sigma^{SORT} = \langle S, F, R, A \rangle$ be a sort signature. A <u>universe</u> U for Σ^{SORT} is a triple $U = \langle D, F_U, R_U \rangle$ where*

- *D is a family of sets $D = \langle D_{se} \rangle_{se \in SE(\Sigma^{SORT})}$, the <u>domain</u> of U with*
    - $D_{se} \subseteq D_\top$ *for any sort expression se*
    - $D_\perp = \emptyset$
    - $D_{int} = \mathcal{Z}$
    - $D_{bool} = \mathcal{B}$
    - $D_{se \sqcap se'} = D_{se} \cap D_{se'}$
    - $D_{se \sqcup se'} = D_{se} \cup D_{se'}$
    - $D_{\neg se} = D_\top \setminus D_{se}$
    - $D_{\{a_1, ..., a_k\}} = \{a_{1_U}, ..., a_{k_U}\}$
    - $D_{\text{with-feature } f \text{ in } ese} = \{x \in dom(fp_U) \mid f_U(x) \in D_{ese}\}$
    - $D_{\text{with-role } r \text{ in } ese} = \{x \in dom(r_U) \mid (x, y) \in r_U \Rightarrow y \in D_{ese}\}$
    - $D_{\text{some } r} = \{x \in dom(r_U) \mid \text{ there is a } y \text{ with } (x, y) \in r_U\}$
    - $D_{\text{agree } fp\, fp'} = \{x \in dom(fp_U) \cap dom(fp'_U) \mid fp_U(x) = fp'_U(x)\}$
    - $D_{\text{disagree } fp\, fp'} = \{x \in dom(fp_U) \cap dom(fp'_U) \mid fp_U(x) \neq fp'_U(x)\}$

where

- $dom(fp_U) = D_s$ for $fp \in FP_{s \to ese}$
- $dom(r_U) = D_s$ for $r \in R_{s,se}$
- $(f;p)_U(x) = p_U(f_U(x))$

- *the set forming extended sort expressions are interpreted as follows*
  - $D_{se^*} = \{\oplus_{i=1}^n m_i \mid m_i \in D_{se}, n \geq 1\}$
  - $D_{se^+} = \{\oplus_{i=1}^n m_i \mid m_i \in D_{se}, n \geq 2\}$

  Note that the sets interpreting se^* and se^+, respectively, are not subsets of D_T but of D_{T^*}.

- F_U *is a set of functions containing a total function* $f_U : D_s \longrightarrow D_{ese}$ *for each feature* $f \in F_{s \to ese}$

- R_U *is a set of relations containing a relation* $r_U \subseteq D_s \times D_{ese}$ *for each role* $r \in R_{s,ese}$

- *an element* $a_U \in D_s$ *for each atom* $a \in A$, *such that* $a \neq a'$ *implies* $a_U \neq a'_U$

The interpretation of the sorts, features, roles and atoms meets the intuition behind these concepts as syntactic means for speaking about sets, one-place functions, two-place relations and objects within certain domains, respectively. These basic notions are used to define those sets within the universe which are described by arbitrary extended sort expressions.

The set-forming operators * and + are idempotent in the following sense:

$$(se^*)^* = se^* \text{ and } (se^+)^+ \subseteq se^+.$$

I.e., these set-forming operators cannot form sets of sets of sets ..., but only do this once.

An important notion in relation to sort expressions is that of subsumption.

Definition 2.4 *Let* Σ^{SORT} *be a sort signature and* se *and* se' *be sort expressions over* Σ^{SORT}. se' Σ^{SORT}-*subsumes* se *iff for any universe* $U = \langle D, F_U, R_U \rangle$ *for* Σ^{SORT} *we have* $D_{se} \subseteq D_{se'}$. *We then write* $se \ll_{\Sigma SORT} se'$.

The notion of subsumption between sort expressions can be extended in a straightforward way to extended sort expressions. We obtain the following relationships:

$$se \ll_{\Sigma SORT} se' \text{ implies } se^* \ll_{\Sigma SORT} se'^* \text{ and } se^+ \ll_{\Sigma SORT} se'^+.$$

The subsumption relation can even be used in a 'mixed style' to compare sort expressions and extended sort expressions. Then we obtain the relationships depicted in the diagram below

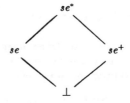

as they are seen in [Link 83]. In addition to the above diagram we have that se and se^+ will always be interpreted by disjoint sets.

Since universes are semantic structures for sort signatures, we can ask the standard question: 'Does there exist a universe for a sort signature?'. Technically speaking we introduce a first concept of consistency.

Definition 2.5 *A sort signature Σ^{SORT} is <u>consistent</u> iff there is a universe U for Σ^{SORT}.*

Considering the sort signature

> **sort** s
>> **features** $f : s' \sqcap \neg s'$
>> **atoms** a
>
> **sort** s'

we observe that sort signatures may be inconsistent. This holds because the sort s has to be interpreted by a non-empty set, since $a_U \in D_s$ must hold in any universe U. Then, however, we cannot interpret the feature f of the sort s by the empty function (where both the domain and the range are the empty set) which is the only possibility to assign a semantic object to f satisfying the requirement that its range has to be the empty set.

3 Sort Hierarchies

In the previous section we have seen that sort signatures can be inconsistent, i.e. there are sort signatures for which we cannot find a universe.

The idea behind the sort hierarchies to be introduced next is to influence the subsumption ordering $\ll_{\Sigma^{SORT}}$ on sort expressions by imposing so-called sort constraints. These constraints allow us to require a sort to be interpreted by the same set as a sort expression, i.e. we may enforce certain identities between the sets of a universe; and these identities should then hold in any universe for a sort signature.

Imposing such sort constraints for sort signatures which are inconsistent doesn't make much sense since these sort signatures don't have a universe and thus will satisfy any sort constraint. Therefore we make the following

General Assumption: Any sort signature has to be consistent.

Definition 3.1 *Let $\Sigma^{SORT} = \langle S, F, R, A \rangle$ be a sort signature. The set of <u>sort constraints</u> $SC(\Sigma^{SORT})$ over Σ^{SORT} is defined as follows*

- $s \doteq se \in SC(\Sigma^{SORT})$, for $s \in S$ and $se \in SE(\Sigma^{SORT})$

Using sort constraints of the form $s \doteq se$ we can define other constraints which will be of interest. For example we can express the constraint $s \leq se$ by $\neg se \sqcap s \doteq \bot$, or $s \doteq s \sqcap se$, or $s \doteq se \sqcap s'$ for a new sort name s'. The disjointness condition for two sorts can be stated directly as $s \sqcap s' = \bot$.

Definition 3.2 *A sort hierarchy $SH = \langle \Sigma^{SORT}, SC \rangle$ consists of a sort signature Σ^{SORT} and a set of sort constraints $\overline{SC}$ over Σ^{SORT}.*

Sort hierarchies are a generalized form of the partially ordered set of sorts which usually describes the set structure of the universe of discourse in knowledge bases defined in order-sorted logic, see [Walther 87]. Looking at the KL-ONE world, sort hierarchies correspond to (restricted) T-Boxes, c. f. [Brachman et al. 83].

As already mentioned, sort hierarchies correspond to the KL-ONE concept of a T-Box. The purpose of sort hierarchies is to impose constraints on the sorts by formulating sort constraints which enforce that a certain sort is interpreted by a special set, namely the set defined by the sort expression forming the righthand side of the sort constraint.

Returning to our example on vehicles, we may want to define that *sportscars* are those vehicles having 4 wheels, 2 doors and which are of type *cabrio*; this can be done by making the following sort constraint part of the sort hierarchy:

> sort *sportscar* = **with-feature** *wheels* **in** $\{4\}$ $\sqcap$
> **with-feature** *doors* **in** $\{2\}$ $\sqcap$
> **with-feature** *type* **in** $\{$ *cabrio* $\}$

Now we want to define the meaning of a sort constraint and introduce the concept of a universe satisfying the constraints of a sort hierarchy. This is straightforward and the definition below is what we expect intuitively.

Definition 3.3 *Let* $\Sigma^{SORT} = \langle S, F, R, A \rangle$ *be a sort signature and* $U = \langle D, F_U, R_U \rangle$ *a universe for* Σ^{SORT}. *U* *satisfies the sort constraint*

- $s \doteq se$ *iff* $D_s = D_{se}$; *we then write* $U \models s \doteq se$.

All these concepts are put together in order to define whether a universe is a semantic structure satisfying the sort constraints of a sort hierarchy.

Definition 3.4 *Let* $SH = \langle \Sigma^{SORT}, SC \rangle$ *be a sort hierarchy and* $U = \langle D, F_U, R_U \rangle$ *a universe for* Σ^{SORT}. *U* *is a* <u>*universe*</u> *for SH iff U is a universe* Σ^{SORT} *and* $U \models sc$ *for any* $sc \in SC$.

The sort constraints can be considered as axioms formulated over the symbols declared in the sort signature. These axioms are expressive enough to make some sort hierarchies inconsistent, i. e. there are sort hierarchies for which no universe exists. The following sort hierarchy is an example of this kind. The sort s has to satisfy the contradicting requirements of being interpreted by the empty set on the one hand, while on the other hand it has to contain at least one element: the object interpreting the atom a.

> sort $s \doteq \perp$
> **atoms** a
>
> sort s'

On the basis of the concept of a universe for a sort hierarchy we can extend the notion of subsumption as follows:

> $se \ll_{SH} se'$ *iff* $D_{se} \subseteq D_{se'}$, *for all universes* $U = \langle D, F_U, R_U \rangle$ *for SH*.

According to this definition we obtain $s \ll_{SH} s'$ for our sort hierarchy above, because it is inconsistent. The sort signature alone is still consistent, but we do not obtain $s \ll_{\Sigma^{SORT}} s'$.

4 Signatures

The previous two sections introduced the syntactic and semantic concepts for describing the universe of discourse of a knowledge base in terms of a sort hierarchy. What we have seen there is a generalization of the usual concept of a partially ordered set of sorts which is used in ordinary order-sorted logic to define sort hierarchies in order to integrate some ideas known from KL-ONE and feature term languages.

The next step is to introduce signatures containing the operator symbols and predicate symbols over which terms and formulas can be formed in order to provide the axioms of a knowledge base.

Being guided by order-sorted logic where the sort hierarchies of the signatures are always consistent we make the

General Assumption: Any sort hierarchy is consistent.

Definition 4.1 *A* L_{LILOG} *signature* $\Sigma = \langle SH, O, RO, P \rangle$ *consists of*

- *a sort hierarchy* $SH = \langle \Sigma^{SORT}, SC \rangle$
- *a family of sets of operators* $O = \langle O_{w \to ese} \rangle_{w \in ESE(\Sigma^{SORT})^*, ese \in ESE(\Sigma^{SORT})}$
- *a family of sets of reference objects* $RO = \langle RO_{ese} \rangle_{ese \in ESE(\Sigma^{SORT})}$
- *a family of sets of predicates* $P = \langle P_w \rangle_{w \in ESE(\Sigma^{SORT})^*}$.

L_{LILOG} signatures define the alphabet of symbols over which terms and formulas will be constructed. We already know about the sort hierarchy being part of a L_{LILOG} signature and shall discuss the next components below.

The operators are those function symbols which we use for the term construction. Reference object identifiers form a distinguished class of constants within our logic constituted by L_{LILOG} . The predicates play the standard role as in any logic. Features, atoms and roles of a sort signature are also operations and predicates, respectively, of a signature.

Definition 4.2 *Let* $\Sigma^{SORT} = \langle S, F, R, A \rangle$ *be a sort signature.*

- *A family of operators* O *over* Σ^{SORT} *is a family of sets*
  $$O = \langle O_{w \to ese} \rangle_{w \in ESE(\Sigma^{SORT})^*, ese \in ESE(\Sigma^{SORT})}$$
  such that

  - $F_{s \to ese} \subseteq O_{s \to ese}$, *for* $s \in S, ese \in ESE(\Sigma^{SORT})$, *i.e. features may be used as operators.*
  - $A_s \subseteq O_{\epsilon \to s}$, *for* $s \in S$, *i.e. atoms may appear as constants.*
    (ϵ *stands for the empty string)*
  - $a_i \in O_{\epsilon \to \{a_1, \dots, a_n\}}$, *for* $1 \leq i \leq n$, *i.e. the atoms of an enumerated sort are operators of that enumerated sort*
  - *the following arithmetic operations are available*

    $$
    \begin{array}{lll}
    +, -, *, / & : int, int & \longrightarrow int \\
    - & : int & \longrightarrow int \\
    \dots, -2, -1, 0, 1, 2, \dots & : & \longrightarrow int
    \end{array}
    $$

    where the numbers are atoms of the sort int.

  - *the following operations on sets are available for any sort expression* se

    $$
    \begin{array}{lll}
    \oplus & : se^*, se^* & \longrightarrow se^+ \\
    card & : se^* & \longrightarrow int
    \end{array}
    $$

    The Plural Logic of [Link 83] has introduced a richer reservoir of operations on sets which we have not built into L_{LILOG} *in its current state.*

- A family of <u>reference objects</u> RO over Σ^{SORT} is a family of sets
$$RO = \langle RO_{ese} \rangle_{ese \in ESE(\Sigma^{SORT})}$$

- A family of <u>predicates</u> P over Σ^{SORT} is a family of sets
$$P = \langle P_w \rangle_{w \in ESE(\Sigma^{SORT})}.$$

such that

- $R_{s,ese} \subseteq P_{s,ese}$ for any $s \in S, ese \in ESE(\Sigma^{SORT})$, i.e. roles appear as binary predicates.

- the sort predicate
$$ese \quad : \top$$
is available for each extended sort expression $ese \in ESE(\Sigma^{SORT})$, i.e. each extended sort expression may be used as a unary predicate.

- Predicates
$$<, \leq, >, \geq \quad : int, int$$
comparing integers are available.

- the equality predicate
$$\doteq \quad : \top^*, \top^*$$
is available.

- A predicate for testing whether an object of sort $\top^*$ is a sub-object of an object of sort se^*. This notion has to be understood in the sense of [Link 83] where all objects of a sort are considered as elements of a lattice of sets.
We use a slightly modified version of Link's ideas in that we don't consider non-set objects as singleton sets. However, semantically the $\leq$ predicate behaves as proposed in [Link 83] and can be understood as a uniform test of equality, membership, and inclusion.
$$\leq \quad : \top^*, se^*$$

- a predicate for testing whether an object of sort se^* belongs to the sort se
$$atomic \quad : \top^*$$
is available.

The idea of signatures is to introduce further constants, function symbols, and predicate symbols for forming logical axioms which are not atoms, features, or roles. While atoms, features, and roles are symbols of a signature which may be used for defining sorts this is not possible for the other symbols of the signature.

We will now continue our story of traveling people and discuss an important predicate that should be part of any signature of a knowledge base on traveling:

predicate travel : who :: person*, from :: location, to :: location, with :: vehicle

Thus, in the concrete syntax we may name the argument positions in order to allow for free placing of the arguments when forming literals with a predicate. Although we didn't speak about literals up to now, it should be clear what they look like; thus with the above predicate as part of a signature we may express that 'John and Mary travel with their Porsche' by the literal

travel(who :: John ⊕ Mary, with :: Porsche, from :: LA, to :: SF)

In this literal the positions of the arguments of the predicates *travel* are interchanged. This is allowed due to the explicit naming of the argument positions. Formally, the above literal requires that the persons John and Mary, the cities LA and SF, and the Porsche have been introduced as objects of the knowledge base and should thus be mentioned in its signature; they would typically be introduced as

reference object *John : man*

reference object *Mary : woman*

reference object *LA : location*

reference object *SF : location*

reference object *Porsche : sportscar*

For reasons of completeness we should introduce the sorts *man* and *woman* into our sort hierarchy over which the signature is formed and thus have the following sort definitions introducing the subsorts *man* and *woman* of the sort *person* and declaring *woman* to be disjoint from the sort *man*.

sort *man* < *person*

sort *woman* < *person*
      disjoint *man*

The syntactic concept of a signature will now be given a meaning by introducing the notion of a model for a signature Σ.

On the basis of the interpretation of the sortal information that will be part of a L_{LILOG} knowledge base we now define the meaning of the operator and predicate symbols and the reference object identifiers appearing in a signature.

Definition 4.3 *Let* $\Sigma = \langle SH, O, RO, P \rangle$ *be a* L_{LILOG} *signature where* $SH = \langle \Sigma^{SORT}, SC \rangle$ *with* $\Sigma^{SORT} = \langle S, F, R, A \rangle$.
A Σ-*model is a triple* $M = \langle U, F_M, R_M \rangle$ *where*

- *U is a universe for SH*

- *F_M is a set of functions containing for each operator $o \in O_{w \to ese}$ a function $o_M : D_w \longrightarrow D_{ese}$ such that $o_M = o_U$, if o is a feature or an atom.*
  Moreover, the built-ins have to be interpreted in the intended way. This should be clear for the arithmetic. The two set-oriented operations $\oplus$ and card are given respectively as the union/set-forming function $\oplus$ of section 2 and the cardinality function on sets.

- *R_M is a set of relations containing for each predicate $p \in P_w$ a relation $p_M \subseteq U_w$ such that $p_M = p_U$ if p is a role.*
  Moreover, the built-ins have to be interpreted in the intended way. This should be intuitively clear for the equality and the arithmetic comparators. The sort predicates, the $\leq$ predicate, and the predicate for testing the atomicity of an object have the following semantics:

  - *$ese_M = D_{ese}$*
  - *$\leq_M \subseteq D_{T^*} \times D_{ese}$ and*

$$
x \leq_M y \Leftrightarrow \begin{cases} x = y & \text{, if } x, y \in D_T \\ x \in y & \text{, if } x \in D_T, y \in D_{T^*} \setminus D_T \\ x \subseteq y & \text{, if } x, y \in D_{T^*} \setminus D_T \end{cases}
$$

$-\ atomic_M \subseteq D_T.$ and

$$atomic_M(x) \Leftrightarrow x \in D_T$$

- *For each reference object* $ro \in RO_{ese}$ *there is an element* $ro_M \in D_{ese}$.

Apart from the involved definition of the universe of a model we have given before, this notion is nothing but the straightforward extension of classical notions of models for a signature to our setting where sorts are not just names but complex expressions.

Again we face the problem that models need not exist for any L_{LILOG} signature which is in contrast to ordinary order-sorted logic.

We close this section by introducing a further version of the subsumption relation. Let Σ be a L_{LILOG} signature and se and se' be two sort expressions.

$se \ll_\Sigma se'$ *iff for any* Σ-*model* $M = \langle U, F_M, R_M \rangle$ *we have* $D_{se} \subseteq D_{se'}$ *for* $U = \langle D, F_U, R_U \rangle$.

We then say that se' $\underline{\Sigma\text{-subsumes}}$ se.

5 Terms

The next step in the definition of L_{LILOG} is to define the construction of terms.

As with all the concepts introduced so far, term construction for L_{LILOG} is a generalization of what is known from order-sorted logic. Instead of the standard *syntactic* condition, namely that the sort s of the argument term t to which the operator o is applied in order to form the term $o(t)$ has to be smaller than the argument sort s' of o, we will impose the *semantic* condition $s \ll_\Sigma s'$. As one can easily see, this semantic condition is also present in the term construction for order-sorted logic: if $s \le s'$ holds, where $\le$ is the reflexive and transitive closure of the user-formulated order $<$ on the sorts, we have $s \ll_\Sigma s'$.

As terms will contain variables let us first say what they are.

Definition 5.1 *Let* $\Sigma^{SORT} = \langle S, F, R, A \rangle$ *be a sort signature. A family of* $\underline{variables}$ V *over* Σ^{SORT} *is a family of sets* $V = \langle V_{ese} \rangle_{ese \in ESE(\Sigma^{SORT})}$.

Definition 5.2 *Let* $\Sigma = \langle SH, O, ROI, P \rangle$ *be a* L_{LILOG} *signature where* $SH = \langle \Sigma^{SORT}, SC \rangle$ *and* V *a family of variables over* Σ^{SORT}.

The $\underline{terms}$ *over* Σ *and* V *form a family of sets*

$T_\Sigma(V) = \langle T_\Sigma(V)_{ese} \rangle_{ese \in ESE(\Sigma^{SORT})}$

such that $T_\Sigma(V)_{ese}$ *contains*

- $V_{ese'}$, *for* $ese' \ll_\Sigma ese$
- $O_{\epsilon \to ese}$, *for* $ese' \ll_\Sigma ese$
- RO_{ese}, *for* $ese' \ll_\Sigma ese$
- $o(t_1, ..., t_k)$, *for* $o \in O_{ese_1, ..., ese_k \to ese}$, *and* $t_i \in T_\Sigma(V)_{ese'_i}$ *such that* $ese'_i \ll_\Sigma ese_i$

where $ese, ese', ese_i, ese'_i \in ESE(\Sigma^{SORT})$.

Since our semantic condition $ese' \ll_\Sigma ese$ always holds for inconsistent signatures, the term construction will then generate any term that could be obtained for an unsorted logic. That is, $T_\Sigma(V)$ would contain terms that are *syntactically ill-typed* in the sense that the term t may be of a sort greater than the source sort of the operator o when forming the term $o(t)$. But since inconsistent signatures don't have any model at all, forming ill-typed terms doesn't make much sense since they can't be evaluated anyway.

The conclusion we draw from these considerations is:

Term construction only makes sense for consistent signatures!

The concept of a term should be clear enough not to have to present further examples beyond those of the previous section.

Term evaluation can be defined along the standard lines.

Definition 5.3 *Let* $\Sigma = \langle SH, O, ROI, P \rangle$ *be a* L_{LILOG} *signature where* $SH = \langle \Sigma^{SORT}, SC \rangle$ *with* $\Sigma^{SORT} = \langle S, F, R, A \rangle$, V *a family of variables over* Σ^{SORT}, *and* $M = \langle U, O_M, P_M \rangle$ *a* Σ-*model.*

- *A variable assignment* α *is a family of total functions*
$$\alpha = \langle \alpha_{ese} : V_{ese} \longrightarrow D_{ese} \rangle_{ese \in ESE(\Sigma^{SORT})}.$$

- *A variable assignment* α *induces a term evaluation*
$$\overline{\alpha} = \langle \overline{\alpha}_{ese} : T_\Sigma(V)_{ese} \longrightarrow D_{ese} \rangle_{ese \in ESE(\Sigma^{SORT})}$$
  where
$$\overline{\alpha}(t) = \begin{cases} o_M & , if\ o \in O_{\epsilon \to ese} \\ \alpha(v) & , if\ v \in V_{ese} \\ o_M(\overline{\alpha}(t_1), ..., \overline{\alpha}(t_k)) & , if\ t = o(t_1, ..., t_k)\ and\ o \in O_{w \to ese} \end{cases}$$

6 Formulas and Knowledge Bases

The final step in the definition of L_{LILOG} is to introduce formulas and knowledge bases and to define their semantics.

While the terms have to be well-typed, we will be more liberal concerning the well-typedness of literals. This is motivated by the natural language application we have in mind. The fundamental idea is that in the framework of natural language understanding, well-typed objects - represented as terms - are okay. However, one wants to be freer in the use of terms in literals leading to *syntactically ill-typed* literals. So for the construction of literals we will not impose our (also the usual) semantic condition that $p(t)$ is a syntactically well-formed literal if and only if in any Σ-model t evaluates to an object of a set to which the relation for p can be applied.

Declaring a reference object ro to be of sort s, and allowing only literals $p(ro)$ to be formed where the source sort s' of p is always interpreted as a superset of the set interpreting s would be too restrictive. The point is that the sort of ro will in general not be the declared sort s (which is determined upon the first mention of ro), but after further text analysis, the conjunction of several sort assertions made for c when processing the text. Thus, the declaration $ro : \longrightarrow s$ only reflects the sort information about ro when ro first appears in the text to be analysed. Further occurrences of ro in the text may force us to attach completely different sort information to ro. This will not be done, however, in terms of overloaded declarations of the constant ro, but rather in terms of sort formulas $s'(ro)$ which appear in the set of axioms of a knowledge base.

Syntactic sort-restrictions for term construction have been imposed in order to avoid semantically irrelevant terms already at the level of their definition. For literals one can be more liberal since *semantic well-typedness*, which is of course what we expect for any literal (and term), can be assured via the definition of satisfiability, due to the availability of sort literals.

The construction of formulas gives us (at the moment) a restricted version of the standard set of formulas of first order predicate logic: we allow only general clauses as axioms of a knowledge base.

Definition 6.1 *Let* $\Sigma = \langle SH, O, RO, P \rangle$ *be a* L_{LILOG} *signature where* $SH = \langle \Sigma^{SORT}, SC \rangle$ *and* V *a family of variables over* Σ^{SORT}.

- *The* <u>literals</u> *over* Σ *and* V *form the set*
  $$L_\Sigma(V) = \{p(t_1, ..., t_k), \neg p(t_1, ..., t_k) \mid p \in P_{ese_1, ..., ese_k} \text{ and } t_i \in T_\Sigma(V)_{T^*}\}$$

- *The* <u>conjunctions</u> *over* Σ *and* V *form the set*
  $$CON_\Sigma(V) = \{l_1 \wedge ... \wedge l_k \mid l_i \in L_\Sigma(V), k > 0\}$$

- *The* <u>disjunctions</u> *over* Σ *and* V *form the set*
  $$DIS_\Sigma(V) = \{l_1 \vee ... \vee l_k \mid l_i \in L_\Sigma(V), k > 0\}$$

- *The* <u>premises</u> *over* Σ *and* V *form the set*
  $$PREM_\Sigma(V) = \{d_1 \wedge ... \wedge d_k \mid d_i \in DIS_\Sigma(V), k \geq 0\}$$

- *The* <u>conclusions</u> *over* Σ *and* V *form the set*
  $$CONC_\Sigma(V) = \{c_1 \vee ... \vee c_k \mid c_i \in CON_\Sigma(V), k > 0\}$$

- *The* <u>axioms</u> *over* Σ *and* V *form the set*
  $$A_\Sigma(V) = \{c \leftarrow p \mid p \in PREM_\Sigma(V), c \in CONC_\Sigma(V)\}$$

- *The* <u>goals</u> *over* Σ *and* V *form the set*
  $$G_\Sigma(V) = \{\leftarrow p \mid p \in PREM_\Sigma(V)\}$$

For our predicate *travel* we may now introduce an axiom expressing that any member of a group of people traveling with some vehicle uses the same vehicle as the entire group.

> **axiom** $M : person,\ G : person^*,\ F : location,\ T : location,\ V : vehicle$.
> $travel(who :: M, from :: F, to :: T, with :: V)$
> $\leftarrow travel(who :: G, from :: F, to :: T, with :: V),\ M \leq G$

With all these notions we can define what we want to consider as a L_{LILOG} knowledge base.

Definition 6.2 *A* L_{LILOG} *knowledge base* $KB = \langle \Sigma, AX \rangle$ *consists of*

- *a* L_{LILOG} *signature* Σ

- *a set of* <u>axioms</u> *AX such that an axiom* $ax \in AX$ *is an element of* $A_\Sigma(V)$ *where* V *is some family of variables* V *over* Σ^{SORT}, *the sort signature of* Σ

Finally let us collect all the pieces of knowledge introduced in the previous sections within one knowledge base about *traveling* shown in Figure 1:

knowledge base *traveling*

    sort *person*

    sort *man* < *person*

    sort *woman* < *person*
    disjoint *man*

    sort *vehicle*
    features *wheels* : [*2 .. 16*]
            doors : [*0 .. 4*]
            type : { *bike, sedan, cabrio, truck* }

    sort *vehicle-type*
    atoms *bike, cabrio, sedan, truck*

    sort *sportscar* = **with-feature** *wheels* **in** {4} ⊓
          with-feature *doors* **in** {2} ⊓
          with-feature *type* **in** { *cabrio* }

    predicate *travel : who :: person*, from :: location, to :: location, with :: vehicle*

    reference object *John : man*

    reference object *Mary : woman*

    reference object *Porsche : sportscar*

    axiom *M : person, G : person*, F .: location, T : location, V : vehicle .*
        travel(who :: M, from :: F, to :: T, with :: V)
            ← travel(who :: G, from :: F, to :: T, with :: V), M ≤ G

end knowledge base

Figure 1: A complete knowledge base on *traveling*

For these last syntactic concepts of L_{LILOG} we now define their semantics in terms of the satisfaction relation between formulas and models and finally we introduce a loose semantics for L_{LILOG} knowledge bases.

Definition 6.3 Let $\Sigma = \langle SH, O, RO, P \rangle$ be a L_{LILOG} signature where $SH = \langle \Sigma^{SORT}, SC \rangle$ and V a family of variables over Σ^{SORT}. Let $M = \langle U, O_M, P_M \rangle$ with $U = \langle D, F_U, R_U \rangle$ be a Σ-model and $\alpha : V \longrightarrow D$ a variable assignment.
M satisfies

- *the literal* $p(t_1, ..., t_k)$ *w.r.t.* α *iff* $\overline{\alpha}(t_i) \in D_{ese_i}$, *for* $1 \le i \le k$ *and assuming* $p \in P_{ese_1, ..., ese_k}$, *and* $(\overline{\alpha}(t_1), ..., \overline{\alpha}(t_k)) \in p_M$; *we then write* $M \models_\alpha p(t_1, ..., t_k)$

- *the literal* $\neg p(t_1, ..., t_k)$ *w.r.t.* α *iff* $\overline{\alpha}(t_i) \in D_{ese_i}$, *for* $1 \le i \le k$ *and assuming* $p \in P_{ese_1, ..., ese_k}$, *and* $(\overline{\alpha}(t_1), ..., \overline{\alpha}(t_k)) \notin p_M$; *we then write* $M \models_\alpha \neg p(t_1, ..., t_k)$

- *the conjunction* $l_1 \wedge ... \wedge l_k$ *w.r.t.* α *iff* $M \models_\alpha l_i$ *for all* i; *we then write* $M \models_\alpha l_1 \wedge ... \wedge l_k$

- *the disjunction* $l_1 \vee ... \vee l_k$ *w.r.t.* α *iff* $M \models_\alpha l_i$ *for some* i; *we then write* $M \models_\alpha l_1 \vee ... \vee l_k$

- *the premise* $d_1 \wedge ... \wedge d_k$ *w.r.t.* α *iff* $M \models_\alpha d_i$ *for all* i; *we then write* $M \models_\alpha d_1 \wedge ... \wedge d_k$

- *the conclusion* $c_1 \vee ... \vee c_k$ *w.r.t.* α *iff* $M \models_\alpha c_i$ *for some* i; *we then write* $M \models_\alpha c_1 \vee ... \vee c_k$

- *the axiom* $c \leftarrow p$ *iff for all variable assignments* α *we have* $M \models_\alpha p$ *implies* $M \models_\alpha c$; *we then write* $M \models c \leftarrow p$.

- *the goal* $\leftarrow p$ *iff there is a variable assignment* α *where* $M \models_\alpha p$; *we then write* $M \models \leftarrow p$.

Note that the definition of satisfaction for literals $p(t_1, ..., t_k)$ and $\neg p(t_1, ..., t_k)$ requires that the terms t_i appearing as arguments have to be evaluated to objects in the sets interpreting the source sorts of the predicate p. Literals for which this is not possible will always be false.

Now we can define the notion of a model of a L_{LILOG} knowledge base.

Definition 6.4 Let $KB = \langle \Sigma, AX \rangle$ be a L_{LILOG} knowledge base and M a model for Σ. We call M a model of KB iff M satisfies all the axioms in AX.

We choose a loose semantics for a knowledge base according to

Definition 6.5 Let $KB = \langle \Sigma, AX \rangle$ be a L_{LILOG} knowledge base. The *semantics of KB is* $\mathbf{mod}(KB)$, the class of all models of KB.

On the basis of this definition we define when a knowledge base KB implies a goal g as follows:

$$KB \models g \text{ iff } M \models KB \text{ implies } M \models g \text{ holds for any } M \in \mathbf{mod}(KB)$$

Consistency of knowledge bases becomes a standard notion:

Definition 6.6 Let $KB = \langle \Sigma, AX \rangle$ be a L_{LILOG} knowledge base. KB is *consistent* iff there is a model for KB.

Finally we give a last definition of subsumption of sort expressions:

$$se \ll_{KB} se' \text{ iff } D_{se} \subseteq D_{se'} \text{ for any model } M = \langle U, F_M, R_M \rangle \text{ of KB}$$

where $U = \langle D, F_U, R_U \rangle$.

7 Structuring Knowledge Bases

This section is concerned with structuring concepts for L_{LILOG} knowledge bases. The concept of knowledge packets presented here is adapted from [Wachsmuth 88] and constitutes a simple form of modularizing logical theories as discussed in [Goguen, Burstall 85] or [Ehrig, Mahr 85]. We expect that the structuring principle offered by L_{LILOG} supports knowledge engineers in the improved modeling of large application domains in much the same way as module concepts in programming languages enable and support the development of structured software systems.

7.1 Syntactic Aspects

The basic idea is to identify so-called knowledge entities within knowledge bases and to assign them to knowledge packets - forming the modules of a knowledge base - thus achieving a separation of a knowledge base into different parts. These different parts of the knowledge base will be related by an accessibility relation linking the different knowledge packets. Since the accessibility relation between knowledge packets will be given in terms of a partial ordering, a situation $kp \prec kp'$ for two knowledge packets kp and kp' expresses that all the knowledge being available in the knowledge packet kp' can also be used in the knowledge packet kp, i.e. the knowledge of kp' can be *imported* into kp. The concept of accessibility of knowledge from other knowledge packets is based on the notion of visibility and reachability defined for arbitrary partial orders.

Definition 7.1 *Let* $\langle M, \prec \rangle$ *be a partial order. For any* $m \in M$ *we define the following two sets*

- *visible(m)* $= \{m' \in M \mid m' \prec m\}$
- *reachable(m)* $= \{m' \in M \mid m \prec m'\}$

A knowledge base is partioned by assigning knowledge entities of the knowledge base to knowledge packets. Thus we shall now define what we consider as a knowledge entity.

Definition 7.2 *Let KB be a knowledge base. The set of knowledge entities KE(KB) of KB contains: the sorts, the features, the roles, the atoms, the sort constraints, the operators, the reference object identifiers, the predicates, and the axioms of KB.*

Structuring a knowledge base KB can now be achieved by relating its knowledge elements to knowledge packets connected by an accessibility relation. The accessibility relation will be a partial ordering containing a top element. The idea behind this accessibility relation is to express which knowledge can be reached from each part of the knowledge base. In order to allow a knowledge element to occur in multiple knowledge packets, we model the distribution of knowledge elements over the various knowledge packets by means of a relation.

Definition 7.3 *Let* $KB = \langle \Sigma, AX \rangle$ *be a knowledge base and* $\langle KP, \prec \rangle$ *a partial order with a top element* $\top$.
A knowledge packet assignment for KB w.r.t. $\langle KP, \prec \rangle$ *is a relation*
$KPA \subseteq KE(KB) \times KP$
relating all the built-in knowledge elements of KB to the knowledge packet $\top$ *and satisfying the following condition:*
KPA(s, kp) for a sort s and a knowledge packet kp holds if and only if

- *KPA(a, kp), for all atoms* $a \in A_s$
- *KPA(f, kp), for all features* $f \in F_{s \rightarrow ese}$
- *KPA(r, kp), for all roles* $r \in R_{s,ese}$

- $KPA(s \doteq se, kp)$, for all sort constraints of the form $s \doteq se$ in SC

Given a knowledge packet assignment $KPA \subseteq KE(KB) \times KP$ we want to identify all those knowledge elements visible from a certain knowledge packet, which we then call the <u>focused</u> knowledge packet. A focused knowledge packet $kp \in KP$ induces the pair

$$KPA^{-1}(kp) = \langle \Sigma', AX' \rangle$$

of knowledge elements visible from kp where

- Σ' is the quadruple $\Sigma' = \langle SH', O', ROI', P' \rangle$ with
  - $SH' = \langle \Sigma^{SORT'}, SC' \rangle$ such that
    * $\Sigma^{SORT'} = \langle S', F', R', A' \rangle$ has the following components
      - $S' = \{s \in S \mid KPA(s, kp') \Rightarrow kp' \in visible(kp)\}$
      - $F' = \langle F_{s \to ese} \rangle_{s \in S', ese \in ESE(\Sigma^{SORT'})}$ with
        $F'_{s \to ese} = \{f \in F_{s \to ese} \mid KPA(f, kp') \Rightarrow kp' \in visible(kp)\}$
      - $R' = \langle R_{s,ese} \rangle_{s \in S', ese \in ESE(\Sigma^{SORT'})}$ with
        $R'_{s,ese} = \{r \in R_{s,ese} \mid KPA(r, kp') \Rightarrow kp' \in visible(kp)\}$
      - $A' = \langle A_s \rangle_{s \in S'}$ with
        $A'_s = \{a \in A_s \mid KPA(a, kp') \Rightarrow kp' \in visible(kp)\}$
    * $SC' = \{sc \in SC \mid KPA(sc, kp') \Rightarrow kp' \in visible(kp)\}$
  - $O' = \langle O'_{w \to ese} \rangle_{w \in ESE(\Sigma^{SORT'})^*, ese \in ESE(\Sigma^{SORT'})}$ with
    $O'_{w \to ese} = \{o \in O_{w \to ese} \mid KPA(o, kp') \Rightarrow kp' \in visible(kp)\}$
  - $RO' = \langle RO'_{ese} \rangle_{ese \in ESE(\Sigma^{SORT'})}$ with
    $RO'_{ese} = \{ro \in RO_{ese} \mid KPA(ro, kp') \Rightarrow kp' \in visible(kp)\}$
  - $P' = \langle P'_w \rangle_{w \in ESE(\Sigma^{SORT'})^*}$ with
    $P'_w = \{p \in P_w \mid KPA(p, kp') \Rightarrow kp' \in visible(kp)\}$
- $AX' = \{ax \in AX \mid KPA(ax, kp') \Rightarrow kp' \in visible(kp)\}$

With the definition of how to spread the knowledge entities of a knowledge base over various knowledge packets and the formal definition of which knowledge elements are accessible when focusing on a particular knowledge packet, we now introduce the concept of a knowledge base structured into knowledge packets.

Definition 7.4 *A structured* L_{LILOG} *knowledge base is a pair*

$$SKB = \langle KB, KPA \rangle$$

where KPA is a knowledge packet assignment for KB over $\langle KP, \prec \rangle$ *such that* $KPA^{-1}(kp)$ *is a* L_{LILOG} *knowledge base for any* $kp \in KP$.

The property of $KPA^{-1}(kp)$ being a L_{LILOG} knowledge base assures us that within a structured knowledge base SKB all the knowledge packets visible from kp form an unstructured knowledge base. That is to say, all the parts of a knowledge base which can be selected by focusing on a knowledge packet have to form a knowledge base, i.e. we impose a closure condition that the focusing mechanism has to satisfy.

Before we discuss the semantic aspects of structured knowledge bases, let us take our example knowledge base and show how to exploit the structuring mechanism introduced above for separating the knowledge base into modules. Although the knowledge base on vehicles is rather small, we observe that L_{LILOG} offers several built-in knowledge entities for handling arithmetic and sets. The structuring mechanism requires us to have at least one knowledge packet which contains the built-ins.

Up to now we have considered our knowledge about travelers and their vehicles to be of the same nature as the built-ins. This would lead to a trivial structuring of our knowledge base into just one knowledge packet *top* which can be depicted as follows

top

It might be more appropriate to introduce a second knowledge packet *travelers-and-vehicles* containing our knowledge referring to travelers and vehicles and place this knowledge packet beneath the knowledge packet *top* leading to the following structure of the knowledge base

This reflects one of the motivations of [Wachsmuth 88] for introducing the concept of knowledge packets: to place knowledge on different topics into different knowledge packets.

To be a bit more precise, the structured version of our knowledge base *traveling* could look as shown in the Figure 2 below:

knowledge base *traveling-with-kps*
knowledge packet structure *travelers-and-vehicles* < *top*
knowledge packet *travelers-and-vehicles*

    sort *person*

    sort *man* < *person*

    sort *woman* < *person*
    disjoint *man*

    sort *vehicle*
    features *wheels* : [*2 .. 16*]
            doors : [*0 .. 4*]
            type : { *bike, sedan, cabrio, truck* }

    sort *vehicle-type*
    atoms *bike, cabrio, sedan, truck*

    sort *sportscar* = **with-feature** *wheels* **in** {4} ⊓
           with-feature *doors* **in** {2} ⊓
           with-feature *type* **in** { *cabrio* }

    predicate *travel* : *who* :: *person**, *from* :: *location, to* :: *location, with* :: *vehicle*

    reference object *John* : *man*

    reference object *Mary* : *woman*

    reference object *Porsche* : *sportscar*

    axiom *M* : *person, G* : *person**, *F* : *location, T* : *location, V* : *vehicle* .
        travel(who :: *M, from* :: *F, to* :: *T, with* :: *V)*
           ← *travel(who* :: *G, from* :: *F, to* :: *T, with* :: *V), M* ≤ *G*

end knowledge packet
end knowledge base

Figure 2: The structured version of the knowledge base on *traveling*

7.2 Semantic Aspects

Before we define the semantics of a structured knowledge base, let us briefly discuss the ideas behind the concept of a focused knowledge packet and its semantic implications. Having focused on kp one 'sees' all those knowledge packets which are predecessors of kp with respect to the accessibility relation given in terms of the partial ordering on the set of knowledge packets. This implies that the semantics of a structured knowledge base should not be identical to the semantics of the entire knowledge base where the structuring has been forgotten. Instead, we intend a 'relative semantics' which is dependent on the focused knowledge packet. This results in the idea that we only interpret parts of a knowledge base, those parts which are visible from a certain knowledge packet, by means of our semantics defined for unstructured knowledge bases. Proceeding this way implies that the semantics of certain knowledge entities varies depending on the knowledge packet being focused.

A kind of monotonicity of the interpretation of the knowledge entities can be observed because of the following relationship:

$$kp' \prec kp \text{ implies } \mathrm{KPA}^{-1}(kp) \subseteq KPA^{-1}(kp').$$

Thus, shifting the focus from a knowledge packet kp to a smaller knowledge packet kp' increases the amount of information on the knowledge elements visible in the focus given by kp. For inferential purposes this means that we have access to more knowledge and might thus produce more answers to certain goals when we are in the focus kp' than we can find in the focus kp.

These considerations lead to

Definition 7.5 *Let $SKB = \langle KB, KPA \rangle$ be a structured L_{LILOG} knowledge base with the knowledge packet assignment $KPA \subseteq KE(KB) \times KP$. Further, let $kp \in KP$ be a knowledge packet.*
The semantics of SKB w.r.t. kp is the class of all models of the unstructured knowledge base $KPA^{-1}(kp)$.

8 Conclusions and Prospects

We have described two central aspects of the knowledge representation language L_{LILOG} : its sort concept and its structuring mechanism for knowledge bases.

The type concept of L_{LILOG} is given in terms of a sophisticated sort description language integrating concepts of feature term languages into the framework of an order-sorted logic. The definition of the language exhibits a closer junction between the level of syntax and the semantic level than e.g. ordinary order-sorted logic. This results from the rich sort description language for which some semantic properties influencing the definition of the syntactic constructs of L_{LILOG} have to be mentioned explicitly, while for simpler sort description languages as present in, ordinary order-sorted logic, say, allow these properties to be guaranteed by simple syntactic means; see [Pletat 89] for a more detailed discussion of the phenomena.

Such a rich sort description language may lead to problems when implementing a language like L_{LILOG} by an inference component. The sort language includes (parts of) an entire paradigm for representing knowledge: basically the idea of KL-ONE and also feature term languages where all knowledge is represented as sorts and relations between the sorts. Thus the typical inference task for a KL-ONE system, namely performing a subsumption test for two sort expressions, is just part of a L_{LILOG} inference task of solving a goal: subsumption tests will have to be performed all the way along during unification within a resolution based inference engine. Apart from completeness problems (see [Schmidt-Schauss 88]), this may cause performance problems.

Looking at L_{LILOG} from the KL-ONE point of view, the language would be called a hybrid one, see [Brachman et al. 85]. The various members of the KL-ONE family of languages differ mainly in the expressiveness of their A-Boxes. In the A-Boxes, logical axioms can be formulated; what can be expressed ranges from propositional logic offered by KL-TWO ([Vilain 85]) or the BACK system ([Luck et al. 87]) to full first order predicate logic as in KRYPTON ([Brachman et al. 85]). L_{LILOG} can be found at the upper end of this scale so that, referring to expressiveness, KRYPTON would be of comparable power. A more detailed discussion of the relationship between L_{LILOG} and KRYPTON may be found in [Beierle et al. 89].

The structuring mechanism offered by L_{LILOG} should be of interest for knowledge engineers to determine a reasonable module structure for a large knowledge base. Thus knowledge bases will no longer be large piles of knowledge entities but collections of knowledge entities organized in a structured way. This will also lead to a greater 'mental hygiene' when developing a knowledge base.

Acknowledgement: The authors would like to thank Bernd Mahr for his critical comments on an earlier version of the paper.

References

[Beierle et al. 88a] C. Beierle, J. Dörre, U. Pletat, C. Rollinger, P.H. Schmitt, R. Studer: *The Knowledge Representation Language L_{LILOG}* . IBM Deutschland, Wissenschaftliches Zentrum, LILOG-Report Nr. 41, Stuttgart 1988

[Beierle et al. 88b] C. Beierle, U. Pletat, H. Uszkoreit: *An Algebraic Characterization of STUF* Proc. Symposium Computerlinguistik und ihre theoretischen Grundlagen, I.S. Batori, U. Hahn, M. Pinkal, W. Wahlster (eds), Informatik Fachberichte, Vol. 195, Springer-Verlag, Berlin 1988

[Beierle et al. 89] C. Beierle, U. Hedtstück, U. Pletat, J. Siekmann: *An Order Sorted Predicate Logic with Closely Coupled Taxonomic Information.* To appear, 1989

[Bouma et al. 88] G. Bouma, E. König, H. Uszkoreit: *A flexible graph-unification formalism and its application to natural-language processing.* IBM J. Res. Develop. **32** 2, 1988, 170-184

[Brachman et al. 83] R.J. Brachman, R.E. Fikes, H.J. Levesque: *KRYPTON: Integrating Terminology and Assertion.* Proc. of AAAI-83 1983 31-35

[Brachman, Schmolze 85] R.J. Brachman, J.G. Schmolze: *An Overview of the KL-ONE Knowledge Representation System.* Cognitive Science **9** (2) April 1985, 171-216

[Brachman et al. 85] R.J. Brachman, V. Pigman, Gilbert, H.J. Levesque: *An Essential Hybrid Reasoning System.* Proc. IJCAI-85, 1985, 532-539

[Cohn 87] A. G. Cohn: *A More Expressive Formulation of Many Sorted Logic.* Journal of Automated Reasoning, 3:113-200, 1987

[Ehrig, Mahr 85] H. Ehrig, B. Mahr: *Foundations of Algebraic Specifications I.* Springer-Verlag, Berlin 1985

[Goguen, Burstall 85] J. A. Goguen, R. M Burstall: *Institutions: Abstract Model Theory for Computer Science.* SRI International Report 1985

[Goguen, Meseguer 87] J. A. Goguen, J. Meseguer: *Order-Sorted Algebra I.* SRI International Report 1987

[Hendrix 79]   G.G. Hendrix: *Encoding Knowledge in Partitioned Networks.* in: Associative Networks, Findler (ed.), Academic Press, 1979, 51-92

[Herzog et al. 86] O. Herzog et al.: *LILOG – Linguistische und logische Methoden für das maschinelle Verstehen des Deutschen.* IBM Deutschland GmbH, LILOG-Report 1a, 1986

[Kasper, Rounds 86] R. T. Kasper, W. C. Rounds: *A Logicical Semantics for Feature Structures.* Proc. 24th ACL Meeting, 257-265, Columbia University 1986

[Link 83]   G. Link: *The Logical Analysis of Plurals and Mass Terms.* in: Bäuerle et al. (eds): Meaning, Use, and Interpretation of Language, de Gruyter Verlag, Berlin 1983, 302-323

[Luck et al. 87] K. v. Luck, B. Nebel, C. Peltason, A. Schmiedel *The Anatomy of The BACK-System.* TU Berlin, KIT-Report No. 41, Jan. 1987

[Luck, Owsnicki-Klewe 89] K. v. Luck, Bernd Owsnicki-Klewe: *Neuere KI-Formalismen zur Repräsentation von Wissen.* in: Christaller (Hrsg.), Künstliche Intelligenz, Springer-Verlag, Berlin 1989, 157-187

[Oberschelp 62] A. Oberschelp: *Untersuchungen zur mehrsortigen Quantorenlogik.* Mathematische Annalen, 145:297-333, 1962

[Patel-Schneider 84] P.F. Patel-Schneider: *Small can be Beautiful in Knowledge Representation.* Proc. IEEE Workshop on Principles of Knowledge-Based Systems, 1984, 11-19

[Pletat 89]   U. Pletat: *Aspects of Consistency of Sophisticated Knowledge Representation Languages.* Proc. IBM Symposium Natural Language and Logic, 1989. To appear in Lecture Notes in Artificial Intelligence, Springer-Verlage, Berlin.

[Schmidt-Schauss 88] M. Schmidt-Schauss: *Subsumption in KL-ONE is Undecidable.* Universität Kaiserslautern, SEKI-Report SR-88-14, 1988

[Shieber 86]   S. Shieber: *An Introduction to Unification-Based Approaches to Grammar.* CSLI Lecture Notes 4, Stanford University 1986

[Smolka 88]   G. Smolka: *A Feature Logic with Subsorts.* Wissenschaftliches Zentrum der IBM Deutschland, LILOG-Report 33, 1988

[Sussman, McDermott 72] G.J. Sussman, D.V. McDermott: *From PLANNER to CONNIVER.* Proc. of Fall Joint Comp. Conf. 1972, 1171-1179

[Vilain 85]   M. Vilain: *The Restricted Language Architecture of a Hybrid Representation System.* Proc. IJCAI-85, 1985, 547-551

[Wachsmuth 88] I. Wachsmuth: *Zur intelligenten Organisation von Wissensbeständen in künstlichen Systemen.* Habilitationsschrift, Universität Osnabrück, 1988

[Walther 87]   C. Walther: *A Many-Sorted Calculus Based on Resolution and Paramodulation.* Research Notes in Artificial Intelligence, 1987

Structure and Control of the L-LILOG Inference System

K.H. Bläsius[1]
C.-R. Rollinger[2]
J.-H. Siekmann[3]

Abstract

L-LILOG II is a knowledge representation language based on order-sorted predicate logic. It is used in a text understanding system to represent the meaning of natural language texts and to represent the background knowledge. An inference system for L-LILOG II has been developed including a control component which restricts the possible inferences and guides the search for proofs. This control of reasoning is based to some extend on meta knowledge, i.e. knowledge about knowledge, which is also represented in L-LILOG II. This paper provides an overview of the inference system and describes the knowledge based control of reasoning in more detail.

1   Fachhochschule Dortmund, FB Informatik, Sonnenstr. 96, D-4600 Dortmund
2   IBM Germany GmbH, Institute for Knowledge Based Systems, P.O. Box 800880, D-7000 Stuttgart 80
3   University of Kaiserslautern, FB Informatik, D-6750 Kaiserslautern

Essential parts of this work have been done while the authors jointly were at IBM

1. Introduction

The aim of the LILOG Project (LInguistic and LOGic methods) is to investigate linguistic and logical tools and methods for the computational understanding of texts. In 1987 a prototype was implemented capable of understanding a German text (taken from a tour guide that describes a hiking tour in the Alsace) and answering questions about the domain and the text. For example, once the text was read in, the system answered successfully questions such as: "Wo beginnt die Wanderung?" or "Welche Sehenswürdigkeiten gibt es?" ("Where does the tour start?" "What objects are worth seeing?")

The prototype was based on the knowledge representation language L-LILOG I, which represented the different kinds of knowledge necessary for processing the text as well as to represent the semantic description of the input text. A knowledge processing component was implemented as interpreter for L-LILOG I that solved problems and extended the knowledge base by deriving new facts.

The experience with this first prototype led to new demands for the representation and processing of knowledge. These new demands asked for more expressive power of the knowledge representation language, as well as for new problem solving capabilities of the inference engine. For example, the linguistic component to construct a semantic representation from a given input text should be directly supported by the inference engine to solve linguistic problems such as disambiguation and anaphora resolution.

Additionally, there were several requests by the system engineers for an extension of the knowledge representation language L-LILOG I such as to utilize "full first order predicate logic" (instead of just Horn logic as in the first prototype), "equality reasoning", "a more powerful sort description language", "default reasoning", "reason maintenance", "vagueness", "uncertainty", "arithmetic", "sets" and "generalized quantifiers".

Most of these requests were accepted and have been integrated to some extend into the new knowledge representation language L-LILOG II, which is essentially based on order-sorted predicate logic. However, the sort hierarchy is not necessarily statically fixed, but L-LILOG II contains a sort description language SORT-L-LILOG that allows the dynamic construction of sorts.

Knowledge based systems (KBS) usually have very large knowledge bases, consisting of problem specific, domain specific and general world knowledge. In this paper, these kinds of knowledge are called "object knowledge". For a given unit of object knowledge there may be some specific control knowledge that specifies how the object knowledge is used during the inferencing task. This kind of control knowledge is also called meta-knowledge, since it is knowledge about knowledge. We also use L-LILOG II to represent this meta knowledge in order to guide the search for a proof at the object level, i.e. L-LILOG II is also used as a control language for this "meta reasoning" task.

The language L-LILOG II and the processing of sort expressions are described elsewhere in this volume: (Pletat, v. Luck 1989) and (Hedtstück, Schmitt 1989). This paper gives an overview of the knowledge processing component (the inference engine), and describes the relationship between reasoning and meta reasoning, both based on L-LILOG II.

In section 2, we give a short and informal description of L-LILOG II, for more details see (Pletat, v. Luck 1989). Section 3 describes the kinds of problems (tasks) to be solved by the inference engine. In section 4, we present the knowledge processing component for the second prototype currently under development, and section 5 gives an account of control language and meta reasoning.

We expect the reader to be familiar with fundamentals of artificial intelligence, especially knowledge representation formalisms, as well as predicate logic and the resolution calculus.

2. The language L-LILOG II

We shall now give a short and informal introduction to the knowledge representation language L-LILOG II, for more details see (Pletat, v. Luck 1989).

The language L-LILOG II is essentially based on order-sorted predicate logic with equality (see e.g. (Cohn 1987), (Walther 1987), (Schmidt-Schauss 1988)). However, there are not only constant sorts, but complex sort terms as described in (Pletat, v. Luck 1989) and (Hedtstück, Schmitt 1989).

The concept of "entrypoints", used in L-LILOG I to restrict the possible deduction steps, has been removed from the knowledge representation language. Instead, a new method to specify and handle control information has been developed (see section 5).

Just as in L-LILOG I (Beierle et al. 1988), arguments of predicates and functions are not identified by a fixed position in the argument list, but by argument identifiers which are called roles. While this is only "syntactic sugar" from the logical point of view, roles may have a certain importance for the linguistic processing, mapping of words to L-LILOG II expressions. Furthermore, we again have reference objects, being logical constants, which are treated in a special way motivated by linguistic requirements.

Literals and clauses in L-LILOG II are assigned attributes containing information which will be used during inferencing. So far, the following attributes have been used:

- default-value, specifying whether a clause is a default clause or not
- consistency-value, specifying whether a clause is consistent with the knowledge base
- vagueness, a value specifying the degree of validity of a literal

Default- and consistency-values have also been used in L-LILOG I. However, reasoning with these attributes will be improved in the second prototype, exploring default reasoning with priority. We also have developed a method to handle vagueness, which is described in (Gemander 1989).

In the second prototype, a restricted treatment of arithmetic expressions will be possible, and L-LILOG II includes certain special predefined symbols for arithmetic expressions.

They are

- a special Sort-Symbol: NUMBER with the property BOTTOM < NUMBER < TOP
- special function symbols: +, *, - : NUMBER x NUMBER --> NUMBER
- special predicate symbols: <, >, ≤, ≥ : NUMBER x NUMBER

Furthermore, each PROLOG number may be used as a logical constant with the sort NUMBER. Arithmetic expressions are immediately evaluated, using PROLOG. This means that evaluation of arithmetic expressions is only possible if the latter are sufficiently instantiated.

L-LILOG II also includes a concept of sets, such that, for example, constants can be used to represent sets of objects. The declaration of sorts is extended to allow the declaration of a symbol to be of sort set. The following special symbols are used:

- constant: empty_set
- function symbols: build_set, union, cardinality
- predicate symbols: is_empty, is_element, is_subset

In order to reason about set expressions, an ACI-matching algorithm and special algorithms to evaluate ground-expressions are used.

3. Inferencing Tasks

In the first LILOG-prototype the inference engine was used for two tasks: first, to derive new facts on the basis of the knowledge extracted from the input text (text knowledge) and second to answer questions on the basis of the textual knowledge and the background knowledge (world knowledge, common sense knowledge). These tasks are going to be performed in the second protopype as well.

Extending the Knowledge Base: from the given input text, knowledge elements are extracted, most of which will be facts. From these, new facts can be derived using the actual knowledge base, which contains general knowledge and domain specific knowledge. These new facts can be seen as an interpretation of the text in relation to the background knowledge, which is why these facts are inserted to the textual knowledge. Extending the knowlegde base at "read time" reduces the search effort during "question time", when questions are answered. However, the process of extending the knowledge base has to be severely restricted, e.g. by a given maximum processing time or derivation depth.

Answering questions: generally speaking, questions about the domain and the input text cannot be answered directly, but need some derived facts which are to be drawn by the inference engine.

In the second prototype, the task to answer questions has been improved by taking the expectations of a user into account. In many cases, the user expects to receive more information than is explicitly asked for in a question. Take, for example, the following question:

Is there a cheap Italian restaurant in Hamburg?

In this case, it will not be sufficient to just answer with "yes". Instead, the answer should be supplemented by giving the name of a concrete restaurant. Hence, the inference engine has to handle such "yes/no-questions" as "which-questions", i.e. not only one, but all solutions are to be found and for each solution a resulting substitution is to be computed which provides for this kind of over-answering.

Another example is

Does Hamburg celebrate the 500th anniversary of its harbour?

The answer should not simply be "no", but "no, the 800th anniversary". Hence, if no proof can be found for the given goal, the goal is generalized and the inference engine tries to prove this generalized goal.

There are plans for further applications of the inference engine during the execution of the components "semantic construction" and "text generation", e.g. to support disambiguation and anaphora resolution. In order to resolve anaphorae, the inference engine should determine all reference objects that have certain properties. To reduce ambiguities, certain readings can be excluded in case an inconsistency is discovered by the inference engine. To enable a fast reaction, a modular and flexible architecture of the inference component is required. This aspect is covered by the subsequent section.

4. The Inference Engine

Within this paper, we cannot fully describe each component of the inference engine. Instead, we would like to present a more general overview and then describe the control aspects in more detail.

4.1. Overview

The knowledge processing component (inference engine) is subdivided into three main modules:

"Selection Strategies", "Inference Rules" and "Basic Operations"

The module "Selection Strategies" contains the selection strategies and heuristics which guide the search for a proof; it determines the sequence of operations. The selection strategies depend on the different tasks at hand like "Extending the Knowledge Base" or "Answering Questions".

The module "Inference Rules" contains the inference rules, and is divided into several submodules that contain the special rules for the treatment of extensions like arithmetic, equality, sets etc. Different calculi may be built by these various inference rules, depending on the task to be performed. For example, the process of extending the knowledge base uses resolution rules which differ from those used in the process of answering questions. The following submodules are currently developed:

"Resolution-rules", "Equality-handling", "Arithmetic" and "Sets"

The different selection functions and inference rules use certain basic operations, which are contained in the third main module. The latter is also divided into several submodules:

"Unification and Matching", "Sort-operations", "Reason Maintenance" and "Vagueness"

Since unification is the central operation for most inference rules, appropriate unification and matching (one side unification) algorithms are required. Besides standard unification algorithms, we also use theory unification (commutativity) and theory matching (commutativity, associativity and idempotence). Furthermore the unification and matching algorithms have to consider sort conditions. For the treatment of sorts see (Hedtstück, Schmitt 1989).

The use of defaults requires a mechanism to represent the proof history in order to be able to retract certain default assumptions and all consequences thereof, which led to an inconsistency. Therefore a reason maintenance system has been implemented and was integrated into the inference engine. The proof history is also the basis for an explanation component.

A method to handle vagueness has been developed as well. For this purpose, the inference rules have been extended to check certain vagueness conditions, and some operations comparing vagueness values have been incorporated into the "basic operations" module. This is described in (Gemander 1989).

4.2. Inference Rules

Knowledge processing is mainly based on resolution and factorization, which are known to be sound and complete for first order predicate logic. Since resolution presupposes a clausal normal form, all formulae of the knowledge base are normalized initially. Besides the general resolution rule, there are special inference rules as well which support forward and backward chaining between horn clauses. These special inference rules, called "forward-resolution" and "backward-resolution" work similar to the respective rules of the first prototype (Bollinger et al. 1988). "Forward-resolution" is only used in the task "extending the knowledge base", i.e. it is used to draw the immediate conclusion at read time. Problem solving and question answering is based on backward-resolution between horn clauses and resolution between two clauses, at least one of which is not a horn clause.

The treatment of sort formulae has been improved and is presented elsewhere in this volume (Hedtstück, Schmitt 1989).

L-LILOG II contains the equality relation as a special predicate. Depending on the actual knowledge base, different principles are used to handle it: the RUE-calculus of Digricoli (Digricoli 1979) and paramodulation (Robinson, Wos 1969).

The RUE-calculus (Resolution by Unification and Equality) is a generalization of resolution incorporating equality. Whereas the usual resolution rule is applicable only if the corresponding subterms of the respective literals are unifiable, this condition is weakened for RUE-resolution: the two literals to be resolved upon only need to have the same predicate and different sign. If there are pairs of subterms which are not unifiable, negated equations for these pairs of terms are generated and added to the resolvent as new subgoals, which are still to be solved.

RUE-resolution is only used in case the knowledge base contains many equations. If there are only a few, then there is a high probability that most of the subproblems are unsolvable: the number of generated inequations increases and the search space becomes too large.

In such a case, we use paramodulation where terms are replaced by equal terms. Paramodulation may produce large search spaces if there are many equations in the knowledge base and most of the paramodulants will never be used.

In other words, these two methods are somehow complementary: RUE-resolution works best if there are many equations in the initial set of clauses but deteriorates rapidly otherwise, whereas paramodulation is difficult to control where there are many equations, but works well if there are just a few equality literals (relative to the initial clause set). Hence, the system checks the quotient nonequality literals over equality literals and sets its options accordingly.

4.3. Selection Strategies

At each step of the inference process, there are several operations applicable and a selection function is necessary to select a suitable operation from the applicable ones.

The selection modules have to provide functions to

- select some literal and the appropriate inference rule
- check conditions like equality of predicates, difference of signs, values of vagueness
- unify termlists
- select one unifier among the many most general unifiers (since theory unification is used)
- activate the selected operation
- manage the other possible operations and unifiers as a basis for subsequent selection processes.

The first prototype was based on a simple breadth first search with backward- and forward-chaining. The selection functions of the second prototype will be more complex since there are stronger requirements for the inference tasks, and the knowledge representation language is no longer based on horn logic, but on full first order predicate logic.

However, as experience has shown so far, in most cases the knowledge can be naturally modelled in the form of horn clauses. This fact will be exploited, preferring the usual forward and backward chaining strategies for the extension of the knowledge base and answering questions, respectively. Other resolution steps are not excluded, but only executed when necessary, e.g. if disjunctions or negated literals are involved. Control information is used to guide the search for proofs. The control information is specified in the special control language which is presented in the following section.

5. Knowledge Based Control of Reasoning

5.1. Motivation

Problem solving in knowledge based systems usually relies on very large knowledge bases, consisting of problem specific, domain specific and general world knowledge (object knowledge). An uncontrolled use of this knowledge may lead to a prohibitively large search space during the problem solving process. Heuristics and strategies are required to control the selection of knowledge elements to be used in a given stage of the search process. To use only domain independent heuristics and strategies turned out to be insufficient: For a certain piece of (object) knowledge, (control) knowledge is necessary for specifying how the object knowledge has to be used during inferencing. This kind of control knowledge we shall call meta knowledge, since it is knowledge about the use of knowledge.

The idea of a separate control language was advocated by P. Hayes as early as 1973 in a project called GOLUX (Hayes 1973). He claimed that if predicate logic was ever to develop into a useful programming language, an important prerequisite would be a control language (preferably based on first order predicate logic) by which detailed instructions could be specified to guide the underlying logic inference engine. Such a control language should directly advise the inference engine on what to do next, as opposed to the refinements and filters investigated in the sixties, which restrict the search space in a certain way but have nothing to offer for the remaining space, where search resorts to blind enumeration.

A similar, albeit extremely limited and primitive form of this idea was realized in L-LILOG I: the user could mark some literals in the input clauses by socalled "entrypoints" (see section 5.5). The marked literals where then given a strategical treatment in the deduction process and this turned out to be one of the most useful control mechanisms in our application domain, without which the system would not have worked. For the task of text understanding with those very large knowledge bases, there are just far too many possible inferences that can be potentially drawn and it is paramount to find just those necessary for the respective linguistic components in order to determine the meaning of a given text. In this sense they should be cognitively adequate. Meta knowledge seems to be necessary to restrict the inferences in this sense.

In the LILOG-project, a control language is currently under development which elaborates these ideas and generalizes the notion of "entrypoints". The control language is also based on order-sorted predicate logic, but there are many special symbols, which are interpreted by the inference engine in order to guide the search for a proof. These special symbols are not interpreted at the level of meta reasoning, hence they don't build a special logical theory at the level of the control language. The correlation between the different levels of reasoning and the different knowledge bases will be explained below.

5.2. Levels of Reasoning

As already mentioned, there are essentially two tasks to be solved by the inference engine:

When a new fact, extracted from the input text has to be inserted into the knowledge base, some conclusions are drawn immediately in order to make the knowledge base more coherent. Figure 1 illustrates this process for the LILOG-system. A sentence from the input text is analysed by the "Linguistic Processing" component, which constructs an L-LILOG-formula representing the meaning of this sentence. The L-LILOG-formula is given to the inference engine which performs some consistency checks and then deduces some further facts using the knowledge base. The new formulae are inserted into the knowledge base.

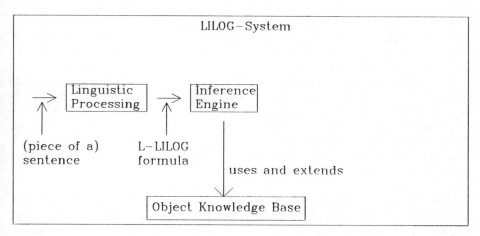

Figure 1

The second main task of the inference engine is to solve problems in order to answer the questions that are posed to the system. Figure 2 illustrates this problem solving process. A question is analysed by the "Linguistic Processing" component, constructing a logical representation (L-LILOG-formula) of the question, which is the input for the inference engine. The inference engine tries to solve this problem by using the knowledge base and, if successfull, gives the answer as an L-LILOG-formula (or simply as the resulting substitution for the variables in the question) to the "Text Generation" component, which constructs the final answer in natural language.

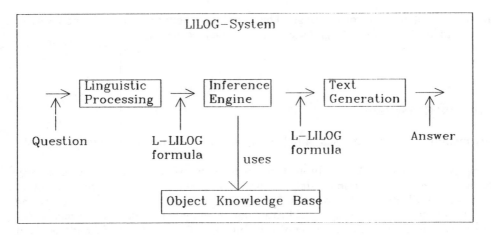

Figure 2

The object knowledge base contains knowledge about the specific domain, general knowledge about the world as far as necessary, and the knowledge extracted from the input text. This knowledge is represented in L-LILOG, and the inference engine is based on a calculus for L-LILOG as explained in section 4. The predicate, function and constant symbols used in the object knowledge base refer to the objects and properties of the specific domain and to the words and phrases used to describe the objects and events in the text. Hence, these symbols are not interpreted by the inference engine, i.e. they don't build a logical theory. They have a special meaning for the linguistic processing and the text generation component, which sets the task for the inference engine or transforms the answer of the inference engine into a natural language sentence, respectively. The linguistic processing and text generation components interpret the symbols used in the logical formulae correlating them to words and phrases.

Knowledge based control of reasoning may be realized by a special knowledge based subsystem of the KBS. The inference engine can be described similar to the LILOG-system above, including an instantiation of the inference engine itself called "meta reasoning engine" (see figures 3 and 4). Since the concept of knowledge based control of reasoning might be applied on several levels, the inference engine might recursively call itself. To simplify matters, we only use two levels of reasoning and use a separate name for the meta reasoning component, although the software-module "inference engine" is also used as the "meta reasoning engine" in the actual implementation.

In the inference engine, the selection module takes the analogous role to the linguistic processing and text generation components in figures 1 and 2. Hence, the selection module sets the tasks and interprets the solutions. Again, we have two different kinds of tasks:

When a certain operation is performed by the inference engine, a logical formula is created and inserted into the control knowledge base. Figure 3 illustrates this process: the selection module which performs the control of the inference engine creates a logical representation for the inference operation and gives it to the meta reasoning engine. The latter inserts this formula into the control knowledge base. Hence, the control knowledge base contains formulae representing information about the actual state of the search process at another level of abstraction.

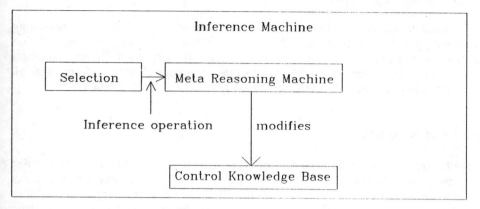

Figure 3

The second task of the meta reasoning engine is to solve problems created by the selection module. Figure 4 illustrates this problem solving process. A control problem is created by the selection module and its logical representation is the input for the meta reasoning engine. The meta reasoning engine tries to solve this problem by using the control knowledge base, and in case of success gives the answer as a logical formula (or simply as the resulting substitution) to the selection module, which interprets the answer and guides the search for a solution of the problem at the inference engine level.

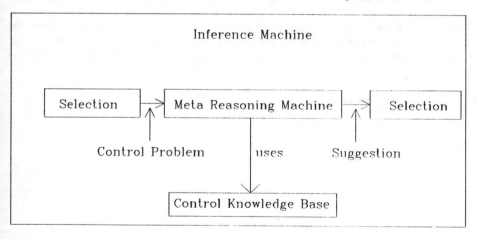

Figure 4

The predicate, function, and constant symbols used in the control knowledge base refer to the objects and properties occurring in a deduction process, such as clauses, inference rules or derivability. These symbols are not interpreted by the meta reasoning engine, i.e. they don't build a logical theory, but they have a special meaning for the selection module, which sets the tasks to the meta reasoning engine and interprets the answer. The selection module interprets the symbols used in the logical formulae by correlating them to clauses of the object knowledge base, to inference operations, or to the actual state of the search process at the inference engine level.

Hence, the control language is simply order-sorted predicate logic. The meta reasoning engine is realized by a subsystem of the inference engine itself, containing a simple version of the selection module. Since the meta reasoning engine is also based on the resolution principle, the control formulae are to be transformed into clausal normal form, and all variables occurring in the control formulae are implicitly universally quantified.

5.3. Predefined Symbols

Arbitrary predicate, function, constant and sort symbols may be defined by the knowledge engineer and may occur in the control language. However, only some predefined symbols can immediately be interpreted by the selection module of the inference engine in order to guide the search for a proof. Other symbols may be defined on the basis of these predefined symbols. In what follows, the predefined symbols of the control language are presented.

Let OSIG = (OS, OF, OP) be the signature for the object knowledge base, and CSIG =(CS, CF, CP) the signature for the control knowledge base. This means that OS is the set of sort symbols, OF the set of function symbols and OP the set of predicate symbols occurring in the object knowledge base, and CS is the set of sort symbols, CF the set of function symbols and CP the set of predicate symbols occurring in the control knowledge base. As usual, 0-ary function symbols are also called constants.

Below we present the predefined symbols of the control language which should be contained in CSIG. First, there are many symbols for the objects in the meta reasoning process. Objects of meta reasoning are the expressions occurring in the object knowledge base as well as the inference rules used by the inference engine.

CS contains as predefined sort symbols:

nat
clause-names
literals
clauses
formulae
terms
fterms
inference-operations

with the subsort relations literals $\leq$ formulae and clauses $\leq$ formulae

CF contains as predefined function symbols:

1, 2, ... : nat
clause: clause-names -- > clauses
literal: clause-names, nat -- > literals
literalpattern: fterms -- > literals
not: fterms -- > fterms
resolution: literals, literals -- > inference-operations
depic-create: formulae -- > inference-operations
depic-prove: formulae -- > inference-operations

Furthermore, each identifier i for a clause in the object knowledge base may be used as 0-ary function symbol of the sort "clause-names" (i.e. it is an element of CF). The function and predicate symbols occurring in the object knowledge base are predefined as follows:

Each n-ary function f from OF, is an n-ary function symbol in CF with f: terms, ..., terms -- > terms
Each n-ary predicate p from OP, is an n-ary function symbol in CF with p: terms, ..., terms -- > fterms

Using these symbols, terms can be created representing the objects of meta reasoning. The function symbols "clause" and "literal" allow the specification of a specific clause or literal in the object knowledge base. The term clause(c1) denotes the clause with name c1, and the term literal(c2, 3) denotes the 3rd literal of clause c2.

The function symbol "literalpattern" allows the creation of patterns for a literal, specifying literals by certain symbols from OS. The meta reasoning engine tries to unify such a pattern with the literals of the object knowledge base. For example let p be an element of OP, a an element of OF and x a variable, then the term literalpattern(not(p(a, x))) represents the literals of the object knowledge base having negative sign, predicate p, the first argument a, and an arbitrary second argument.

The function symbols "depic-create", "depic-prove" and "resolution" allow the specification of certain inference steps. For each inference rule of the calculus for L-LILOG II, a function symbol may be used denoting this rule, in order to specify the knowledge about the activation of this rule. To simplify we only treat the resolution and depiction rules in this paper. The term resolution(literal(c2, 3), literalpattern(not(p(a, x)))) represents the resolution between the 3rd literal of clause c2 and a literal of the object knowledge base having negative sign, predicate p, the first argument a, and an arbitrary second argument. The depiction rules are the interface to a special module performing spatial reasoning (Khenkhar 1989), (Habel 1989). The terms depic-create(clause(c1)) and depic-prove(clause(c1)) represent the inference operations constructing a depiction of the clause c1 and solving the problem specified by c1 using depictions, respectively.

The predefined symbols introduced so far are necessary to specify the objects of meta reasoning. Predefined predicate symbols are required, too, and may be separated into two classes:

- predicates for an active control of the inference engine and
- predicates for the description of the actual state of the search process

Using the first class of predicate symbols, three different kinds of control information may be specified by the knowledge engineer:

- Selection and activation of a certain inference operation or of a sequence of operations
- Prohibition of certain operations
- Allocation of priorities for certain inference operations

Corresponding to these kinds of control information, three predicates are defined, belonging to CP:

activate: inference-operations
prohibit: inference-operations
priority: inference-operations, nat

With these predicates, control formulae can be created in order to guide the search for a proof. For example, the formula activate(resolution(literal(c2, 3), literalpattern(not(p(a, x))))) is to be interpreted by the selection module in such a way that each resolution step specified by the argument of the predicate is to be performed immediately. Analogously a control formula with predicate "prohibit" prevents the execution of the inference steps specified in the argument. The predicate "priority" has two arguments and allows the allocation of certain priority values to certain inference operations. For example, the formula priority (resolution(clause(c1), X), 5) means that any resolution with clause c1 has priority 5. The interpretation of such formulae by the selection module is explained in the subsequent section.

Besides of the above predicates concerning the activation of steps, there are also predicates describing the actual states of the search process. Hence CP also contains the predefined predicate symbols:

kb-extending
problem-solving
executed: inference-operations

"kb-extending" and "problem-solving" are 0-ary predicates indicating the kind of task actually solved. A control formula executed(resolution(literal(c1,3), literal(c4,2)) means that a resolution step between the literals specified has been executed. Further predicates of this class and their interpretation are explained in (Klabunde 1989).

5.4. Interpretation of the control language

We shall now describe the interpretation of control formulae by the selection module. Formulae concerning the actual state of the search process are created by the selection module of the inference engine and are inserted into the control knowledge base. When the processing of some task starts, the selection module gives the control formula "kb-extending" (if the actual task is "extending the knowledge base") or "problem-solving" (if the actual task is "problem solving") to the meta reasoning engine, which inserts it into the control knowledge base as explained in figure 3. In the same way formulae with the predicate "executed" are inserted into the control knowledge base.

The selection module also inherits the control information of the parent clauses to the resolvent after each resolution step. If a resolution step between clauses is executed so that at least one of the clauses is not a horn clause, then for each literal l′ of the resolvent and each control formula f(...,l,...) having the parent literal l of l′ as one argument, the control formula f(...,l′,...) is created and inserted into the control knowledge base.

As shown in figure 4, search problems are created by the selection module which have to be solved by the meta reasoning engine. The answers are again interpreted by the selection module leading to the execution of a certain inference operation at the object level.

In any state of the search process, the selection module determines the next inference step performing the following actions:

1. The goal   activate(x)

   is created and given to the meta reasoning engine.

2. If there are any solutions, then the corresponding inference steps are executed.

3. If no solution has been found, then the goal

   priority(x,y) and not prohibit(x)

   is created and given to the meta reasoning engine.

4. If there are any solutions, then the solution with highest priority is selected and executed.

5. If no solution has been found, all possible operations are determined and for each operation o the goal

   prohibit(o)

   is created. If this goal succeeds, then o is deleted from the list of possible operations.

6. If the remaining list of possible operations is not empty, one operation is selected and executed.

7. If the remaining list of possible operations is empty, no inference rule is applicable, and the proof at object level fails.

The procedure described above is just one method to use the information specified in the control knowledge base and it will be implemented in the second LILOG-prototype as an experimental basis. Experiments are necessary to modify and improve the selection procedure and to adjust the heuristics, especially the priority values.

The selection process may be further improved using the connection graph method (Kowalski 1975). In a connection graph procedure (also called clause graph procedure) the possible operations are determined initially. After each deduction step, the new possible deduction steps are computed by inheritance. The main advantage of using the clause graph method in knowledge based systems with very large knowledge bases is that most of the effort of searching for possible inference steps is done initially, i.e. at "read-time", and the efficiency of the processes at "question-time" (possibly a dialog with a user) can be improved drastically.

The clause graph method seems to be suitable for a partial evaluation of the control knowledge. Many of the actions described in the selection procedure above may be shifted to the construction of an intial clause graph, again saving time for problem solving at question-time. For example, the links for inference steps to be prohibited may simply be deleted from the clause graph.

5.5. Example

The concept of entrypoints was used in the first prototype to express control information. Entrypoints are markers for literals in rules. They are used in order to control the inference process. If a premise of a rule is marked by an entrypoint, and the premise is matched by a fact, then this rule may be used for a forward inference step. If the conclusion is marked by an entrypoint, and a subgoal of a proof process matches the conclusion, then the rule may be used for a backward inference step. A rule may have entrypoints for both premises and the conclusion.

These entrypoints are a special case of what has been proposed above and can easily be realized with the control language: Just prohibit all resolution steps with a literal which has no entrypoint. Suppose the 2nd literal of clause c4 has no entrypoint. This effect can be expressed in the control language by

    prohibit(resolution(literal(c4,2),X))

However, the knowledge engineer could also define a new predicate "entrypoints" by

    not entrypoint(X) implies prohibit(resolution(X, Y))

and specify facts like

    entrypoint(literal(c4,1))

entrypoint(literal(c4,4))

Instead of specifying the literals for which a resolution step is admissible, the knowledge engineer could have also specified corresponding facts for those literals where a resolution step is prohibited. The control knowledge base would then contain the control formulae

entrypoint(X) implies prohibit(resolution(X, Y))
entrypoint(literal(c4,2)).

Additional examples can be found in (Klabunde 1989) where this method of user controlled reasoning is described in more detail.

6. Conclusion

The experience made with the first LILOG-Prototype has led to L-LILOG II, a new formalism to represent knowledge by combining the paradigms of first order predicate logic and KLONE. The knowledge processing component for the second prototype is currently developed. Control of reasoning is performed by interpreting meta knowledge, also represented in L-LILOG II. Hence, two levels of reasoning are used, both of which are based on order-sorted logic.

Meta reasoning will be a main effort for the future work on the inference system. However, experimental results are required in order to evaluate and modify the concepts developed so far.

References

C. Beierle, J. Dörre, U. Pletat, P.H. Schmitt, R. Studer: The Knowledge Representation Language L-LILOG. LILOG-Report 41, IBM Germany, Stuttgart, 1988.

T. Bollinger, U. Hedtstück, C.-R. Rollinger : Reasoning in Text Understanding, Knowledge Processing in the LILOG-Prototype. LILOG-Report 49, IBM Deutschland, Stuttgart, 1988.

A.G. Cohn: A More Expressive Formulation of Many Sorted Logic. Journal of Automated Reasoning 3, 113-200, 1987.

V.J. Digricoli : Resolution by Unification and Equality. Proc. 4th Workshop on Automated Deduction, Texas, 1979.

C. Habel: Zwischen-Bericht. In: C. Habel, M. Herweg, K. Rehkämper (eds.): Raumkonzepte in Verstehensprozessen. Niemeyer Verlag, Tübingen, 1989.

P.J. Hayes: Computation and Deduction. Proc. 2nd MFCS Symp., Czechoslovakian Academy of Sciences, 105-118, 1973.

U. Hedtstück, P.H. Schmitt: A Calculus for Order Sorted Predicate Logic with Sort Literals (this volume).

M.N. Khenkhar: DEPIC-2D: Eine Komponente zur depiktionalen Repräsentation und Verarbeitung räumlichen Wissens. in D. Metzing (eds.): GWAI-89 13th German Workshop on Artificial Intelligence, Informatik Fachberichte 216, Springer Verlag, 1989.

K. Klabunde: Erweiterung der Wissensrepräsentationssprache L-LILOG um Konstrukte zur Spezifikation von Kontrollinformation. Diplomarbeit, EWH Koblenz, 1989.

R. Kowalski: A Proof Procedure Using Connection Graphs, JACM Vol 22, No.4, 1975.

U. Pletat, K. v. Luck: Knowledge Representation in LILOG (this volume).

J. Gemander : Schlußfolgern unter Vagheit und / oder Unsicherhert. Diploma thesis No. 586, Institut für Informatik, Universität Stuttgart, 1989.

G. Robinson, L. Wos: Paramodulation and TP in first order Theories with Equality. Machine Intelligence 4, 135-150, 1969.

M. Schmidt-Schauss: Computational Aspects of an Order-Sorted Logic with Term Declarations. LCNS vol. 395, Springer, 1989.

C. Walther: A Many-Sorted Calculus Based on Resolution and Paramodulation. Research Notes in Artificial Intelligence, Pitman, London, and Morgan Kaufmann, Los Altos, 1987.

A General Characterization of Term Description Languages

Bernd Owsnicki-Klewe

PHILIPS Research Laboratories Hamburg

P.O. Box 540 840

D-2000 Hamburg 54

1 Introduction

In this paper we will give a general, yet short, characterization of knowledge representation systems based on the KL-ONE paradigm. Since many different systems have been built from this approach, we subsume them under the notion of *term description languages*, other possible terms would be *term-forming languages, frame description languages* or *KL-ONE lookalikes*. These languages are used to build *terminological* models of a domain within a knowledge based system.

We will especially focus on the distinction between the (agreed upon) notational and semantic framework of these formalisms and the "instantiation" of a formalism as a complete or planned computer implementation.

One important aspect of these languages is their well-defined semantics on which several useful inference mechanisms are based. This semantics also serves as the base for "hybrid" systems, i. e. systems that embody another knowledge source besides the terminological one. The basic issue in these systems is the question of how to integrate knowledge from these sources in order to maintain an overall consistency of the knowledge base.

From these considerations, we can assess some problems involved in extending the expressive power of these systems.

2 Terminological representation systems

Terminological representation systems constitute a special variant of sorted formalisms — evolved from the paradigm of KL-ONE [2,3] — which mainly concentrate on a structural analysis of *term descriptions*. By assigning a compositional semantics to the language of term descriptions (called a *term description language* or *TDL*) descriptions are mapped into elements of a subset lattice[1] — the *taxonomy* or *TBox*.

The advantage of assigning a extensional semantics to intensional descriptions is to free the user from problems of setting up and maintaining the taxonomic structure which is now done by the representation system itself.

---

[1] Actually, KL-ONE and its descendants only offer language constructs that make the algebraic structure a lower semi-lattice, i. e. a structure closed under intersection. But we will always use the term lattice, since it is less clumsy.

Thus, basic constructs of a TDL are descriptions plus a way to connect a description with a *name*. Algebraic properties of the taxonomy are just derived from the semantic interpretation of a description built from the semantics of the language primitives.

2.1 Syntax and semantics of a small term description language

As an example, we will briefly introduce a small terminological language $\mathcal{L}$ which is a stylized version of the languages embodied in terminological representation systems (see [8,10] for an overview). These TDLs are usually more expressive, but for illustration purposes we will just introduce this quite weak one.

To give a formal semantics of $\mathcal{L}$, we will give a translation into a set of equations on a certain set $\mathcal{D}$. For this, we define a map $\mathcal{E}: \mathcal{L} \to 2^{\mathcal{D}} \cup 2^{\mathcal{D} \times \mathcal{D}}$ (the *extension* map).

Syntax		Semantics
Expr	←	
	Name = Descr	$\mathcal{E}[\![Name]\!] = \mathcal{E}[\![Descr]\!]$
	Name < Descr	$\mathcal{E}[\![Name]\!] \subseteq \mathcal{E}[\![Descr]\!]$
Descr	←	
	ConceptName	$\mathcal{E}[\![ConceptName]\!]$
	any	$\mathcal{D}$
	(and *Descr*$_1$ ... *Descr*$_n$**)**	$\bigcap_{i=1}^{n} \mathcal{E}[\![Descr_i]\!]$
	(all *Role Descr***)**	$\{x \in \mathcal{D} \mid \forall y \in \mathcal{D}: (x,y) \in \mathcal{E}[\![Role]\!] \Rightarrow y \in \mathcal{E}[\![Descr]\!]\}$
	(atleast *n Role***)**	$\{x \in \mathcal{D} \mid \|\{y \in \mathcal{D} \mid (x,y) \in \mathcal{E}[\![Role]\!]\}\| \geq n\}$
	(atmost *n Role***)**	$\{x \in \mathcal{D} \mid \|\{y \in \mathcal{D} \mid (x,y) \in \mathcal{E}[\![Role]\!]\}\| \leq n\}$

An example of terminological knowledge expressed in this language could look like this:

$$
\begin{aligned}
\text{display} \quad &< \quad \textbf{any} \\
\text{monochrome-display} \quad &< \quad \text{display} \\
\text{colour-display} \quad &< \quad \text{display} \\
\text{computer} \quad &< \quad \textbf{(and (all } \text{has display) } \textbf{(atmost } 2 \text{ has))} \\
\text{monochrome-system} \quad &= \quad \textbf{(and } \text{computer } \textbf{(all } \text{has monochrome-display))}
\end{aligned}
$$

2.2 Basic operations

An extensional semantics for intensional descriptions allows one to reformulate terminological questions as algebraic problems that can be solved by formal means. Typical terminological questions that can be solved that way are (cf. [12]):

Subsumption: Is a given description more general or more special than another one, or can no such relation be established? *As an algebraic problem, this is the problem of looking up the connection between two elements of the set lattice.*

Coherence (Consistency): Is a given description logically coherent, i. e. can there be an instance of this term? *The semantic equivalent is to (dis)prove the identity of a lattice element with the lattice bottom element $\perp$.*

Identity: Are two given descriptions identical, even when expressed by different (syntactic) structures? *This is solved by (dis)proving the identity of two lattice elements.*

Compatibility: Can it be shown that two given descriptions allow for common instances? *This is shown by proving that the greatest lower bound (glb) of two lattice elements is not the bottom element $\perp$.*

Common specialization: Given two descriptions, what are the properties of a third one, introduced as the common specialization or sub-term of them? *This is reformulated as the task of constructing the glb of two lattice elements.*

The algebraic interpretation of terminological questions shows a system of dependencies between them, i. e. a procedure that solves some of them yields a procedure that solves the rest:

Subsumption answers identity: Two descriptions are identical, iff there is a mutual subsumption between them.

Common specialization and identity answer subsumption: A term A subsumes another term B, iff their common specialization is identical with B.

Subsumption answers coherence: A term is incoherent, iff it is subsumed by $\perp$.

The task of selecting the primitive operations for a given implementation is usually *not* solved according to some *minimum principle*, i. e. selecting a small set of operations from which to derive the rest. Rather, economic considerations about frequency of use and efficiency of implementation play the dominant role.

For example, defining subsumption (by far the most frequently used operation) by common specialization and identity — though elegant — is too expensive, such that in most systems all three operations are implemented directly. The mechanism for incorporating the correct subsumption relations into the TBox is generally referred to as the *classifier*.

The most prominent question within the scope of terminological representations is that of *tractability*, i. e. •the problem of *decidability* and *complexity* of these basic operations. It has been shown that in any language with high expressiveness, the problem of computing subsumption is intractable [6,12] or even undecidable [15].

KL-ONE based representation *systems* take a more or less pragmatic attitude towards these theoretical questions. Usually they either restrict the terminological language in order to stay in the region of tractability, like KRYPTON [16], or resort to *incomplete inferences*, like NIKL [5], BACK [7], MESON [13] or LOOM [9].

3 Assertional components

While terminological knowledge constitutes a category of vital importance for domain modelling, other categories are needed in order to account for many applications of knowledge based systems.

Thinking of TBox concepts as representatives for domain terms, another category deals with *referents* or *term instances*, i. e. representatives for *domain objects*. A system component storing knowledge about instances is generally called the *assertional component*, or *ABox*.

The main problem within systems working from more than one knowledge source is how to relate knowledge expressed in different portions of the knowledge base. We want a piece of knowledge to be expressed *only once* and we want pieces of knowledge referring to each other to be held complete and consistent, if one of them changes.

In [11] the solution to this problem is called *integration* of a terminological and an assertional component. In the first place, this requires an interpretation of assertions on the same formal base as used for terminological expressions.

At the beginning of work on ABoxes the first problem to be solved was to state a clear distinction between what constitutes terminological or assertional knowledge (still a controversial topic!).

TBox	ABox
intensional knowledge	extensional knowledge
sets	members of sets
analytical knowledge	contingent knowledge

Whatever is appealing in any of these distinctions, reality makes its own rules. Whether the knowledge expressed in the TBox really *is* intensional, seems to be debatable considering the purely extensional semantics underlying the formalism. Assigning to the ABox the task of representing *set elements* only, puts a severe restriction on its expressiveness, stemming from the fact that assertions about *sets of instances* are quite natural in many domains [1,13]. Many contingent facts (like "any vehicle having a combustion engine is also a vehicle using gasoline", where the underlined phrases should denote TBox terms) seems to be misplaced in the ABox; in fact, the LOOM system [9] puts these contingent facts into its *universal box* or UBox.

While the terminological part of the respective systems is more or less fixed (apart from syntactic questions), ABox designers have made their own choices. Requirements for the expressiveness of an actual ABox are different for different applications and implementations,so that we cannot rely on an overall agreement on what an ABox language should be able to express. We find languages that express full first-order logic, like KRYPTON, just some positive fragments of first-order logic plus some extensions towards set theory, like MESON, or languages that can express *negative* or *disjunctive* assertions, like BACK.

3.1 Syntax and semantics of a small assertional language

For illustration, we will define a very small assertional language that is able to express simple positive facts about the world. An assertion is either of the form

- $C(x_0)$, where C is a TBox concept, and x_0 is an individual constant (generally referred to as an *instance name*), which is translated into the sentence $x_0 \in \mathcal{E}[\![C]\!]^2$ or

- $\mathcal{R}(x_0, y_0)$, where $\mathcal{R}$ is a TBox role and x_0, y_0 are instance names, which is translated into the sentence $(x_0, y_0) \in \mathcal{E}[\![\mathcal{R}]\!]$.

Questions to the ABox will be answered according to the two underlying basic principles:

The Unique Name Assumption: Distinct symbols denote distinct world objects, allowing one to count symbols and interpret their number as the number of world objects.

The Open World Assumption: A question is answered with *"false"* only if it would be inconsistent with the actual knowledge base contents.

4 Expanding the expressive power in a hybrid system

In the previous sections we have discussed the basic properties of hybrid representation systems containing a terminological representation language. We pointed out the necessity to *integrate* knowledge from the terminological and assertional component on a common semantical base.

This requirement places a multiple burden on the system designer who wants to enhance the language expressiveness by adding new TBox or ABox language constructs:

- A semantics must be provided that fits into the formal scheme previously established,

- If the extension applies to the terminological side, the basic operations have to be modified to comprise a (more or less) complete and (perfectly) correct treatment of the new construct,

- The new construct must be reflected in the ABox, in order to be integrated into a complete hybrid system.

Examples for extensions on the terminological side are mainly changes in the semantics, in order to cope with the intractability of subsumption [14] or to base more powerful language constructs on it [4].

The above catalogue seems to be a complicated procedure to follow, when working on language extensions in a hybrid environment. We want to sketch this procedure (as far as we can) by studying one of the language extensions proposed in [1].

---

[2] In order to simplify matters, we identify instance names with elements of a term extension. Otherwise, we would have to introduce a *denotation map* (the assertional counterpart of the terminological *extension map*) that maps instance names into world objects that could be members of a term extension.

4.1   A case study: The γ operator

In [1] some generalized quantifiers are proposed for which the catalogue given above has to apply. We pick the γ operator as an example to study the above steps in more detail. The first step (embedding the operator into the formal framework) is quite straightforward:

Semantics of the γ operator. *An expression* $R(x_0, \gamma x. C(x))$, *where* C *is a TBox concept,* R *is a TBox role and* x_0 *is an individual constant (an instance) translates into the formula*

$$\forall x. (x \in \mathcal{E}[\![C]\!] \Rightarrow (x_0, x) \in \mathcal{E}[\![R]\!]).$$

A simple taxonomy to illustrate the consequences of integrating this operator into the inferential framework is the following:

$$
\begin{array}{rcl}
A & < & \text{any} \\
B & < & (\text{all } R \ A) \\
C & = & (\text{and } B \ (\text{atleast } 1 \ R))
\end{array}
$$

Assume the ABox is now filled with two assertions:

1. $B(x_0)$ (asserting x_0's membership in the extension of B)
2. $R(x_0, \gamma x. A(x))$ (establishing R between x_0 and the *complete* extension of A)

Now the task of proving $C(x_0)$ is equivalent to proving $\|\mathcal{E}[\![A]\!]\| \geq 1$[3], which could e. g. follow from one of the following independent assertions

1. $A(z_0)$,
2. $B(y_0) \wedge R(y_0, z_0)$ or
3. $C(y_0)$

which all have the consequence of asserting the existence of *at least one* instance of A. The first two assertions would then also imply the stronger $R(x_0, z_0)$.

We can see that the γ operator — though probably hard to implement completely — might be a source of quite strong inferences. It poses some new problems inside the inference mechanism, but the general problem of *non-locality* of ABox inferences already exists anyway.

5   Conclusions

We have examined the scope of terminological and especially hybrid representation systems from a very general point of view, focussing on the problem of extending the expressiveness of the representation formalism. We presented a three-step procedure that should be followed in order to have a fully integrated system.

This procedure of adding or modifying language constructs is somewhat difficult, but when done properly it pays off with enhanced inference capabilities and flexibility of use.

---

[3]Note that without having asserted $B(x_0)$ this would *not* be sufficient!

References

[1] J. Allgayer, C. Reddig-Siekmann: *What KL-ONE Lookalikes Need to Cope with Natural Language*, in this volume

[2] R.J. Brachman: *What's in a Concept: Structural Foundations for Semantic Networks*, Int. Journal of Man-Machine Studies, Vol. 9 1977

[3] R.J. Brachman: *On the Epistemological Status of Semantic Networks*, in: N.V. Findler (ed.): *Associative Networks*, Academic Press, 1979

[4] J. Heinsohn, B. Owsnicki-Klewe: *Probabilistic Inheritance and Reasoning in Hybrid Knowledge Representation Systems*, in: W. Hoeppner (ed.): *GWAI-88*, Springer-Verlag, 1989

[5] T. Kaczmarek, R. Bates, G. Robins: *Recent Development in NIKL*, Proceedings of the AAAI-86 1986

[6] H.J. Levesque, R.J. Brachman: *A Fundamental Tradeoff in Knowledge Representation and Reasoning*, in: R.J. Brachman, H.J. Levesque (eds.): *Readings in Knowledge Representation*, Morgan Kaufmann, 1985

[7] K. v. Luck, B. Nebel, C. Peltason, A. Schmiedel: *The Anatomy of the BACK-System*, TU-Berlin, KIT-Report No 41, 1987

[8] K. von Luck, B. Owsnicki-Klewe: *Neuere KI-Formalismen zur Repräsentation von Wissen*, in: T. Christaller (ed.): *KIFS-87*, Springer-Verlag, 1989

[9] R. MacGregor, R. Bates: *The LOOM Knowledge Representation System*, Information Science Institute, Marina del Rey (CA), 1987

[10] R. MacGregor: *The Evolving Technology of the KL-ONE Family of Knowledge Representation Systems*, Proceedings of the Workshop on Formal Aspects of Semantic Networks, St. Catalina, 1989

[11] B. Nebel, K. von Luck: *Issues of Integration and Balancing in Hybrid Knowledge Representation Systems*, in: K. Morik (ed.): *GWAI-87*, Springer-Verlag, 1987

[12] B. Nebel: *Computational Complexity of Terminological Reasoning in BACK*, Artificial Intelligence 34, 1988

[13] B. Owsnicki-Klewe: *Configuration as a Consistency Maintenance Task*, in: W. Hoeppner (ed.): *GWAI-88*, Springer, 1989

[14] P.F. Patel-Schneider: *A Four-Valued Semantics for Frame-Based Description Languages*, Proceedings AAAI-86, 1986

[15] P.F. Patel-Schneider: *Undecidability in NIKL*, Artificial Intelligence, 39(1989)

[16] M. Vilain: *The Restricted Language Architecture of a Hybrid Representation System*, Proceedings of the IJCAI-85

Sorts in Qualitative Reasoning

Werner Dilger
Fraunhofer-Institut für Informations- und Datenverarbeitung
Fraunhoferstraße 1, D-7500 Karlsruhe

Hans Voß
Gesellschaft für Mathematik und Datenverarbeitung
Schloß Birlinghoven, D-5205 St. Augustin

Abstract

According to the principles of Qualitative Reasoning physical systems are represented by deep models (component or process oriented) and are simulated on the basis of the models, either by interpretation of the models or by envisioning. Models of Qualitative Reasoning can be conceived as logical theories, but not as arbitrary ones, rather as theories that have models in the model theoretic sense. Sorts are an integral part of Qualitative Reasoning models, although in the existing approaches only slight attention is pasid to this feature. So sorts can be used to guide the construction of composite models from primitive ones and to specify models.

1. Introduction

One of the roots of Qualitative Reasoning (QR) is Hayes' proposal of a *naive physics* [Hayes 79]. The idea of naive physics is to create a formalism for the representation of commonsense knowledge of the physical world. Hayes argues that a formalism for this purpose should make „use of a highly sorted logic" and supposes that it should be a first order logic. In contrast to scientific or philosophical reasoning which uses uniform formalisations, commonsense reasoning deals with a „richly structured collection of entities" involving a lot of sorts and different kinds of relationships between the entities.

The use of sorted logic as a representation language has a long tradition in AI, cf. [McCarthy/Hayes 69], [Sandewall 71], [Hayes 71], [Pople 72], and [Dilger/Zifonun 78]. Sorts are regarded as a natural means to organize the domain of individuals, corresponding to the human tendency to classification. The definition of sorts ranges from a very coarse subdivision of the individual domain into only five sets corresponding to five sorts in [Sandewall 71] to a complete hierarchy of sorts induced by a partial ordering on the set of sort symbols, semantically viewed as set inclusion.

In QR the concept of sorts is closely related to the concept of *views*. In particular, Forbus defines so-called individual views of objects, which can be regarded as sorts [Forbus 84]. For example, a particular substance may occur in solid, fluid, or gaseous form, depending on temperature and pressure. These are three individual views of that substance. Taking the substance as a sort, the three forms of its occurrence can be conceived as subsorts. In [Struss 88] the term *view* is almost synonymous with *sort*, whereas in [Kippe 88] it is used in a more general sense. This will be described in more detail in Section 4.

In order to make this paper self-contained, Section 2 gives a short introduction to QR. The term *model* is used in QR in a sense different from that in logic. This will be explained in Section 3. Section 4 concentrates on the places where sorts and views can occur in different approaches of QR, and in Section 5 the use of sorts and views in these approaches is described and some problems with views are discussed.

2. Principles of Qualitative Reasoning

A major impetus for QR research originated from the Engineering Problem Solving Project at the MIT. It became known to the majority of the AI community via the special volume of the AI Journal on QR [Bobrow 84], as a result of which it has been given a fresh impetus throughout the whole AI world. In that volume the main approaches of QR were presented by de Kleer [de Kleer/Brown 84], Forbus [Forbus 84], and Kuipers [Kuipers 84].

All these approaches have in common that they model physical systems explicitly with respect to their structure and their behavior. The kinds of models built in QR are sometimes called *deep models*. This term is chosen in contrast to the term *surface models*, by which the usual way of modeling in conventional rule based expert systems is described. In surface models, a relation between two properties is established by means of rules if any relationship between them can be observed in the physical system. But this relation in general is not based on an insight into the structure of the physical system and the causal relationships holding within it. Deep models describe the structure and behavior of physical systems explicitly. This has several advantages. Firstly, an observed relationship between two properties can be explained by means of the model. Secondly, the physical system represented by the model can be simulated by means of the behavior description. And thirdly, deep models are highly modular and thus they can easily be changed, improved, or adapted to a reconfiguration of the modeled system.

The works of the authors mentioned above also share another feature: the description of the structure of physical systems is rather rudimentary. This holds even for de Kleer's *component oriented* models - in contrast to Forbus' *process oriented* models - suggesting that structure description plays a prominent role. But whereas de Kleer's ideas were the basis for the work of other authors introducing more extended component oriented modeling features (cf. [Dilger/Kippe 85], [Kippe 88], [Struss 88], and [Voss 86]), he attaches importance to the modeling of behavior by means of *confluences*. Component and process oriented descriptions have become the two main streams in qualitative modeling.

According to the component oriented approach, the basic active unit of a physical or technical system is the *component*. The definition of a component comprises structural and behavioral information. The basic elements of a component are its *physical properties* (or quantities, variables, parameters). A component has slots, defined by sets of physical properties, by which it can interact with its environment; these slots are called *gates* or *terminals*. The behavior of a component is usually defined by means of *constraints*, i.e. relations between the physical properties, which may have special forms like de Kleer's confluences or Kippe's concepts. A set of components may be composed into a more complex unit, calledc an *aggregate*, via the gates, supposing that there are gates with identical sets of physical properties. Figure 1 shows a naive picture of a buzzer, taken from [Voss 86], and Figure 2 a component oriented model of it (omitting the wires). It consists of five components. The edges denote the gates of the respective components.

Each component has its own behavior. For example the behavior of the switch can be described by the rule-like constraints

IF Switch = out   THEN $I = 0$
IF Switch = on   THEN $I = U / R$

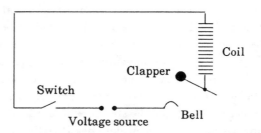

Figure 1: A naive picture of a buzzer

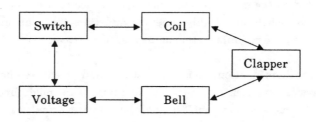

Figure 2: A component oriented model of the buzzer

The voltage source may itself be a complex aggregate, e.g. a battery, with electro-chemical processes going on inside it. The bell is an elastic object that oscillates if some other object hits it. The clapper can move up and down around its suspension point like a lever. The coil builds up a magnetic field if current flows through it, thus attracting metallic objects. The power of the magnetic field is defined by the equational constraint

$$H = n*I$$

where n is the number of turns on the coil. The physical properties occurring in the model are voltage, current, magnetic force, etc.

Component oriented models can be organized hierarchically. That means, an aggregate composed of several components can be regarded on a more abstract level as a single component. On this level it can be composed with other components, that may also be abstractions of aggregates, to build a higher aggregate and so on. This is illustrated for the buzzer example in Figure 3. Here the voltage source is conceived as an aggregate consisting of three components; thus the hierarchy has three levels.

The process oriented approach focusses on the processes going on in a physical system. Processes are the basic active units, and every event occurring in a physical system is directly or indirectly caused by a process. Thus importance is attached to the *behavior* of a physical system, although structural information need not be omitted as can be seen in the process definitions of [Forbus 84] and [Janson-Fleischmann/Sutschet 88]. Processes represent the dynamics of the system explicitly. In particular the relations between physical properties (which are the core of a process definition and can be compared with the constraints in the component oriented approach) do not hold in general, rather they depend on explicitly defined conditions about the activity of the process. Therefore one can say that

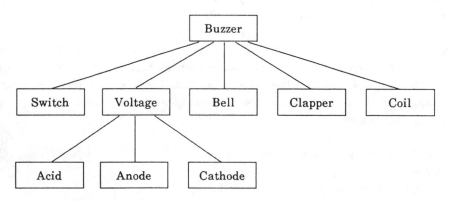

Figure 3: A hierarchical model of the buzzer

the relations between physical properties contained in a process hold *as long as* the process is active.

Here the aspect of *time* comes in. A process is active in some time interval that can be regarded as the domain of validity in a temporal interpretation of the process conditions. In qualitative models this interval is not quantitatively determined, it is just some interval. But sometimes relationships between the activity intervals of processes are known, such as *during, overlaps, meets, before*, etc. They can be denoted, for instance, by the interval calculus of Allen [Allen 83] as is done in [Voss 86] and [Janson-Fleischmann/Sutschet 88].

A process can be defined as having subprocesses, each of which may in turn have subprocesses, and so on. Thus process models can be organized hierarchically like component oriented models. For example the buzzer process, i.e. the oscillation process that causes the ring of the buzzer, can be modeled in CAPAS [Janson-Fleischmann/Sutschet 88] as follows:

> *process*   buzzer-ring
>   *objects*   buzzer
>   *params*   (U (const)), (R (low, infinite)), (I (0, greater0))
>   *subs*   increasing-of-magnetic-force, decreasing-of-magnetic-force,
>   lifting-of-clapper, falling-of-clapper
>   increasing-of-magnetic-force {m, <} decreasing-of-magnetic-force
>   decreasing-of-magnetic-force {m, <} increasing-of-magnetic-force
>   lifting-of-clapper {m} falling-of-clapper
>   falling-of-clapper {m, <} lifting-of-clapper
>   increasing-of-magnetic-force {o} lifting-of-clapper
>   lifting-of-clapper {o} decreasing-of-magnetic-force
>   *conds*   weight(clapper) < H(coil), pos(switch) = on
>   *relations*   R = low ↔ I = greater0, R = infinite ↔ I = 0

A structure model of the buzzer including coil, clapper, voltage-source, switch, and bell is assumed to be defined elsewhere. The physical properties, here called *params*, are defined with qualitative value domains. The symbols m, $<$, and o denote the temporal relations *meets, before*, and *overlaps*. The process increasing-of-magnetic-force is taken as an example for a subprocess:

> *process*   increasing-of-magnetic-force
>   *objects*   coil
>   *params*   (H (0, low, high)), (I (0, greater0))
>   *conds*   I = greater0
>   *influences*   + ∈ Infl(H)

The notation $+ \in \text{Infl}(H)$ means: H is positively influenced by the process. In a similar way the other subprocesses of buzzer-ring can be defined.

Deep models can be used as a basis for simulation. Simulation means computing possible states which the system can be in (defined by value assignments to the physical properties) and possible transitions between the states. This can be done immediately in an interpretative way on the basis of the information encoded in the models, mainly the behavior description, or by some kind of compilation of this information which is called *envisioning*. The first way is used e.g. in [Kuipers 84], the second way is used in [de Kleer/Brown 84], [Forbus 84], [Janson-Fleischmann/Sutschet 88], and [Voss 86].

The main steps of envisioning are: definition of states, computation of states, and computation of state transitions. A state is defined either by a value assignment to a number of selected physical properties or by a set of processes that can be active at the same time. In the latter case the values of the properties in the process conditions must be known; thus it is similar to the former. Having defined a state, the values of the undetermined physical properties are to be computed. This is usually done by constraint propagation. In general, the values of unknown properties cannot be evaluated unambiguously. In this case, the original (undetermined) state is expanded to a set of fully determined states. Finally, the state transitions are computed from the actual values of the physical properties and their trends of change. These trends are parts of the values of the properties. Mathematically, they denote the values of the first derivatives of the properties, considered as functions of time.

Take as an example the process *buzzer-ring*. From the temporal relationships between its subprocesses it can be concluded that there is a state where only the process *increasing-of-magnetic-force* is active, followed by a state where both *increasing-of-magnetic-force* and *lifting-of-clapper* are active. In the first state there is a positive influence on H, which expresses the power of the magnetic field of the coil, and this is the only influence, thus it can be concluded that the value of H increases and will reach the next value after some interval of time, e.g. low if it was 0 before, and perhaps even the next one, namely high. Just this situation could be part of the activity condition of the process *lifting-of-clapper*. Now the second state is reached: the current decreases very quickly to 0 (which should be denoted as an influence in the process *lifting-of-clapper*), and after a short period of time the process *increasing-of-magnetic-force* stops and the process *decreasing-of-magnetic-force* starts, and a new state is reached.

The envisioning process results in a state transition diagram that describes all possible behaviors of the whole physical system. Simulation means then determining a path through the diagram, starting with input values of some of the physical properties. Because the qualitative values are inherently ambiguous, in general not just one path but a number of paths, i.e. a subgraph of the transition diagram, will be found.

3. QR-models and logical theories

In the previous section the terms *model* and *modeling* frequently occurred in the sense they are used in QR-literature. A model in this sense is a set of expressions formulated according to a more or less precisely defined syntax, and modeling means constructing such models. So far a model is just a syntactic entity. This is clearly exhibited by Forbus who shows by an example how his process definitions can be translated into expressions of first order logic [Forbus 84]. As processes have the syntactical form of frames, this translation follows the lines that are given in [Hayes 80]. In the terminology of logic, a set of such expressions forms a *theory*. Thus a logician would regard a model consisting of several components or processes, whether hierarchically arranged or not, as a logical theory. Hayes explicitly takes this view [Hayes 79].

In logic, the term *model* is used in another sense than in the QR-literature, namely in that of model theory. In this sense a model is a mapping of a theory into a semantic domain, consisting of individuals and functions and relations over these individuals, such that the theory becomes true under that mapping.

So far it seems reasonable to distinguish clearly between the two uses of the term *model* in QR and in model theory. However there is also some relationship between them. Hayes argues that the semantic domain of a model (in the model theoretic sense) need not be another formal system as it is usually defined in logic textbooks, rather it may be a section of the real world, e.g. a physical system. He points out that there may be different theories that can be mapped onto the same semantic domain of this kind, such that they are true, and on the other hand there may be different models of one and the same theory. Following Hayes' arguments, the natural semantic domains of QR-theories are the physical systems themselves (or more generally, dynamical systems of any kind). Considering the examples in the QR-literature, it becomes obvious that most QR-researchers assume their theories (i.e. QR-models) to be true with respect to some interpretation that has the modeled physical system as its domain, or in other words, they only formulate or want to formulate theories that have models in the model theoretic sense.

If one regards an interpretation, which is a mapping, as a function graph pairing domain and range values, the entities of the theory become constituents of the interpretation. Thus if the interpretation is a model, the theory itself is part of the model. For this reason it makes sense to speak of QR-theories as models without conflicting with the logical use of this term. In the rest of this paper we will usually speak of *models* in the QR-context and of *theories* in the logical context if no confusion can arise.

4. Sorts and views in QR-models

The basic elements of all kinds of deep models are the physical properties. As the examples of Section 2 suggested, these properties can be conceived as belonging to different *sorts*, e.g. U, R, I are *electrical* properties, H is a *magnetic* property. There are other kinds of physical properties such as mechanical or thermal ones. Thus ELECTRICAL, MAGNETIC, MECHANICAL, or THERMAL can be taken as sorts of physical properties. There are still other classification schemes for physical properties. One of them is to distinguish between pressure-like, flow-like, and friction-like properties. Obviously these terms stem from the mechanics of fluids and gases, but they have analogues in other areas, e.g. electricity. There, voltage is a pressure-like property, current is a flow-like property, and resistance is a friction-like property. These terms form a system of sorts separate from the system of sorts mentioned above. However they can be integrated in that system by taking them as subsorts of e.g. ELECTRICAL and MECHANICAL, thus getting a small hierarchy of sorts which is illustrated in Figure 4. The three subsorts FLOW-LIKE, PRESSURE-LIKE, and FRICTION-LIKE should not be regarded as an exhaustive enumeration of all subsorts of the respective supersorts; there may be still other sorts, e.g. the electrical subsort to which the property *charge* belongs.

In Section 4.1, several QR representation languages are examined with respect to the occurrence of sorts. In Section 4.2, the observations are summarized and classified according to the criteria of sorted logic.

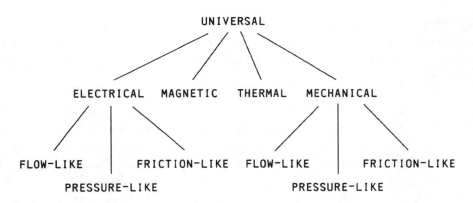

Figure 4: A small hierarchy of sorts

4.1. The use of sorts in QR representation languages

The first approach to be examined is that of Forbus. Although Forbus gives no formal definition of the syntax of the processes and individual views, one can infer such a syntax from his examples. According to this hypothetical syntax, objects associated with processes are always defined as belonging to some sort. Take as an example the entries of the *Individuals*-slot in the process *heat-flow* [Forbus 84]:

> src an object, Has-Quantity(src, heat)
> dst an object, Has-Quantity(dst, heat)
> path a Heat-Path, Heat-Connection(path, src, dst)

Here *object* and *Heat-Path* are sorts, Has-Quantity and Heat-Connection are predicates over the individuals and the physical property *heat*. In QPT (Qualitative Process Theory), physical properties are called *quantities*. A typical phrase in the *Relations*-slot of the process definitions is

> Let x be a quantity

Thus *quantity* is also a sort. Nothing is said about an ordering of these sorts. It can be assumed that *object* and *quantity* are rather general sorts, whereas *Heat-Path* may be a subsort of some other sort. Sometimes terms are used as sorts that are defined as so-called individual views, e.g.

> w a contained-liquid

The individual view *contained-liquid* describes some piece of stuff that is liquid and contained in a container. This individual view can be regarded as the definition of a whole class of entities, i.e. as a sort, which is a subsort of the sort *liquid*. Individual views are a special means to define sorts.

The representation language HIQUAL [Voss 86] is a strictly typed language. The types of HIQUAL are similar to the types of programming languages; they are not types in the sense of typed logic. All syntactical entities have types: variables (the physical properties of HIQUAL), attributes (constant properties), ports (the gates of HIQUAL), models (the components of HIQUAL), and aggregates. Consequently, the functions are defined with respect to the types of their arguments. The basic types are the *simple types* which are sets of numbers or symbols that may or may not be ordered. In HIQUAL only finite simple types are used. For example NAT3 $= \{0, 1, 2, 3\}$ is a simple type. Simple types can be composed into *records*.

The types have a strictly extensional semantics. Thus the type of a variable or a port is just the set of its possible values, cf. the type NAT3 or the type NOISE that is defined by the set {low, zero, high}. In a model definition, the types of all parts of the model have to be explicitly specified, as in a program. Therefore the type of a variable that occurs in several models may vary from model to model, even if it has in some sense the same meaning in all models. For example the type of any variable v might be {zero, greater-zero} in one model and {zero, low, medium, high} in another model, although in both models v is e.g. an electrical variable, say current. The term *electrical* can be conceived as the name of a sort as in figure 4, and compared with the types it has an intensional meaning. Thus the types of variables in HIQUAL specify value domains. The fact that different variables or some variables and ports have intensionally the same meaning cannot be expressed in HIQUAL.

On the other hand, there are two special types in HIQUAL called *model* and *aggregation*. Every model and every aggregate defined in HIQUAL is of type *model* or *aggregation* respectively. This means that it obeys the syntax of the model- or aggregate-definition, or in other words, that it is an instance of the generic object *model-type* or *aggregation-type* respectively. Here the term *type* is used in another sense than above: it is not extensional, rather it is the intensional description of a class of entities and can thus be conceived as a sort.

Sorts in the usual meaning are explicitly used in [Struss 88]. Although Struss does not define a particular syntax for his representation language but represents his models directly as LOOPS-objects, the *views* in the models are obviously sorts. The views are used to express different aspects of components, connections (some kind of dummy-components), and terminals (the gates in Struss' approach), such as electrical, thermal, and mechanical aspects. This is illustrated by the resistor definition. The resistor is firstly defined without regard to different aspects, but later on the views are introduced and the resistor definition is extended by including views for the parameters (the physical properties in Struss' approach) and for the terminals. The extended version reads as follows:

```
Class Resistor
    (Supers Component)
    (InstanceVariables
        (views (electrical thermal)
        (parameters NIL    electrical (resistance)
                           thermal (temperature specificHeat))
        (resistance #(NIL CreateVariable ReplaceMe))
        (temperature #(NIL CreateVariable ReplaceMe))
        (specificHeat #(NIL CreateVariable ReplaceMe))
        (material NIL)
        (terminals NIL electrical (t1 t2) thermal (ht))
        (t1 NIL    type CurrTerminal specifications NIL)
        ...
        (ht NIL    type HeatTerminal specifications NIL)
        ...)
```

There are: one parameter of view *electrical* (resistance), two parameters of view *thermal* (temperature, specificHeat), two terminals of view *electrical* (t1, t2) and one terminal of view *thermal* (ht). The views are different from value domains whose definition is left open.

According to Struss, the views are partially ordered. Figure 5 shows this ordering. In [Struss 88] another view called THERMAL-ELECTRICAL is defined, which is a subview of both, THERMAL and ELECTRICAL. Thus the views do not form a tree-like hierarchy as usual. But here, the definition of the views and the view-hierarchy and their use are mixed. A view THERMAL-ELECTRICAL could at best be defined as a superview of THERMAL and ELECTRICAL, but this would be somewhat artificial because it is hard to find a physical property in naive physics that is both, electrical *and* thermal. In fact, Struss simply means the union of the sets of electrical and thermal physical properties, which can be produced if needed but need not be predefined in the view-hierarchy.

The representation language COMODEL [Kippe 88] defines five types of basic objects that are similar to the types *model* and *aggregate* of HIQUAL. These types form a small hierarchy, which is shown in Figure 6. Only the concepts may have successors of the same kind, i.e. we can build a hierarchy of concepts. The core of a concept is a number of constraints over the *phyprops* (physical properties). They describe the behavior of the modeled system. A COMODEL-model can be transformed into a constraint network composed of the constraints in the concepts of the whole model. Kippe argues that it should be possible to partition this network with respect to different aspects. One of them is called the „functional aspect", and here he means the electrical, mechanical, thermal, ... aspect. Nothing is said about any relationship between these aspects, e.g. a hierarchical arrangement. The parameters for the partitioning of the constraint network are called *views*. Thus the term *view* is used in a broader sense than in [Struss 88] but is similar to it with respect to the functional aspect.

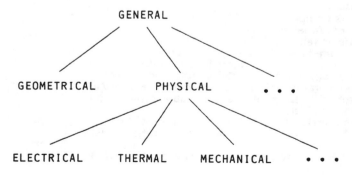

Figure 5: A hierarchy of sorts according to Struss

4.2. Types of sortedness in the QR languages

In Section 4.1, we examined only many-sorted approaches. According to [Oberschelp 89] the simplest form of many-sortedness is restricted quantification. Here a universal variable is quantified by a quantifier that is restricted to a „sort predicate". For example in the formula

$$\forall x \in D_s: P(x)$$

D_s is a sort predicate that may be interpreted by a subsort of the individual domain, and x is restricted to D_s. In this sense, Forbus' QPT can be regarded as a calculus with restricted quantification.

A more elaborated form of many-sortedness is what Oberschelp calls „strictly many-sorted logics". Here each individual constant and each argument slot of relation and function constants belongs to one and only one sort. A term that can be used as an argument of a relation in order to form a formula must be of the same sort as the corresponding argument slot. Only formulas that obey this restriction are well-formed. This demand is reflected in HIQUAL in the definition of functions and relations with respect to the types, though the types are somewhat different from sorts.

The next step in many-sorted logics is to define an ordering on the set of sorts. The main purpose of ordered sorts is the following. Assume s_1 and s_2 are sorts and $s_1 < s_2$. If an argument slot of some relation or function constant has sort s_2, then every term of sort s_2 or s_1 can be put into this slot. Of the QR languages considered in Section 4.1 only Struss' approach contains explicitly ordered sorts. However the strictness of the sorts, in particular the sorts of argument slots, is not reflected. This is perhaps due to the fact that the introduction of sorts in Struss' approach is motivated by the need to view physical systems under different aspects, hence the name *view*.

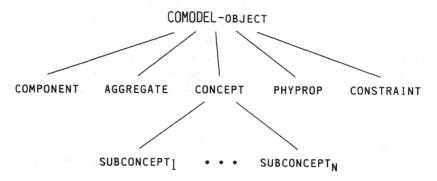

Figure 6: A hierarchy of of basic objects in COMODEL

5. Making use of sorts and views

In deductive systems sorts are used as restrictions to narrow down the search space. For example in theorem proving, the sorts restrict the set of unifiable literals, thus reducing the number of possible inference steps. In QR, the envisioning process or the interpretative simulation of models can be regarded as a kind of deduction which may be influenced by the sorts. Sorts can also be used for the definition of a model, or more precisely, for the specialization of a model from a general model type by instantiation. There is still another place where sorts are used or at least should be used in QR, namely in the construction of models. The process of model construction can also be described as a kind of inference. This is described in Section 5.1, and in Section 5.2 the sort directed instantiation of models will follow.

5.1 Building models

Assume that a number of components are already defined and are stored in a component library. Assume further that a knowledge engineer wants to build a model of a large physical system. If he/she does not want to start from the scratch, he/she will make use of the models in the library. That means, he/she tries to compose these models within a particular context for which they were not originally designed. The models are composed via the gates, and for this purpose it is required that the gates (which are sets of physical properties) are identical. But what does identical mean here?

Clearly, the names of the physical properties involved don't matter and can be arbitrarily changed. Following Voss, the identity of types, in particular of the simple types, is required. To illustrate this, take as an example a system consisting of a power plant that produces electricity and a consumer who wants to receive electricity to have electric light or to watch TV. Assume both the power plant and the consumer are already modeled, and the models have gates consisting of the physical properties U1 and U2 respectively, among others. Of course, it is intended that both properties are of the same kind or sort, namely ELECTRICAL, even of the subsort VOLTAGE. Thus the knowledge engineer can try to compose the two models via their VOLTAGE-gates. But then, the consumer will quickly burn down in the model, because the knowledge engineer did not take into account the value domains of U1 and U2, i.e. their types in the sense of HIQUAL. In fact, the power plant produces voltage of thousands of volts but the consumer can only accept a voltage of, say, 200 volts.

The example shows that identity of the sorts of the physical properties in the gates is not sufficient to connect models. However, the sorts can be used to restrict the possible connections. If there are physical properties in gates of different models with the same sort but different types, then the question is whether the value domain (type) of one physical property is a *refinement* ($D_1 \subseteq D_2$) of the value domain of the other physical property or not (cf. [Dilger 88]). If it is, both properties have the same maximal and the same minimal value and the finer value domain can be taken for both. If it is not, then a special component is needed that transforms one value domain to the other. Such a component can be called a *transformation-component*; in electricity it is just called a *transformer*, in areas where temperature is the relevant physical property it might be some kind of cooling aggregate, and so on.

A transformation-component transports any physical entity (e.g. material, energy, information), and transforms the value domain of at least one relevant physical property of this entity. Corresponding to this definition, other kinds of components could be defined, for example a *production-component* can be conceived as one which produces a new physical entity out of others. Examples of production-components are a generator or a chemical reactor. These different kinds of components can be viewed as subsorts of the sort COMPONENT in COMODEL or MODEL in HIQUAL.

The construction of models can be conceived as an inference process. Considering the models as expressions of first order logic as described in Section 3, the model library forms a theory in this logic. Connecting two components of the model library to an aggregate corresponds to the derivation of an expression for the aggregate from the expression for the two components. For this step a general axiom is needed that can be verbally defined as follows:

If there are two components C1 and C2 with gates G1 and G2 respectively, such that the physical properties of G1 and G2 have the same type,

then an aggregate A can be constructed with C1 and C2 as subaggregates and with a connection consisting of G1 and G2.

Clearly, this axiom is a logical implication. According to its content it can be viewed as a constraint restricting the possible connections between components. The symbols C1, C2, G1, G2, and A play the role of variables. But these variables stand for whole logical formulas; thus the axiom is not first order but rather second order. Therefore the process of building models can only be formulated in second order logic.

The axiom above can be reformulated for sorts instead of types:

If there are two components C1 and C2 with gates G1 and G2 respectively, such that the physical properties of G1 and G2 are of the same sort but have different types and if there is a transformation-component C_T with gates G_T1

and G_T2 such that the physical properties of G1 and G_T1 and the physical properties of G2 and G_T2 have the same types respectively,

then   an aggregate A can be constructed with C1, C2, and C_T as sub-aggregates and with a connection consisting of G1 and G_T1 and of G2 and G_T2 respectively.

This axiom is the logical description of the way sorts can be used to guide the construction of QR models.

5.2. Specifying models

According to [Kippe 88] and [Struss 88], the purpose of views is to focus the interest of the user, to reduce the complexity of a model, and to represent possible variations of the internal structure of a model. In this sense a view can be conceived as a model regarded under a special aspect, therefore Kippe suggests representing a view of a component as an instantiation of that component with respect to the parameters defining the view. One of the parameters in Kippe's more general definition of views, namely functionality, corresponds to the usual meaning of sorts as noticed in Section 4. Now the question is: How can components be instantiated with respect to this parameter, and in other approaches with respect to sorts in general?

A possible solution would be to mark all parts of the model definition with special symbols denoting membership into one of the views for each part. This is the way it is done for the physical properties if views are in use. But what about functions and relations and other parts of the model definition? In a strictly sorted approach, functions are sorted according to the sort of their values, but sorts of relations or other syntactical constructs like whole slots cannot be defined this way. Nevertheless, the modeler could assign sorts to these constructs for technical reasons, namely to have clearly defined subdivisions of the model which he/she can treat as views. This would be a static solution to the problem.

Another solution would be to use sorts strictly in the sense of sorted logic, i.e. to assign sorts only to physical properties and functions. But then, if one wants to instantiate a model, for a number of parts of the model definition it is not clear if they should be included in the view. However, this can be determined by means of special rules guiding the instantiation, or even by different sets of rules which allow the production of different instantiations starting from the same sort according to different relevance criteria. For example if there is a relation relating an electrical, two mechanical, and a thermal physical property, one may ask whether this relation is relevant, e.g. for an electrical view. In some respect it might be, in another it might not be. This way to produce views from models would be a dynamical one.

However, the dynamical solution may have unexpected effects. To modify the above example of the power plant and the consumer of voltage, assume the consumer runs an electric motor. The whole system, consisting of the generator in the power plant, the transport of voltage including transformers, and the consumer's motor, has electrical, mechanical and other aspects. In a mechanical view of the system the wires and transformers would probably not be included, hence in this view it cannot be inferred that the rotation of the armature in the motor is caused by the rotation of the armature in the generator. From the standpoint of logic this is not surprising. If the model of the whole system is regarded as a logical theory, say T, the mechanical view is a subtheory T'. It is a well-known fact that from $T' \subseteq T$ it follows that $Cons(T') \subseteq Cons(T)$, where $Cons(T)$ is the set of consequences of T.

This effect is a principal problem with views that cannot be eliminated. But in order to handle views flexibly one can find a better way than just to build a view and to make inference steps, some of which may already have been made in another view. Such a better way might be based on a reason maintenance system, which would take the axioms of a theory as assumptions and protocol all inference steps. For each derived theorem it could then easily recover the assumptions and inference steps on which it depends. Regarding the construction of a new model from a number of given primitive models as an inference process, the primitive models would be treated as assumptions, and the composite models and all derived behaviors as consequences. Building a view may also start from the primitive models as assumptions, annull some of them and keep the rest. The inferences drawn from those assumptions that are still in use are kept as well, whereas the others are rejected, and to manage this is the task of the reason maintenance system.

The use of a reason maintenance system facilitates the specification of views and the changing of views. Take the above example. If the user observes that a desired inference cannot be drawn from the view he/she has selected, he/she may extend this view simply by reinforcing some of the previously annulled assumptions. Thus he/she may create an electro-mechanical view as a union of the electrical and the mechanical view, which allows for the derivation of the causal relationship between generator and electric motor.

Notice that in the current state of discussion the concept of *views* is not yet defined very concisely. For example, views are used to distinguish the different aggregate states of a substance, say water, and they are also used to denote different aspects of the same substance e.g. its thermal, magnetic, or mechanical properties. Semantically, these distinctions are related to their object in very different way. Whereas at any instant of time water can only exist in one of its aggregate states, at any time we may consider any of its thermal, magnetic, or mechanical properties simultaneously. From the point of view of a problem solver, e.g. during simulation, only one view with respect to the aggregate states will be instantiated, whereas arbitrary other aspects might be relevant and be instantiated. Assume, for example, that during simulation the water switches state from liquid to solid, back to liquid and so forth. Then clearly, in the last liquid phase we won't recompute all derivations already made in a former liquid phase. Instead, the reason maintenance system should automatically switch to the former state of reasoning.

6. Conclusions

Sorts naturally come in QR representation languages. As in sorted logic they are used to restrict the search space during the process of model construction, which can be conceived as an inference process. In addition they are used to restrict the set of axioms (or assumptions) and hence the set of possible deductions in specifying views of models. This may cause some problems. Sorts are not yet an explicit topic of QR research, but dealing with sorts in QR could make a helpful contribution to the careful handling of QR representation languages. Sorts could also be useful for precisely defining extensions of such languages

References

[Allen 83]
J.F. Allen: Maintaining knowledge about temporal intervals. CACM 26.11 (1983) 832 - 843.

[Bobrow 84]
D.G. Bobrow (ed.): Qualitative reasoning about physical systems. Special volume of AI.
AI 24 (1984).

[de Kleer/Brown 84]
J. de Kleer, J.S. Brown: A qualitative physics based on confluences. AI 24 (1984) 7 - 83.

[Dilger/Zifonun 78]
W. Dilger, G. Zifonun: The predicate calculus-language KS as query language. In:
H. Gallaire, J. Minker (eds.): Logic and data bases, Plenum Press, New York 1978, 377 - 408.

[Dilger/Kippe 85]
W. Dilger, J. Kippe: COMODEL: A language for the representation of technical knowledge.
In: Proceedings of IJCAI 85, Los Angeles 1985, 352 - 358.

[Dilger 88]
W. Dilger: Composing qualitative models. TEX-B Memo 39-88, FhG-IITB, Karlsruhe 1988.

[Forbus 84]
K.D. Forbus: Qualitative process theory. AI 24 (1984) 85 - 168.

[Hayes 71]
P.J. Hayes: A logic of actions. In: Machine Intelligence 6, Edinburgh University Press 1971, 495 - 520.

[Hayes 79]
P.J. Hayes: The naive physics manifesto. In: D. Michie (ed.), Expert Systems in the Micro-Electronic Age,Edinburgh University Press 1979, 242 - 270.

[Hayes 80]
P.J. Hayes: The logic of frames. In: D. Metzing (ed.): Frame conceptions and text understanding, de Gruyter, Berlin 1980, 46 - 61.

[Janson-Fleischmann/Sutschet 88]
A. Janson-Fleischmann, G. Sutschet: A process oriented approach for qualitative modelling and analysis of dynamical systems. In: H.-W. Früchtenicht et al. (eds.): Technische Expertensysteme: Wissensrepräsentation und Schlussfolgerungs-verfahren, Oldenbourg-Verlag, München 1988, 227 - 247.

[Kippe 88]
J. Kippe: Komponentenorientierte Repräsentation technischer Systeme. In: H.-W. Früchtenicht et al. (eds.): Technische Expertensysteme: Wissensrepräsentation und Schlussfolgerungsverfahren, Oldenbourg-Verlag, München 1988, 155 - 226.

[Kuipers 84]
B. Kuipers: Commonsense reasoning about causality: Deriving behavior from structure. AI 24 (1984) 169 - 203.

[McCarthy/Hayes 69]
J. McCarthy, P.J. Hayes: Some philosophical problems from the standpoint of artificial intelligence. In: Machine Intelligence 4, Edinburgh University Press 1969, 463 - 502.

[Oberschelp 89]
A. Oberschelp: Order Sorted Predicate Logic. This volume.

[Pople 72]
H.R. Pople: A goal-oriented language for the computer. In: H.A. Simon, L. Siklossy (eds.): Representation and meaning, Prentice Hall, Englewood Cliffs, NJ, 1972, 329 - 413.

[Sandewall 71]
E. Sandewall: Representing natural language information in predicate calculus. In: Machine Intelligence 6, Edinburgh University Press 1971, 255 - 277.

[Struss 88]
P. Struss: Assumption-based reasoning about device models. In: H.-W. Früchtenicht et al. (eds.): Technische Expertensysteme: Wissensrepräsentation und Schlussfolgerungs-verfahren, Oldenbourg-Verlag, München 1988, 23 - 54.

[Voss 86]
H. Voss: Representing and analyzing causal, temporal, and hierarchical relations of devices. Ph.D. Thesis, University of Kaiserslautern, 1986.

III. Sorts and Types in Natural Language (Understanding) Systems

Eventualities in a Natural Language Understanding System

Kurt Eberle
Institut für Maschinelle Sprachverarbeitung
Universität Stuttgart
Keplerstr. 17

Abstract

This paper focuses on the discussion of suitable representations of eventualities in a formal language and of the possibilities to draw inferences from representations. We argue in favour of the treatment of eventualities as individuals, which are structured along different lines. When reifying eventualities, there are different possibilities of individualization. This is similarly true for the domain of objects. Thus we investigate these possibilities in parallel with objects, and obtain a rather symmetric structuring of the domain of individuals, i.e. a sort hierarchy which is sensible for different kinds of eventualities and objects respectively.

Next to this classification and the pure temporal ordering, there is another possibility of structuring: eventualities can be partitioned into subevents, and be grouped together to form episodes.

The representation of sets of eventualities is needed to deal with certain plural phenomena correctly. Normally this leads to the introduction of second order variables and, allowing quantification over such variables, to the extension of the language to one of second order type. However there is no complete calculus for second order logic. This is one of the reasons why we favour a lattice approach, which enables us to model sets as structured objects of first order type.

The discussion and integration of the different principles of structurization mentioned above will be the main focus of this paper. [1]

The background of this paper is provided by the *travelling scenario* of the LILOG-project and the language L_{LILOG} [2] which is used in the LILOG-system.

1 Introduction

There are a number of criteria to distinguish different verb types. With respect to the task of inferring facts from the formal representation of a Natural Language-text (NL-text), the linguistic classification of verbs into *Aktionsarten* by (Vendler) is especially interesting. The definition of the Aktionsarten is based on a classification of the facts normally described by the verbs. Vendler distinguishes between *states, activities, accomplishments* and *achievements*. States describe static situations, i.e. the validity of the statement is inherited by all subintervals of an interval for which the statement holds (even by points if

---

[1] A more exhaustive discussion and description of the following proposal - with minor (notational) deviations - can be found in (Eberle89).

[2] Compare the paper of Bläsius, Hedtstück, Pletat in the same volume.

they are allowed by the logic). This is - with minor restrictions - equally true for activities. Achievements and accomplishments do not have this quality: they describe changes between different states. These changes are temporally extended in the case of accomplishments, in the case of achievements they are not. Because of this quality of change, achievements and accomplishments are also called *telic events*. In contrast to this, states and activities are *atelic (events)*. The origins of such a classification can be traced back to Aristotle. In the AI-tradition, similar approaches can be found: Allen distinguishes between *properties, processes* and *events* (cf. (Allen84)), (McDermott) between *facts* and *events*. Roughly speaking, one can identify properties and facts with states, processes with activities, and events with either achievements or accomplishments. In order to disambiguate between *events* in the strict sense (accomplishments and achievements) and *events* in the wider sense (subsuming in addition states and activities), (Bach) has introduced the notion of *eventuality* for the latter.

Thus we get the following picture:

Eventualities:

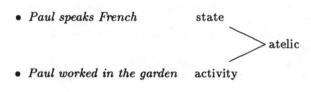

- *Paul speaks French*   state
- *Paul worked in the garden*   activity

                 atelic

       Events:

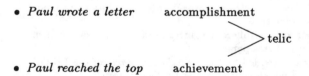

- *Paul wrote a letter*   accomplishment
- *Paul reached the top*   achievement

                 telic

It should be emphasized that the type of the values of the subcategorized functions contributes to the determination of the Aktionsart. For instance *writing* describes an activity, *writing a letter* an accomplishment and *writing letters* again describes an activity. In addition, temporal and local adverbials can contribute to the determination of the Aktionsart. The problem of computing the correct Aktionsart of an eventuality will be dealt with in section **10**.

2   Eventualities as Individuals

In order to represent eventualities in the framework of classical temporal logic of Priorian style, verbs can be assigned predicates whose arguments are the values of the different thematic roles of the verb in the sentence. In approaches of this kind, the influence of tense morphemes is represented by operators applied to formulae. Thereby

P requires the evaluation at a time (a point or an interval depending on the underlying tense logic) in the past, and

F requires the evaluation at a time in the future.

In order to obtain correct representations for certain natural language phenomena, other operators were introduced step by step (for different shiftings of the evaluation time (cf. (van Benthem)), which in the long run led to rather unintuitive extensions of the language.

To avoid such unintuitive extensions occurring, a suggestion was put forward to introduce temporal objects, (i.e. constants and variables for time intervals or time points) directly on the syntactic level, instead of introducing operators and associated evaluation constraints with respect to tense structures, thus making such temporal objects available only on the semantic level. These times can be introduced as fillers of an additional argument place of the verb-predicates. This *Method of Temporal Arguments (MTA)* was argued for by (Haugh) recently.

In the AI-tradition, the proposals of Allen (cf. (Allen84)) and (McDermott) are prominent. In addition, they suggest the reification of different *eventuality-types*. These types are assigned times by means of particular validity predicates.

In the notation of Allen the sentence:

John was ill

is rendered by

holds(ill(john),i),

where $i \, B \, i'$ (" i before i' ") and $i' = speech \ time$.

Note that the interpretation of *ill(john)* in a model would be an individual of the domain of the model (since *ill* is interpreted as a function over the domain)!

It has been objected repeatedly that interval (or point) semantics are too weak to model natural language phenomena.

Parsons:
"*There remains a substantive question, however, of whether or not a "pure" interval semantics can be adequate for various phenomena of ordinary speech without appeal to eventualities.*" [3]

As he points out, one of these phenomena is the possible coincidence of two eventuali-

---

[3] Cited by (Bäuerle), p.18.

ties of the same type. His example:

Miles was wounded by a bullet twice yesterday

could be rendered in Allen's notation as follows:

occur(wounded-by(miles,bullet),i)
occur(wounded-by(miles,bullet),i′)

where *occur* relates event-types and intervals (whereas *holds* relates property-types and intervals). If $i = i'$ holds, it is no longer distinguishable that the eventuality occured twice.

Of course, in the conflict case, we assume that there is exactly one bullet which caused the two injuries. The conflict does not exist if there are two different bullets, since then we get different types of eventualities. By means of the proper reification, functions over the domain of the eventuality-types like *twice* or *composite* are allowed and defined by Allen. However, the axioms associated with these functions require in the first case that the two occurrences of the event type do not coincide, and in the second case, that the occurrence of the composite type is equivalent to the coinciding occurrences of the two event types of which the composite is composed. Thus in the case of identity of the two event types, the composite type is absolutely meaningless or, even worse, it corresponds to the simple type. Of course, one could remedy this shortcoming by turning the equivalence into a simple implication. However, in this case, the composite type is no longer sufficiently characterized. Thus the approach of Allen succeeds in such cases only if he provides an argument place for each possible thematic role (except the temporal trace) in the functions which return eventuality types (the *wounded-by* function in the example above) and existential quantification over the variables of the non-instantiated slots (outside the *occur* predicate). This is not very practical, however.

It may well be also for this reason that Allen allows in the 1983 paper (Allen83) for coinciding but not identical intervals, thus anticipating the reification not only of eventuality types but also of eventuality tokens.

Be that as it may, it seems that examples such as the one given make it plausible to argue for the reification of eventuality tokens, a reification which, on the philosophical level, has been especially motivated by (Davidson).

On the text level, anaphora, nominalizations and definite descriptions respectively present further evidence:

Last week I gave a lecture. It was in Rome.
Yesterday John boxed with Peter. The fight took place in hall C.

In this case, *it* refers to the event of *giving a lecture* which is attributed the additional quality of *taking place in Rome* by means of this reference. *The fight* designates explicitly the event of *boxing*, and serves as argument of the local statement.

The examples mentioned above describe events and, within the linguistic discussion, they put the suggestion forward to ascribe a specific, non-propositional type to events. However, is it plausible to ascribe the same quality to states ? (Bäuerle) argues against (Löbner) who takes the view that states are propositions. In this case, under the premise that negation is an operation which, when applied to propositions, returns propositions, negated states should be accepted in places where unnegated states are accepted. However, as Bäuerle points out, there are test cases where states are accepted but not their negation:

Seit ich bei Bosch arbeite, wohne ich in Degerloch.
(Since I have been working at Bosch, I have been living in Degerloch.)

Sobald die Sonne scheint, holen wir die Wäsche rein.
(As soon as the sun shines, we will bring the wash in.)

* *Seit ich nicht bei Bosch arbeite, wohne ich in Degerloch.*
(* Since I have not been working at Bosch, I have been living in Degerloch.)

* *Sobald die Sonne nicht scheint, holen wir die Wäsche rein.*
(* As soon as the sun doesn't shine, we will bring the wash in.)

In such cases, pure negation is unacceptable . An indirect reformulation of the negation with an inchoative reading (denoting the end of the described state) seems to be requested. As it stands, it seems necessary to make a distinction between the things in the world (which cannot be negated) and the statements about the world. [4] And like other eventualities states are things of the world whereas for instance negated nominalizations of eventualities (*the non-arrival of the train*) are statements about the (non-)existence of eventualities in the world.

We agree with this position and adopt it for the formal representation of natural language texts: all eventualities are rendered as individuals. Splitting off the arguments which are values of thematic roles of the eventuality in question by using accordingly named features allows for a representation which uses only unary predicates with respect to verbs (i.e. "*the running of Peter*" is rendered as $running(e) \land agent(e)=peter$ instead of $running(e,peter)$). Except for the use of unary predicates, this is the way it is done in L_{LILOG}. Instead of unary predicates, L_{LILOG} uses sort symbols.

Below, we will use a sorted first order language for representations, which is related, but not identical to L_{LILOG}. This language is used only for notational convenience, and serves to give an idea of how the axioms formulated in the following could be represented in L_{LILOG}. We do not aim at computational efficiency as one does with respect to the design of L_{LILOG}.

Our language utilizes sorts, which are defined using primitive sorts and features along the lines of the recursive definition of sort expressions as stated by (Smolka) in his paper on *feature logic*. In our approach, primitive sorts are nothing else than unary predicates. They are highlighted as sorts because they are central predicates. Features are central functions (they describe the thematic roles). We use sort expressions only to obtain abbreviations. Below, we will use the following abbreviations in particular:

---

[4] With regard to this, compare the discussion in (Bäuerle) where the distinction between eventualities in the world and statements about the world is also attributed to Vendler's point of view.

Sort-Expr1 $\leq$ Sort-Expr2   for $\forall x \in$ Sort-Expr1: $x \in$ Sort-Expr2

$\forall x \in$ Sort-Expr: $(\ldots)$ for $\forall x :$ $(x \in$ Sort-Expr $\rightarrow (\ldots))$

We frequently have sort expressions such as

WRITING $\bigcap object = c \bigcap agent$:HUMAN.

In this case, $e \in$ WRITING $\bigcap object = c \bigcap agent$:HUMAN is an abbreviation for

WRITING$(e) \wedge object(e) = c$ $\wedge$ HUMAN$(agent(e))$.

(Here WRITING and HUMAN are assumed to be primitive sorts, i.e. unary predicates.)

We will not go into further detail, i.e. defining the exact syntax and semantics of the language used, since this is rather straightforward. On the basis of the given hints, the reader should see the possibility of "syntactic desugaring".

Below, we sometimes use the notion *Discourse Referent (DRF)* for constants in formulae. This notion comes from the so called *Discourse Representation Theory (DRT)* developed by Kamp (Kamp81). Using this framework, the LILOG-system translates NL-texts first into so called *Discourse Representation Structures (DRSs)*, which are especially suited to represent discourse phenomena. DRSs are then translated into L$_{LILOG}$-formulae, where constants of this language are assigned discourse referents. Even if we call constants *discourse referents*, it is this level of representation that we aim at with our sorted language.

3   Reification of Sets

In the previous section, we decided to reify eventualities. This means that discourse referents for eventualities in the representation are mapped onto individuals in corresponding models. Here, we will argue for the reification of sets, i.e. discourse referents for sets (of eventualities or objects, etc.) should also be mapped onto individuals in corresponding models. This is motivated by quantificational phenomena.

In addition to the universal and existential quantification involved in such sentences as:

All Germans like Bach

where the translation into a first order language is no problem:

$\forall x \in$ HUMAN: $(german(x) \rightarrow \exists e \in$ EVENTUALITY:
$\qquad\qquad (like(e) \wedge protagonist(e) = x \wedge object(e) = bach))$

natural language provides so called *generalized quantifiers* like *most, many, few, some*. In this case, the direct translation into a first order language is impossible. Normally such sentences are translated into second order formulae where quantification over sets is possible; for example, the sentence:

Most Germans like Bach

may be translated into the second order formula:

$Y :=$ the set of all Germans
$$\exists X(X \subset Y \wedge most(X,Y) \wedge \forall x(x \in X \to \exists e \in \text{EVENTUALITY:}$$
$$(like(e) \wedge protagonist(e) = x \wedge object(e) = bach)))$$

Since such phenomena are not restricted to objects in the narrow sense but can be observed also for other kinds of objects, such as times or eventualities, a more general solution seems to be necessary.

On more than half of the days last year the sun was shining.

$Y :=$ the set of all days of the last year
$$\exists X(X \subset Y \wedge 2 \times |X| > |Y| \wedge \forall x(x \in X \to \exists e \in \text{EVENTUALITY:} (x \subseteq e \wedge sunshine(e))))$$

Many of the games took place in the rain.

$E' :=$ the set of all games
$$\exists E(E \subset E' \wedge many(E, E') \wedge \forall e(e \in E \to \exists e' \in \text{EVENTUALITY:} (e \subseteq e' \wedge rain(e'))))$$

Such a solution is provided by a (semi-)lattice approach where sets are reified, i.e. where they are modelled as objects. Such "set-objects" have the property that they are the least common upper bound of the elements of the set with respect to the partial order which is defined by the lattice operation.

A positive side effect of the lattice approach is that there is no notational difference between constants which denote "atomic" objects in a model, and constants which denote set-objects in the model. Thus, if knowledge about the cardinality of a discourse referent is lacking, we get the desired underspecified representation, which allows for different models, i.e. for different readings. In some models, the discourse referent denotes a plural entity, in other models it denotes a singular entity:

Last week John bought five books

translates in the following assertion of literals:

$t \in \text{WEEKS}, e \in \text{EXCHANGES}, u \in \text{BOOKS}$

$buy(e), agent(e) = john, object(e) = u$
$|u| = 5, |t| = 1, e \subseteq t$

Note that the sample sentence and its representation describe the situation incompletely.

There is no indication there allowing one to decide whether there was one event of buying five books or whether there were several events of buying books such that the sum of the books bought amounts to five.

It should be stressed here that the extensions of sorts like BOOKS, WEEKS, EXCHANGES, etc. contain atomic objects as well as sums of atomic objects of the described kind. We use the corresponding starred sorts to denote atomic objects, where S* is defined from S by means of the cardinality feature $|\cdot|$ (which will be introduced in section **5**).

4 The Upper Structure of a Sort Hierarchy for an NL Understanding System

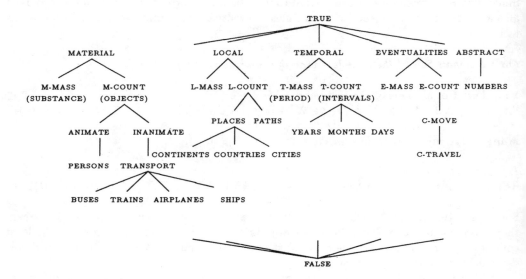

TRUE and FALSE are the maximum and minimum of **S** respectively, which is the set of primitive sorts used in the system partially ordered by means of $\leq$. TRUE always denotes the domain of a model, and FALSE the empty set. (Therefore, sorts S and S' are disjoint in a model M, if $S \cap S' =$ FALSE holds in M).

MASS and COUNT are disjoint primitives which are used to define subsorts of the disjoint sorts MATERIAL, EVENTUALITIES, LOCAL and TEMPORAL.

5 The Sort MATERIAL

5.1 Individualization of Objects and Corresponding Portions of Matter

There are several choices open to us if we want to represent a sentence such as

The ring is not old, but the gold the ring is made of is old

in our first order language. We can adopt a more realistic point of view and create exactly one constant for an object, one DRF, which is qualified as being old gold and being a not-old ring. In order to avoid inconsistency, we are thus forced not to interpret adjectives like "old" intersectively, i.e. as predicates in their own right, but as modifiers of predicates. It is frequently argued, on the basis of many similar examples, that the latter is more adequate. [5] However if we continue the story by:

The gold comes from South Africa. It was brought to Europe in 1922. The ring was produced in 1923 in London

it becomes clear that we have to construct rather unintuitive representations if the DRF for the gold is identified with the DRF for the ring. In addition, it would be difficult to formulate suitable DRS construction algorithms for such representations.

Examples like this led Link in his 1983 paper (Link83) to make an ontological distinction between objects and the substance they consist of. The associated predicates also illustrate a difference which motivates mental discrimination: Two portions of gold mentally brought together form a new entity in the extension of *gold*, whereas the sum of two rings is not in the extension of *ring*, i.e. "gold" is a mass noun and "ring" is a count noun. We have to take this difference into account and structure the sorts accordingly.

In (Link83), Link sets up conditions for the material entities of model domains for NL-representations, which the following structurization of the sort MATERIAL heavily relies upon.

M-MASS := MATERIAL $\cap$ MASS

M-COUNT := MATERIAL $\cap$ COUNT

It is perhaps mnemonically better to call M-MASS "SUBSTANCE" and M-COUNT "OBJECTS". Thus we define:

SUBSTANCE := M-MASS

OBJECTS := M-COUNT.

Now let the extensions of SUBSTANCE and OBJECTS be complete semilattices with respect to the interpretation of the lattice operation symbols $\sqcup_s$ (defined on SUBSTANCE) and $\sqcup_o$ (defined on OBJECTS) respectively.

This means that in the OBJECTS case the following set of axioms hold:

$\forall x, y \in$ OBJECTS: $(x \sqcup_o y = y \sqcup_o x)$	commutativity
$\forall x \in$ OBJECTS: $(x \sqcup_o x = x)$	idempotence
$\forall x, y, z \in$ OBJECTS: $(x \sqcup_o (y \sqcup_o z) = (x \sqcup_o y) \sqcup_o z)$	associativity
$\forall x, y \in$ OBJECTS: $\exists z \in$ OBJECTS: $(x \sqcup_o y = z)$	completeness
$\forall x, y \in$ OBJECTS: $(x <_o y \rightarrow \exists^{=1} z \in$ OBJECTS$:(\neg x \circ_o z \wedge x \sqcup_o z = y))$	(strict) complementarity

---

[5] Compare, for instance, (Cresswell).

where the proper part $<_o$ and the overlap o_o are defined in the following way out of $\leq_o$ which is the partial order given by the semilattice operation:

$$\forall x, y \in \text{OBJECTS}: \quad (x \leq_o y \leftrightarrow x \sqcup_o y = y) \qquad\qquad\qquad \text{part}$$
$$\forall x, y \in \text{OBJECTS}: \quad (x <_o y \leftrightarrow x \leq_o y \wedge \neg x = y) \qquad\qquad \text{proper part}$$
$$\forall x, y \in \text{OBJECTS}: \quad (x \, o_o \, y \leftrightarrow \exists z \in \text{OBJECTS}: \quad (z \leq_o x \wedge z \leq_o y)) \quad \text{overlap}$$

In the case of SUBSTANCE, an analogous set of axioms holds whose formalization we omit here.

The substantialization function *subst* relates the two sorts. On OBJECTS it is a homomorphism with values in SUBSTANCE:

$$\forall x, y \in \text{OBJECTS}: \quad (subst(x) \sqcup_s subst(y) = subst(x \sqcup_o y))$$

On SUBSTANCE it is the identity:

$$\forall x \in \text{SUBSTANCE}: \quad (subst(x) = x)$$

By means of this function we can define a relation $\leq_m$ which extends $\leq_s$ on MATERIAL:

$$\forall x, y \in \text{MATERIAL}: \quad (x \leq_m y \leftrightarrow subst(x) \leq_s subst(y))$$

$\leq_m$ then is a partial order on SUBSTANCE and a preorder on OBJECTS. Intuitively it describes the *material "partof"*.

5.2 Countability

The intuitive meaning of $\sqcup_o$ is the set forming operation, i.e. if a is a ring and b is a ring then $a \sqcup_o b$ is an object which intuitively denotes the set consisting of the two rings. In this sense, OBJECTS denotes plural and singular objects whereas SUBSTANCE denotes portions of matter, say on a pure "singular" level, and thus $\sqcup_s$ models the material fusion of portions of matter.

Taking this into account, it makes sense to define a function $|\cdot|$ on OBJECTS with values in INT(EGERS) ($\leq$ ABSTRACT) which models cardinality.

$$\forall x, y \in |\cdot| : INT : \quad (x \leq_o y \rightarrow |x| \leq |y|)$$

$$\forall x, y, z \in |\cdot| : INT : \quad (x \sqcup_o y = z \wedge \neg x \, o_o \, y \rightarrow |x| + |y| = |z|)$$

$$\forall x, y, z, z' \in |\cdot| : INT : \quad (x \sqcup_o y = z \wedge z' \leq_o x \wedge z' \leq_o y$$
$$\wedge (\forall z'' \in |\cdot| : INT : \quad z'' \leq_o x \wedge z'' \leq_o y \rightarrow z'' \leq_o z')$$
$$\rightarrow |x| + |y| - |z'| = |z|)$$

Each individual in OBJECTS consists of atoms:

$$\forall x \in \text{OBJECTS}: \quad \exists y \in \text{OBJECTS}: \quad (y \leq_o x \wedge o - atom(y))$$

where an object-atom is defined as nondivisible:

$$\forall x \in \text{OBJECTS}: \quad (o - atom(x) \leftrightarrow \neg(\exists y \in \text{OBJECTS}: \quad y <_o x))$$

Atoms have cardinality 1:

$$\forall x \in OBJECTS: \quad (o - atom(x) \leftrightarrow |x| = 1)$$

As it can be shown, $| \cdot |$ is determined (in the finite case) by the values of the singular individuals, the atoms of OBJECTS (where the atoms of a sort expression ST are denoted by $\text{ST}^* := \text{ST} \cap | \cdot | = 1$).

This setting influences the ordering of the subsorts of OBJECTS. For instance, we cannot say that a train is an object consisting, in the sense of $\sqcup_o$, of several cars and an engine, since then a train would be of cardinality > 1. However, when we are counting trains we do not count parts of trains. This means that the sort TRAINS and the sort which subsumes sums of cars and engines have to be disjoint. Link points out that it is cognitively adequate to make a distinction between the structured object and its set of parts, say between the DRF for the train and the DRF for the set consisting of its cars and its engine. In his argument he uses the example of some decks of cards which are not numbered but which consist of numbered cards: "Being numbered" thus serves as a distinguishing criterion. A lot of similar examples can be found making this distinction plausible. If you do not accept the predicative distinction, having in mind what has been said above, think of the same committee with changing membership or different committees with the same members.

This means that being an atom in the sense of a subsort is equivalent to being an atom in the sense of OBJECTS, and being a plural object of some subsort means that all parts in the sense of $\leq_o$ of that object are individuals of the same subsort.

What we are arguing for can be set up as conditions for primitive subsorts of OBJECTS:

Each S in S with $S \leq \text{OBJECTS}$ ($S \neq \text{FALSE}$) is a complete sub-semilattice of OBJECTS with respect to $\leq_o$ such that

$$\forall x \in S, y \in \text{OBJECTS}: \quad (y <_o x \rightarrow y \in S)$$

By means of this we get the quality of strict complementarity transfered from OBJECTS to all primitive subsorts of OBJECTS (and to their intersections as can be shown). In addition, the overlap between two objects is subsort-independent within OBJECTS, i.e if a and b overlap in OBJECTS and if they are of the same subsort S, a and b overlap in S [6].

---

[6]Our primitive OBJECTS subsorts denote singular and plural objects whereas Link's predicates denote only singular objects. By the * operator applied to predicates, he gets what we call (primitive) sorts. The difference becomes relevant when we want to define sorts for mixed objects, e.g. objects which consist of trains and buses. The set of such objects is not described by TRAINS $\bigcup$ BUSES, since here combinations of buses and trains are excluded. However on the basis of our semilattice sorts - containing plural objects - it is possible to define a feature, such that the desired supersort containing mixed objects is describable by

We can straightforwardly define subrelations of the material part relation $\leq_m$, *elementof* and *memberof*, which relate, in the case of *elementof*, atoms with entities which are built up by such atoms, and, in the case of *memberof*, (not necessarily singular) objects with entities of which they are "members", e.g. cards with decks of cards.

For *memberof*, it is useful to require in addition the "cumulative reading". If we render

The persons a, b and c are members of the commitees d, e and f

by:

$memberof(a \sqcup_o b \sqcup_o c, d \sqcup_o e \sqcup_o f)$,

we want to infer that for each person there is at least a commitee, and that for each commitee there is at least a person, such that the person is a member of the commitee. The detailed formalization of this axiom and other similar ones can be found in (Eberle89).

Normally, counting means that the things counted are not part of each other. Unfortunately, there are cases where the atoms of a sort materially overlap. Think for instance of twigs: There might be parts of a twig which can be regarded also as twigs. In this case, counting should be restricted to the same level: when counting the "sub-twigs", one should not count the entire twig as a twig. We cannot exclude such unintuitive sets in our lattice approach, but can at least mark them as non-normal sets by means of the material part relation:

$\forall x \in$ OBJECTS: $(normalset(x)$
         $\leftrightarrow (\forall y, z \in$OBJECTS: $(elementof(y,x) \wedge elementof(z,x) \rightarrow \neg y \circ_m z)))$

where $\circ_m$ is straightforwardly defined by means of $\leq_m$.

As with OBJECTS, all primitive subsorts of SUBSTANCE are complete subsemilattices with respect to the corresponding lattice operation $\sqcup_s$. However, in contrast to the OBJECTS-case, we neither require that complementarity or overlap is inherited from SUBSTANCE nor do we require that there is a notion of counting.

With the first assumption we follow (Krifka). On the one hand, if $\leq_m$ really is meant to model the intuitive material part relation, as indeed it is, then there are still cases where parts of portions of matter do not have the same substantial property as the entire portion. Think for instance of mixtures of fluids. In addition, if we speak for instance of portions of gold we seldom mean pure gold. Impurities are allowed. Thus, in both cases: no inheritance of complementarity. On the other hand, if such portions of gold overlap in SUBSTANCE

---

means of feature terms of Smolka's feature logic. This is not the case for Link's starred predicates P*. The are not definable by feature terms from the primitive predicates P. and therefore $*(P \bigcup Q)$ is no featur term. Since in our approach all sorts should be feature terms, the Link solution is not suitable. With regar to the definition of the feature-term-equivalent for $*(P \bigcup Q)$ compare (Eberle89).

it is possible that they overlap exactly in their non-gold parts and therefore do not overlap in GOLD. The naive physics approach that we envisage should take such phenomena into account.

Concerning the second assumption, "no counting", we do not agree with (Link83). We think that each act of counting presupposes the mental conception of objects which are counted. In the case of portions of matter, this means that there is a formation, a creation of objects, which precedes the counting and it means that counting takes place with respect to these objects and not with respect to the substance they are "made of". For instance if we say that

German gold is stored at the Bank of Switzerland and French gold is stored at the Bank of Switzerland

and if we then want to count, we have to form objects. That these objects differ on the mental level from their substance can be seen from the fact that they can consist at different times of different substance:

Both quantities of gold have changed in the last ten years.

Such definite descriptions thus act like functions from times into substances; they are *fluids* in the terminology of (McDermott), but nevertheless have to introduce DRFs on the discourse representation level.

Of course, much closer connections between the substance and the corresponding object may be formulated. Especially in deictic contexts, the mental image of the substance - it is in this sense that we have used and will use in the following the notion "substance" - and the related object-DRF may be connected in such a way that the substantialization function remains constant in time. However, it seems that in the mental act of conceiving the substance as a whole, as a closed entity, a transformation in aspect is carried out which is responsible for different degrees of acceptability with respect to predications. "**This** water is warm." seems to be acceptable whereas "**This** quantity of water is warm" sounds quite strange (since it is not the quantity that is warm but the substance it consists of).

Two final remarks:

Link's portions of matter in the object sense can be made available in this system. In the case of gold, for example, they are **these** elements of *subst*:GOLD $\cap$ OBJECTS which are nothing else than objects, i.e. which are not elements of unions and/or intersections of primitive subsorts of OBJECTS.

The discussion above has shown that the correct modelling of a cognitively adequate ontology would require a temporal indexing of the substantialization function. In addition, the property of being an element of a sort should be indexed too. For instance grades can change: a former *major* can become a *general*. What is even worse, the relation between sorts can be changed (if such relations are used as a means to describe a world stage). A train, for instance, can lose the property of being a means of transport (in the strict sense) without losing the property of being a train (a train in a museum say), i.e. the relation TRAINS $\leq$ TRANSPORTS is no longer valid. In order to avoid notational complications (and, within the LILOG-framework, a decrease of computational efficiency), we do without models which are "tensed" in this sense. However it should be emphasized that, without loss of correctness, this is only possible if we envisage *stages* of objects in the sense of (Carlson) - which is for the moment not considered in the LILOG-project - or if the eventualities allowed do not change the sortal properties of the individuals of the thematic roles. In this sense, a model can be regarded as such a time slice of a world or of a partial world where the eventualities contained do not change sortal properties. We think that with respect to the LILOG scenario this is justifiable, at least if the behavior of the eventualities allowed serves as a criterion for making a predicate a sort or not.

6 The Sort EVENTUALITIES

6.1 Individualization of Events and Corresponding Portions of "Activity"

It is often argued that with respect to the domain of eventualities there is a similar distinction as in the case of material individuals, i.e. the mass/count-distinction. While it holds that (at least some) material parts of a quantity of gold are gold and that material parts of a ring are not rings, it is equally true that "material parts" of a running are "runnings" and that material parts of a crossing are not crossings. Above all, criteria of this kind have led to the distinction of Aktionsarten: Accomplishments and achievements do not have the "mass-quality" of "downward-heredity" (Shoham). For achievements this is excluded already by the assumption that they are unstructured "points in time". States do have this quality and activities have it to a certain extent. They have it down to a certain lower bound of perceptibility of the quality in question. In this sense, "gold" and "running" behave similarly: there are such boundaries.

However, it seems to be the case that, within an approach which uses reified eventualities, this classification is a classification of **predicates** rather than a classification of individuals, stating, in the case of states for instance, that the extension of a predicate qualified in this way is closed with respect to a corresponding material part relation. Thus, it could be possible that there are eventualities for which a mass- and a count-predicate hold, an activity and an accomplishment predicate say, for instance *running(e)* and *crossing(e)*. This is similar to the MATERIAL case. However, there, in addition, we have argued for an ontological differentiation which is dependent on such predications. It seems that, in the case of eventualities, there is a similar ontological differentiation besides this classification into states, activities, accomplishments and achievements which aims at the properties of predicates. This differentiation is related to the notion of the *nucleus* (Moens/Steedman). In Moens and Steedman's approach, the assumption is that there are different aspects of the same denoted "physical" eventuality which in discourse can be highlighted and which the nucleus provides. The event of *crossing the street* for instance can be seen as a whole, as an accomplished event, as an accomplishment in the ontological sense. However, in addition, we can stress the activity it consists of, for instance the running which leads to the completion of the event, which in this sense is the *preparatory process* of the event. Furthermore, there is the *culmination* of the event, the arriving at the other side of the street. By means of the implicitly given boundaries of the event, we can see the event from the inside and see it as an *event in progress*, i.e. we can stress the *progressive state* of the event, or, by means of the culmination, we can emphasize the consequences of the event, the *perfective* or *consequent state*, for instance the state of being at the other side of the street. Of course, the distinction between these aspects is closely related to the distinction between the properties of the predicates by which the aspects can be described. It may well be the case that it is this relatedness which has sometimes led to the notational confusion we want to disentangle.

Here, we retain the Vendler classification for the different aspects of the nucleus. As well as this, we use a notion defined by (Krifka), namely *quantized* for "count"-predicates and non-*quantized* for "mass"-predicates, where being quantized or having quantized reference means with respect to a predicate P classified in this way, that it is not *downward-hereditary*. In other words, for each eventuality e with $P(e)$ there are no proper subparts e' of e with $P(e')$. Being non-*quantized* means being downward-hereditary (at least to a certain limit). Using the criterion of heredity, (Shoham) has defined a more fine-grained classification of predicates which we can use to rename the notions "state", "activity", "accomplishment" and "achievement" if they are used in the sense of the predicate-classification.

Now, what about reification of the different aspects? With respect to the progressive state of an eventuality, it seems that there would be no need for reification if it were not for the so-called *imperfective paradox*. We cannot deduce the existence of a complete event, for instance *John crossed the street*, from the existence of the corresponding progressive state *John was crossing the street* (it may be that John had an accident before arriving at the other side of the street). Without this problem we could do without the reified progressive state, since its semantical impact would consist only in asserting that the corresponding event exists and that it is in progress at the evaluation time. Thus, constructing the temporal structure of an NL-text, it would suffice to localize this event correctly with respect to the evaluation time. There are several approaches to deal with this problem. Without going into detail, it seems that the suggestions for relating the truth conditions of a progressive state of an accomplishment to the truth conditions of the corresponding event or to the truth conditions of the preparatory process of this event are, in the first case, not practicable and, in the second case, deficient. (A perceived process of *going in the direction of the other side of the street* might not suffice for truly uttering *he is crossing the street*.) At least in cases such as the example above where there is a reading of an intentionally acting agent, it seems that we cannot do without making reference to the belief state of this agent when we want to represent correctly the semantic contribution of a truly asserted progressive state. However, such belief states are not available in our first order representation-language. For this reason, since we do not see suited decompositional solutions within our first order framework, we allow discourse referents for progressive states which are identified as such by means of the value of an aspect feature *asp*. [7] (e is the DRF for a progressive state iff *asp(e)=prog*.)

A similar (since more technically motivated) solution is suggested for perfective states. In general, it is very difficult to decide what consequences of an accomplishment are in focus when the corresponding perfective state is emphasized (viz. *He has just eaten something*), since this appears to be highly dependent on the context (*He is not hungry* vs. *He is recovering* in the context of a disease). Thus in the absence of a highly intelligent inference component, it seems preferable to have discourse referents for perfective states available which are related to the corresponding events and for which no additional predication is

---

[7] For the problem of the imperfective paradox and a solution on the basis of belief states compare also (Eberle88).

asserted. These DRFs are identified as such by means of the aspect feature $(asp(e)=perf)$.

Culminations can be made available as subevents of the corresponding accomplishments.

With regard to the aspectual difference between the event and the activity it consists of, it seems to me that, at the level of discourse representation, the argument for reifying these different aspects of the same denoted "physical" entity, when valid for material individuals, is equally valid for eventualities. It may be that the difference is felt less strongly than in the case of material individuals. The reason for this may be due to the fact that in the material case, there is a certain liberality in time. The gold of the ring existed before the ring came into being and it may exist for longer than the "lifespan" of the ring. On the other hand, the idea of the ring, the pure form, the planned ring say, existed without necessarily being destined to be materialized later in this portion of gold. It is, in a way, accidental that the both entities coincided. Another possible world is conceivable where this relation never holds. In contrast, in the case of eventualities, the two aspects are not independent in such a way. The same activity cannot form different events at different times. Thus, it is not possible to tell stories about the event and the activity which are similarily independent as in the case of material individuals (remember the stories about the gold and the ring in the last section).

However, as mentioned in the previous section, the relevant distinction, which relies on seeing something as a whole and as a closed entity, or, in contrast to that, as pure substance regardless of any criterion of form or separation, without highlighting the boundaries, is strong enough to legitimate the ontological differentiation even in the case of eventualities. Of course the distinction can be more or less explicit in the case of material individuals (gold/ring vs. gold/quantity) as in the case of eventualities: It is easier to distinguish between the crossing and the activity of running it consists of, than between "the run" and the corresponding activity of running. (Note that running in this case is not meant as one of the subevents of crossing; in this example we assume a homogenous crossing consisting only of running.) It has to be stressed that the distinction does not rely on the fact that we can say *"that was a slow crossing but a fast run"* - this is the argument for classifying adjectives as predicate modifiers. The distinction relies on the fact that we mean different "things" when we say *"he enjoyed the running but not the run"*. What the runner enjoyed was the pure activity, the being in motion in each phase of the event, whereas the run as a whole compared to other runs may have been unpleasant.

In other words, what we do is subdivide the domain of eventualities in a similar way as in the case of material individuals:

E-MASS := EVENTUALITIES $\cap$ MASS

E-COUNT := EVENTUALITIES $\cap$ COUNT

with the obvious localization of events in E-COUNT and the corresponding reified "portion of matter"-aspects, including states, in E-MASS.

We require the same structural properties for E-MASS and E-COUNT (and their subsorts) as in the case of M-MASS and M-COUNT respectively. The corresponding lattice operations are $\sqcup_{e_m}$ and $\sqcup_e$ which give rise to the partial relations $\leq_{e_m}$ and $\leq_e$. We extend the substantialization function *subst* to EVENTUALITIES, which behaves on EVENTU-ALITIES in a similar way as on MATERIAL. This allows for the straightforward extension of $\leq_m$, the "material part" relation onto the domain of EVENTUALITIES.

It has to be stressed that it is not the behaviour of the attributed sort alone which decides about the aspect, event or corresponding activity: the sort RUN for instance has non-empty intersections with MASS and with COUNT: M-RUN and C-RUN. Of course, M-RUN is non-quantized. But it is possible to imagine also C-RUN events which consist of material subparts which can be regarded as C-RUN events. On the other hand, we could say that each running has at least an *agent*, a local *source* where this running begins, a local *path* where the running takes place and a local *goal* where the running ends. Such thematic roles are features, as we have said. And it wouldn't be cognitively adequate to say that they are only partially defined on RUN, for instance only on C-RUN. Compare the similar SUBSTANCE case, where, of course we would say that for each portion of matter the feature *weight* is defined as well as for the corresponding objects. This means that for each running, be it element of C-RUN or M-RUN, there are values of such features. Perhaps we do not know them, because they are not provided by the (local) context in which the eventuality in question is introduced or we are not interested in knowing them, but there are such values. If we would accept the definition, that an eventuality e is an event if its parts do not have the same property as e, then all running eventualities (and not only running eventualities) would be events. Summarizing we can say that, within this approach of partially ordered sorts, eventualities can be in the extension of different sorts with different inheritance properties. In other words, these sorts cannot decide the character of the eventuality. It is decided by the inheritance property which *that* predication which introduces the eventuality on sentence level normally has. Deduced sort memberships have no influence on this determination.

C-RUN should be regarded as quantized in the normal case. There is a striking similarity between C-RUN events and objects like "twigs" however. In both cases we can sometimes delimit a repetition of the form in proper parts of the entities. Conversely, there is, at least for certain cases, *cumulative reference*: conjoint runs of the same agent can be regarded as one event of running, one "run". This means that we have to deal with the problem of "anomalous sets" not only in the OBJECTS-case but also within the E-COUNT domain.

A predicate P has *cumulative reference* in the sense of (Krifka) if and only if the *fusion* of some individuals has the property P if these individuals have the property P, where *fusion* stands for a lattice operation with a similar intuitive meaning to our $\sqcup_{e_m}$. In contrast to $\sqcup_{e_m}$, the Krifka-fusion is defined for all kinds of eventualities. However, making a fusion operation defined in this way available for the atomic level of the E-COUNT domain could be a sort of hyper-modelling: If we think of highly separated events like

Yesterday John drank a beer in Canada. (e_1)
Last week Mary conquered Mount Everest. (e_2)

it is difficult to group them together in **one** event (note that this does not prevent the forming of a "set" e_3, element of the non-atomic level of E-COUNT, consisting exactly of e_1 and e_2).

Perhaps, even in the case of E-MASS, one should be more restrictive. Do we really conceptualize the fusion of the activity of drinking by John and the activity of climbing by Mary which the corresponding events e_1 and e_2 respectively consist of, as an activity of some more general sort? Still, there are many more examples in the MASS-case than in the COUNT-case where this is plausible: Think for instance of the "portions" of activities of running in two spatio-temporally separated "runs" by the same agent, which form a sub-semilattice with respect to fusion. It seems that at least individuals of this kind can be conceptualized as a portion of activity (e.g.: *In both runs he enjoyed the running from a to b.*)

Of course, we could do without the fusion operation on E-MASS, having available only the partial order $\leq_{e_m}$ and the corresponding preorder on E-COUNT via *subst*. However, the notion of "substantialization" - as in the case of MATERIAL - means in the assignment of "unformed" entities - pure substances and pure activities respectively - a designation of more "realistic" entities (in the sense of naive physics). It seems that we are aware of precisely this difference between mental objects being closer to the imagined actual world and others being more dependent of cultural forming. It may well be this intuitive difference which we use to get the aspectual difference. In this sense we can regard MASS as a kind of imagined spatio-temporal denotation of the entities of COUNT and in this sense the domain of E-MASS is closed with respect to the fusion operation. (Note that the pure temporal denotation of such eventualities is not necessarily an interval. We will come back to that point in the next section.) Of course, in the general case, there is no lexical primitive which subsumes the corresponding activity of events. Running is in this sense an exception. Conquering a mountain consists of climbings and other activities whose fusion is not necessarily an element of a lexicalized predicate.

There is another problem of individualization which, in contrast to the case above, depends on the specific formalism of the representation. Eventualities which are symmetric with respect to the partaking agents allow for different representations which cannot be made identical:

John exchanged his orange for Mary's apple.
The deal took place yesterday in the market place.

Clearly by means of the definite description the exchange is conceptualized as one event. We agree with (Kowalski/Sergot) who deal with this example in assuming that "an act of exchanging has two actors and two objects." In our language with unary predicates or sorts these individuals must be values of corresponding thematic roles or features; call them *actor, coactor, object, coobject* (as (Kowalski/Sergot) do). It is natural to say then that *actor* will be instantiated by *John* and *object* by *orange*, thus taking into account the topic of the sentence. Assume a genuinely symmetric eventuality in the "real" world of which the example is a description. Depending on a suitable context, the description *Mary exchanged her apple for John's orange* is then perfectly acceptable. However, in this case, we get a second representation which doesn't allow for the identification of its discourse referent for the eventuality with the one of the first representation. Note that in the case of n-ary predicates this problem does not arise, since we could require that:

$$\forall e \in \text{E-COUNT}^*, x, x' \in \text{PERSONS}, y, y' \in \text{OBJECTS:}$$
$$(exchange(e, x, x', y, y') \leftrightarrow exchange(e, x', x, y', y))$$

In other words, we require that the values of the actor and coactor slot are replaced by each other iff the same is the case for the object and coobject slot, thus making use of the relational, not (necessarily) functional behaviour of the slots.

In the literature the discrimination of different events is sometimes argued for in such cases (cf. (Bäuerle)), since there seems to be no identical behaviour with respect to causes and effects. (Often this is used as a criterion of identification: if some events have the same causes and effects they are identical.) With respect to the example above there is a change of meaning if we say:

John exchanged his orange for Mary's apple (e_1), since he wanted to have an apple. (e_3)
or:
Mary exchanged her apple for John's orange (e_2), since he wanted to have an apple. (e_3)

However, it seems that such a phenomenon is not a sufficient reason for the reification of two events. It is not adequate to say that e_3 is the cause for e_1 but not (necessarily) for e_2 (i.e.: *cause(e_3, e_1)*, $\neg$ *cause(e_3, e_2)*) thus distinguishing e_1 and e_2. Such a treatment of causality would be similarly superficial as the treatment of (some) adjectives as predicates instead of predicate modifiers. e_3 is neither the cause for e_1 nor for e_2. It describes the motivation for John to partake in such an event of exchange. The motivation for Mary to partake in the **same** event thus can be completely different. Nevertheless, there might be other reasons for such a refined discrimination of events, getting suited representations in such cases without a great amount of deeply structured semantic analysis for instance. Link, for example in his *Algebraic Semantics* paper takes over such a fine-grained structuring of the atomic level of events. In his approach, even if more general predicates are used, he assumes different events, viz. an event of *kissing* **is not** a *touching* event but **presupposes** an event of *touching*.

Such a micro granulation is not possible in our approach, as long as we want to structurize and use the sort hierachy efficiently (each kissing then would be a touching). However, we are free to keep subsorts of the same sort disjoint, for instance SELL ($\leq$ EXCHANGE) and BUY ($\leq$ EXCHANGE) - where in both cases the *differentia specifica* is the restriction of the value of the *coobject*-feature to individuals out of MONEY. Also, we are free to introduce - in the kissing-example - *another* individual in TOUCH which is no kissing. The then necessary relation of material equivalence is provided by the *material part* relation, where $e =_m e'$ iff $e \leq_m e'$ and $e' \leq_m e$. Note that it is only in the case of MASS-entities that $=_m$ corresponds to $=$, since e_m is only a preorder on E-COUNT. Thus in the example above, we get two events of exchange which are materially equivalent. The pure $\leq_m$ on E-COUNT* is used to mark the relation of subevent.[8]

---

[8]Note however that we get into problems with regard to corresponding progressive and perfective states, where we have to postulate identity for the differently topicalized DRFs, since they are introduced as elements of E-MASS. We see this problem and for the time being can solve it only approximately in a rather technical manner: by decomposing the eventualities in question or by assuming *maximal phases* of these states which can be conceptualized as events and thus can be localized in E-COUNT. Then the initially

6.2 Countability

It is clear by now that the intuitive meaning of $\sqcup_{e_m}$ and $\sqcup_e$ is the same as in the case of $\sqcup_s$ and $\sqcup_o$, i.e material fusion and set union respectively. Thus we extend $|\cdot|$ onto E-COUNT.

Yesterday he worked three times.
He hated her exactly two times in his life. The first time, when she took his money and the second time when she came back with the money.

Clearly in both cases there is quantification over eventualities which are described by non-quantized predicates; in the terminology of Vendler by an activity predicate and by a lexical state predicate respectively. Of course the resulting predication is quantized: $\lambda e(e \in WORK \wedge agent(e) = c \wedge |e| = 3 \wedge e \subseteq yesterday)$ in the first case. But this presupposes that the atomic eventualities which are counted are elements of E-COUNT; they have to be regarded as events and they are described by non-quantized predicates. This shows once more that the behaviour of a basic predicate alone is not sufficient for the localisation of eventualities described in such a way in E-MASS or E-COUNT. Why do the eventualities described in this way behave like events? By means of the downward-heredity of such predicates there are parts for correspondingly described eventualities which have the same property. That means: if we have n such entities we have in general even more than n entities. At least in the second example above, this is not the intended reading. So, when counting takes place the notion of a form must be present, an implicit predication which prevents the downward- (and upward-!) heredity. It seems to me that in such cases of quantification this implicit predication is the *maximality* of a such described eventuality with respect to the (convex) temporal extension. In the previous example, we are counting two maximal, separated phases of hating and three maximal, separated phases of working respectively. This implicit stressing of the boundaries makes the eventualities in question events. In short, a similar argument as in the object/portion of matter case of counting is valid in this case. Of course, similar to the case of twigs or portions of matter, it depends on the implicitly contextually given focus, what kind of portions are given a form such that they can be regarded - or better: such that the corresponding aspects can be reified - as singular objects or events respectively.

Having reached the end of this section, we can now summarize the structuring of EVEN-TUALITIES as follows:

- EVENTUALITIES = E-MASS $\bigcup$ E-COUNT

- complete, strict complementary semilattice E-MASS:

  - lattice operation $\sqcup_{e_m}$
  - corresponding partial relation $\leq_{e_m}$
  - all ST $\leq$ E-MASS are complete sub-semilattices (not necessarily complementary, where ST element of the $\bigcap$-closure of **S** ($:=$ int**S**) with ST $>$ FALSE)

---

introduced states are material parts of these events and must not be described further. Compare the nex section on such event-aspects of states and activities.

- complete, strict complementary semi lattice E-COUNT:

  - lattice operation $\sqcup_e$
  - corresponding partial relation $\leq_e$
  - all ST $\leq$ E-COUNT are complete complementary sub-semilattices with $e \leq_e e'$ $\wedge e' \in$ ST $\rightarrow e \in$ ST for all $e, e' \in$ E-COUNT, ST $\leq$ E-COUNT such that ST $\in$ intS (ST $\neq$ FALSE)
  - cardinality function $|\cdot|$ (straightforward extension to a (functional) subrelation of (M-COUNT $\bigcup$ E-COUNT) $\times$ INT)

- substantialization homomorphism *subst* (straightforward extension to a (functional) subrelation of (MATERIAL $\times$ M-MASS $\bigcup$ EVENTUALITIES $\times$ E-MASS)

- material part relation $\leq_m$ (straightforward extension to a subrelation of MATERIAL $\times$ MATERIAL $\bigcup$ EVENTUALITIES $\times$ EVENTUALITIES)

- the atoms of E-COUNT: E-COUNT* := E-COUNT $\bigcap |\cdot| = 1$ (= extension of the predicate *e-atom*) structured by means of $\leq_m |_{E-COUNT* \times E-COUNT*}$

7 The Sort TEMPORAL

TEMPORAL serves to anchor temporal subsorts like YEARS, MONTHS, DAYS, etc. In addition, it provides the values for the temporal trace *temp*. *temp* is a homomorphism defined for EVENTUALITIES such that:

temp: INTERVALS = E-COUNT

temp: PERIOD = E-MASS

TEMPORAL is structured as MATERIAL and EVENTUALITIES, i.e.:

- TEMPORAL = T-MASS $\bigcup$ T-COUNT

- INTERVALS := T-COUNT, PERIOD := T-MASS

- PERIOD: complete, strict complementary semilattice:

  - lattice operation $\sqcup_p$ ("fusion")
  - corresponding partial relation $\leq_p$

- INTERVALS: complete, strict complementary semilattice:

  - lattice operation $\sqcup_i$ ("set union")
  - corresponding partial relation $\leq_i$

- all ST $\leq$ INTERVALS are complete complementary sub-semilattices with $i \leq_i i'$ $\wedge i' \in$ ST $\to i \in$ ST for all $i, i' \in$ INTERVALS, ST $\leq$ INTERVALS such that ST $\in$ intS (ST $\neq$ FALSE).
- cardinality function $|\cdot|$ (straightforward extension to a (functional) subrelation of (M-COUNT $\bigcup$ E-COUNT $\bigcup$ INTERVALS) $\times$ INT

- substantialization homomorphism *subst* (straightforward extension to a (functional) subrelation of (MATERIAL $\times$ M-MASS $\bigcup$ EVENTUALITIES $\times$ E-MASS $\bigcup$ TEMPORAL $\times$ PERIOD)

- material part relation $\leq_m$ (straightforward extension to a subrelation of MATERIAL $\times$ MATERIAL $\bigcup$ EVENTUALITIES $\times$ EVENTUALITIES $\bigcup$ TEMPORAL $\times$ TEMPORAL which is a partial order on MASS and a preorder on COUNT)

- the atoms of INTERVALS: INTERVALS* := INTERVALS $\bigcap |\cdot| = 1$ (= extension of the predicate *i-atom*) structured by means of the preorder $\leq_m |_{INTERVALS* \times INTERVALS*}$

The subdivision of MATERIAL and EVENTUALITIES is psychologically motivated whereas the similar structuring of TEMPORAL is not: we can regard periods as unions of intervals of numbers (where numbers are provided by ABSTRACT). Thus the conception of *subst* should be suitably organized in order to provide a means of calculating the duration of tensed individuals such as intervals and eventualities.

Making plural entities available for intervals is motivated by examples such as in section **3**, where generalized quantifiers are used with regard to intervals. Since the temporal trace is a homomorphism, in addition, such plural entities serve as values of *temp* for "sets" of eventualities.

Up to this point the temporal order is still lacking. There are several different possibilities open to remedy this shortcoming. With regard to the atoms of the tensed sorts, we can demand that they be ordered with respect to the (liberal) temporal relations $\prec$ (precedes) and $\oslash$ (overlaps) of the so called *"Event Structures"* (Kamp79) or with respect to the 13 temporal relations of the so called *"Interval Structures"* (Allen83), which subclassify $\prec$ and $\oslash$ respectively.

Definition: Event Structure Let E be a $\prec, \oslash$-structure.
$$E \text{ is an } \textit{event structure} \text{ iff } E \models \Phi_{ev}$$
where Φ_{ev} is the conjunction of the following 7 axioms:

E_1	$\forall e \forall f :$	$(e \prec f \to \neg f \prec e)$
E_2	$\forall e \forall f \forall g :$	$(e \prec f \wedge f \prec g \to e \prec g)$
E_3	$\forall e :$	$(e \oslash e)$
E_4	$\forall e \forall f :$	$(e \oslash f \to f \oslash e)$
E_5	$\forall e \forall f :$	$(e \prec f \to \neg e \oslash f)$
E_6	$\forall e \forall f \forall g \forall h :$	$(e \prec f \wedge f \oslash g \wedge g \prec h \to e \prec h)$
E_7	$\forall e \forall f :$	$(e \prec f \vee e \oslash f \vee f \prec e)$

We require:

$$< EVENTUALITIES*, \prec, \oslash > \models \Phi_{ev}$$
$$< INTERVALS*, \prec, \oslash > \models \Phi_{ev}$$

If necessary we can strengthen the structuring of the temporal domain, requiring that EVENTUALITIES* and INTERVALS* be *Interval Structures* in the sense of (Allen83), meaning that they have to be linearly ordered with respect to the more precisely subclassifying 13 relations.

These "refined" relations can be defined by means of $\prec$ and $\oslash$.

We omit such definitions and depict the corresponding relation of subsumption in an informal way only:

$$\prec \left\{ \begin{array}{l} meets \\ before \end{array} \right. \qquad \oslash \left\{ \begin{array}{l} identical \\ starts \\ during \\ finishes \\ overlap \end{array} \right\} \text{temporal inclusion: } \subseteq$$

The inverse relations (*before-inverse* etc.) are straightforwardly defined from the basic relations.

By means of *subst* we get a homomorphic ordering on a subset of PERIOD and E-MASS respectively. We now claim that *temp* is a structure preserving mapping with respect to the temporal relations, and require for *temp* and *subst* that:

$$\forall e \in EVENTUALITIES : \quad temp(subst(e)) = subst(temp(e))$$

Notice that PERIOD and E-MASS are, in the normal case, not event- or interval-structures. We are free to define period relations for these domains such as *synchro-overlaps*, *alternation* or *multiple* in the style of (Kandrashina).

Since there is no relation between the temporal and the semilattice-order yet, we claim that for singular intervals i and eventualities e the following holds:

$$i_1 \leq_m i_2 \iff i_1 \subseteq i_2$$

$$e_1 \leq_m e_2 \implies e_1 \subseteq e_2$$

$$i_1 \circ_m i_2 \iff \exists i_3(i_3 \leq_m i_1, i_2) \iff i_1 \oslash i_2$$

$$e_1 \circ_m e_2 \iff \exists e_3(e_3 \leq_m e_1, e_2) \implies e_1 \oslash e_2$$

Note that in the case of atomic eventualities, there holds no equivalence between semilattice and temporal inclusion and overlap respectively, but only implication. This is because eventualities may overlap in time without overlapping in the sense of having a common subevent or subactivity.

On the basis of the sketched axioms, we can deduce the intuitively correct *convexity* of intervals:

$$\forall i, i', i'' \in \text{INTERVALS*}: \quad (i', i'' \leq_m i \wedge i' \prec i'' \rightarrow (\forall \bar{i} \in \text{INTERVALS*} :$$
$$(i'\{overlaps, meets, before\}\bar{i}\{overlaps, meets, before\}i'' \rightarrow \bar{i} \leq_m i)))$$

We proceed modelling the intuitive conception of intervals by claiming "completeness" of the level of singular intervals, i.e. we require the existence of a semilattice operation $\oplus$ defined for the domain of singular intervals, which makes INTERVALS* a complementary $\oplus$-semilattice where the corresponding partial order $\leq^+$ equals $\leq_m |_{INTERVALS* \times INTERVALS*}$.

Notice that this semilattice is not strictly complementary:

Example:

$$i = i_1 \oplus i_2 = i_1 \oplus i_3$$

The atomic level of INTERVALS is isomorphic to the subset in PERIOD of convex periods with respect to the corresponding temporal and semilattice relations respectively.

Also, each period in PERIOD can be understood as the equivalence class of those sets of intervals, which are identical in the sense of pure temporality (modulo cardinality). The following "sets of intervals" are equivalent:

In contrast to the case of TEMPORAL, we do not require a similar lattice operation on E-COUNT*, as mentioned in the last section, since there is no need to conceptualize the material fusion of two events of E-COUNT* as an atomic event.

We can depict the structural situation as follows:

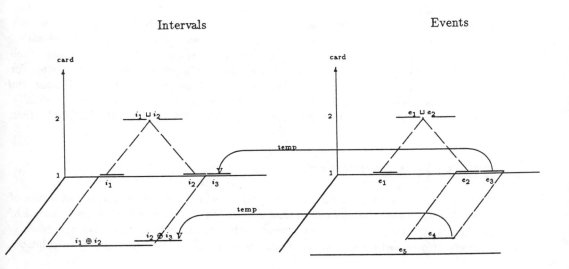

In each case we get "sets" of atomic tensed entities and it may be that there are cases where two events are grouped together and conceptualized as one atomic event (viz. e_2 and e_3), but in general this is not the case (although it is the case for INTERVALS). This does not exclude the possibility that events are regarded as subevents of a "bigger" episode (viz. the episode e_5).

8 The Sort LOCAL

LOCAL is structured similarly to the other main sorts. It is subdivided into L-MASS and L-COUNT, with corresponding lattice operations and relations. As in case of TEMPORAL, there is no psychological motivation for L-MASS. We can regard L-MASS as consisting of sets of "space-points" or sets of elements of a three-dimensional number space. Thus as in the case of PERIOD, the substantialization function *subst* should provide a means to compute local extensions of L-COUNT objects. Of course the domain of *subst* therefore has to be extended accordingly to LOCAL (such that the property of being a homomorphism is preserved). We can regard the reachability of L-MASS by means of *subst* also as an interface to a depictional component. LOCAL provides the values of the local trace, i.e. of the homomorphism *loc* defined on EVENTUALITIES. (Normally one does not assign a local trace to (lexical) states. If we want to keep *loc* totally defined on EVENTUALITIES we can assign the total set of space points or the maximal element of L-COUNT respectively to such states.) In addition, LOCAL provides the values of other local features such as *source, goal, path* which return with respect to move-actions the place of the start, the place of the end and the path over which the eventuality takes place as a function of time. All these features are homomorphisms. Note that the value of *path* is not a pure place. For correct representations we need a notion of directionality here . Thus, we get the subsorts PLACES and PATHS in L-COUNT. Paths can be regarded as ordered "sets" of places. A more elaborate discussion - especially about LOCAL - can be found in (Eberle89), where, in addition to the corresponding semilattice structurings of LOCAL, the relations *sub-path* and *path-specification* are defined for PATHS. *sub-path* relates paths with subpaths, for instance the path described by "*from France to Portugal*" with the path described by "*from France to Spain*". *path-specification* relates paths with more specific paths, for example "*from France to Spain*" with "*from Paris to Malaga via Bordeaux, Burgos and Madrid*". I do not want to go into further detail. Of course, LOCAL - or more specifically, L-COUNT - serves to anchor local subsorts like CONTINENTS, COUNTRIES, CITIES, etc.

9 The Sort ABSTRACT

ABSTRACT subsumes NUMBERS and other things which are not central to the topics of this paper. Note however that we have a choice here. We can require that desired models of the sketched set of axioms be structures which subsume the real or the rational numbers, or we can possibly do with a special set of axioms which allows for rudimentary calculation.

10 Inferring Facts from Eventualities in Narrative Texts

10.1 Inferring the Aspect and the Internal Properties of Eventualities

Different aspects (whether an event is in progress or not for instance) can be indicated by morphological markings or other contextually given parameters. Recently (Moens/Steedman) suggested an *aspectual net* in which the possible transitions are described (e.g. from the process of *travelling* to the culminated process of *travelling to Spain*). Thus we acquire a means for the calculation of the aspect of the eventuality described by the sentence: We use the Aktionsart entry for the verb (in the sense of the Vendler-classification with respect to predications) and then process this value along the lines of an associated aspect calculus. The impact of such an aspect calculus for the formal representation of a text is twofold:

(a) its output enables us to get insights into the internal structure of a described eventuality and
(b) this output, the aspect, contributes to the correct reconstruction of the temporal structure of the text. In other words, it contributes to the determination of the temporal relations between the eventualities. This reconstruction of the temporal structure is often called *temporal resolution*. [9]

Example (a)
John played the sonata for three hours.

---

[9]For the temporal resolution on the basis of the aspect or on the basis of the classification into Aktion sarten respectively compare, for instance, (Hinrichs), (Partee), (Kamp/Rohrer), (Eberle/Kasper).

Playing as such is a non-quantized predication, i.e. the eventuality in question is in the extension of a downward-hereditary sort PLAY. (In the case of *-sorts, we normally assume that singular eventualities are introduced if there is no explicit quantification). However, *playing the sonata* makes reference to a subsort of C-PLAY* which is quantized. Now, it is well known that in contrast to the so-called *in-adverbials* the so-called *for-adverbials* accept only eventualities of the non-quantized type. We therefore need a reinterpretation. There are two possibilities. We can assume an iterative reading: the eventuality describes the fact that there are several events of playing the sonata, or we can assume that John played only parts of the sonata for three hours, i.e. "material portions" of the sonata. In both cases, we arrive at downward-hereditary predications (downward-hereditary to a certain limit): C-PLAY $\cap | \cdot | > 1 \cap$ *object=sonata* and M-PLAY $\cap$ *object*: M-SONATA respectively. (Of course a sonata is not "material" in the strict sense. For such cases we have either to extend the notion of MATERIAL or to provide suitable subsorts in ABSTRACT).

This may suffice to exemplify the one dimension of a suitable aspect calculus. The other can be illustrated by the following examples:

Examples (b)
Early in the morning Peter went to his office.

(i) *He wrote a letter. Then he called Jim.*
vs.
(ii) *He was writing a letter and that caused him trouble.*

These examples motivate a view often taken in the literature that events shift the reference time forward in time whereas states or activities do not. (i) is understood as describing a chain of three events whereas (ii) is understood as describing one event which is temporally overlapped by states. (In addition,these states motivate the action of going to the office.)

10.2 Temporal Resolution

For temporal resolution the mere computation of the aspect on the sentence level is deficient. One reason for this is because there are languages such as German which - in the absence of the morphological marking of progressivity (like in English) - do not provide a clear distinction between an event and the corresponding progressive state reading. Often, this distinction is only provided by background knowledge:

Hans fuhr nach Frankreich (e). An der Grenze stellte er fest, daß er seinen Ausweis vergessen hatte. Also kehrte er um.
(Hans travelled (was travelling) to France (e). At the border he discovered that he had forgotten his passport. So he turned back.)

We should leave the value of the aspect feature of the eventuality in question uninstantiated in such cases, and rely on getting it instantiated by means of an intelligent inference component (which deals with incompatibilities of eventualities) on the basis of the representation of the whole text.

Notice that this problem arises mainly when we have to deal with so-called *elaboration -*

as in the example above. *Elaboration* describes a relation between discourse segments. That an eventuality is elaborated means that it is described by subevents. There is a switch to a more fine-grained level of description. There are other such discourse relations such as *continuation, résumé, contrast* or *causality* which influence the temporal resolution. Only in the case of continuation - as in examples (a) and (b) - the so called *Temporal Discourse Interpretation Principle (TDIP)* (Dowty) is valid. TDIP, roughly speaking, requires that the reference time is shifted forward in time by events unless there are explicit indications (in the form of adverbials for instance) which require other localizations. States and activites have to be resolved with respect to the existing reference time. In order to detect the correct discourse relation, once again, an intelligent inference component is necessary which deals with knowledge about the normal grouping of event types or, in other words, which deals with scenic frames (cf. (Bartsch), (Schank/Abelson)). This means that we have to know that "normal" events of *"travelling from Germany to France by car"* contain subeventualities such as *being in Germany, driving, being at the border ,being in France*. If there is no state of *being in France* then the described eventuality of *"travelling to France"* cannot be a completed event. It has to be interpreted as a progressive state as in the example above (i.e. $asp(e)=prog$).

In the *contrast* case, the temporal ordering is frequently not specified. Furthermore, it is often not clear how many eventualities are described:

Last year John and Peter travelled to Denmark and Holland (e), Mary to Ireland and Italy, Mike to Poland and Jane to Portugal.

It is possible that e consists of one event of travelling to Denmark and Holland or it could be that the agents travelled separately; separately to Denmark and Holland or separately to Denmark and separately to Holland and so on. Here our lattice approach supplies us with a means for unspecified representations which subsume, when interpreted in models, all the different readings. Instead of $e \in$ C-TRAVEL* we introduce:

$$e \in \text{C-TRAVEL} \land agent(e) = john \sqcup_o peter \land goal(e) = denmark \sqcup_l holland$$

Note that this possibility depends on the fact that the features *agent* and *goal* are defined as homomorphisms!

10.3 Temporal Deductions on the Basis of the Temporal Structure of a NL-Text

Besides the well known monotonic and non-monotonic reasoning systems for temporal structures which can be made available for our approach (for instance the Allen-algorithm for the computation of the transitive closure with respect to interval relations (cf.(Allen83) or the event calculus of (Kowalski/Sergot)), we can infer sub- and superevents from the assertion of events by using the inheritance property of certain sorts and suitable axioms respectively.

Following one of (Krifka's) suggestions, we think that there are features which have the so-called *mapping to events*-property or the so-called *mapping to objects*-property with respect to certain sorts, such as for instance *path* which has both properties with respect to TRAVEL*. This means that for each singular travelling event e, the corresponding path p and a subpath p' of p there is a travelling event $e' \leq_m e$ over p' (at least down to a certain lower bound for the size of p'). On the other hand, for each subevent e'' of e which is a travelling by the same agent there is a (more specific) subpath p'' of p which is the path of e''. By means of such properties, we obtain a certain "multi-dimensionality" of the constitution of subevents. (Each feature which has such a property determines the corresponding dimension.)

A slightly generalized version of the mapping-property with respect to *path* can be rendered as follows:

$$\forall e \in \text{C-MOVE*}: \forall w, w' \in \text{PATHS*} : \quad (path(e) = w \wedge w' \leq^+ w$$
$$\rightarrow \exists e' \in \text{C-MOVE*}: \quad (e' \leq_m e \wedge path(e') = w'))$$

$$\forall e, e' \in \text{C-MOVE*}: \forall w \in \text{PATHS*} : \quad (path(e) = w \wedge e' \leq_m e \wedge agent(e') = agent(e)$$
$$\rightarrow \exists w' \in \text{PATHS*} : \quad (w' \leq^{s+} w \wedge path(e') = w'))$$

To get singular superevents from the assertion of singular events we have to state suitable axioms for sorts with specific cumulative reference. For instance, travelling events can be amalgamated if they are temporally overlapping and if they have the same agent. The path of the superevent is then the least common upper bound with respect to the *more specific subpath*-relation which is defined as a combination of *subpath* and *path-specification*.

We can generalize such statements for suitable (lexicalized) sorts S_i, S_j, S_k with $S_i, S_j \leq S_k < \text{E-COUNT}$. Such a generalization could be rendered for instance as follows:

$$\forall e \in S_i * \forall e' \in S_j * \forall x \in TRUE : \quad (agent(e) = x = agent(e') \wedge e \oslash e'$$
$$\rightarrow \exists e'' \in S_k* : \quad (agent(e'') = x \wedge e \leq_m e'' \wedge e' \leq_m e''))$$

11 Conclusion

We have developed a rather symmetric structurization of the sort hierarchy for a NLU-system. We have tried to motivate this structuring, which is based on the global mass/count-distinction, psychologically. By means of the lattice structures, we have modelled "sets" in such a way that the quantification over "set"-objects is available within first order logic. In addition, the definition of lattice operations allows for suited unspecified representations. With respect to eventualities, the approach permits in particular:

- the sorting of eventualities according to a hierarchy of primitive sorts,

- the sorting of eventualities according to feature values,

- the discrimination of subevents according to different dimensions,

- the formalization of conditions constraining the construction of superevents from subevents.

Within this approach the inference systems of Allen and Kowalski/Sergot can be made available for representations of NL-texts if they are slightly modified.

The problems of this approach are that a suitable decompositional treatment of progressive and perfective states is lacking. However, on the one hand a lot of empirical work has still to be done to understand the semantic impact of such aspects with regard to the variety of cases. On the other hand, we need the hierarchical structuring of our representations in the DRT-style to render the beliefs of the agents.

Furthermore, in the long run, dynamic sort membership is required to represent temporal phenomena of NL-texts correctly.

References

[1] Allen, J.(1983): *Maintaining Knowledge about Temporal Intervals*. In: Comm.ACM 26, 1983, pp.832-843

[2] Allen, J.(1984): *Towards a General Theory of Action and Time*. In: Artificial Intelligence 23, 1984, pp.123-154

[3] Bach, E.(1986): *The Algebra of Events*. In: Lingustics and Philosophy 9, 1986, pp.5-16

[4] Bäuerle, R.(1987): *Ereignisse und Repräsentationen*. Habilitation. Universität Konstanz, 1987

[5] Bartsch, R.(1987): *Frame Representations and Discourse Representations*. ITLI Prepublication Series 87-02. University of Amsterdam

[6] van Benthem, J.(1983), The Logic of Time. Dordrecht : Reidel

[7] Carlson, G.(1980): *Reference to Kinds in English*. In: Hankamer, J. (ed.) Outstanding Dissertations in Linguistics. Harvard University

[8] Cresswell, M.J.: *Adverbial Modification*. Dordrecht: Reidel.

[9] Davidson, D.(1967): *The Logical Form of Action Sentences*. In: Davidson, D.: Essays on Actions and Events. Oxford: Clarendon Press

[10] Dowty, D.(1986): *The Effects of Aspectual Class on the Temporal Structure of Discourse: Semantics or Pragmatics?*. In: Linguistics and Philosophy 9, 1986, pp.37-61

[11] Eberle, K.(1988): *Partial Orderings and Aktionsarten in Discourse Representation Theory*. In: Proceedings of Coling, Budapest 1988

[12] Eberle, K.,Kasper, W.(1989): *Tenses as Anaphora*. In: Proceedings of E-ACL, Manchester 1989

[13] Eberle, K.(1989): *Quantifikation, Plural, Ereignisse und ihre Argumente in einer mehrsortigen Sprache der ersten Stufe*. LILOG-Report 67. IBM Deutschland, WT LILOG, Stuttgart

[14] Haugh, B.(1987): *Non-Standard Semantics for the Method of Temporal Arguments*. In: Proceedings of the Tenth International Joint Conference on Artificial Intelligence.

[15] Hinrichs, E.(1986): *Temporal Anaphora in Discourses of English*. In: Linguistics and Philosophy Vol.9,No.1 (1986) pp.63-82

[16] Kamp, H.(1971): *Formal Properties of Now*. In: Theoria 37, 1971, pp.227-273

[17] Kamp, H.(1979): *Events, Instants and Temporal Reference.* In: Bäuerle, R.,Egli, U.,von Stechow, A. (eds.) Semantics from Different Points of View, Berlin, Springer-Verlag

[18] Kamp, H.(1981): *A Theory of Truth and Semantic Representation.* In: Groenendijk, J.A.G., Janssen T.M.V., Stokhof, M.B.J. (eds.) Formal Methods in the Study of Language. Mathematical Centre Tract, Amsterdam

[19] Kamp, H.,Rohrer, C.(1985): *Temporal Reference in French.* Ms. Institut für Maschinelle Sprachverarbeitung, Universität Stuttgart.

[20] Kandrashina, E.Yu.(1983): *Representation of Temporal Knowledge.* In: Proceedings of the Eighth International Joint Conference on Artificial Intelligence, Karlsruhe.

[21] Kowalski,R.,Sergot,M.(1986): *A Logic-Based Calculus of Events.* In: New Generation Computing 4(1) (1986) pp.67-95

[22] Krifka, M.(1987): *Nominal Reference and Temporal Constitution: Towards a Semantics of Quantity.* FNS-Bericht 17. Forschungsstelle für natürlich-sprachliche Systeme, Universität Tübingen.

[23] Link, G.(1983): *The Logical Analysis of Plurals and Mass Terms: A Lattice-Theoretical Approach.* In: Bäuerle, R.,Schwarze, C.,von Stechow, A. (eds.), Meaning, Use and Interpretation of Language, Berlin, de Gruyter, pp.302-323

[24] Link, G.(1988): *Algebraic Semantics of Event Structures.* In: Groenendijk et al. (eds.) Proceedings of the Sixth Amsterdam Colloquium, ITLI, University of Amsterdam, pp.243-262

[25] Löbner, S.(1985): *Definites.* In: Journal of Semantics 4.4, 1985, pp.279-326

[26] McDermott, D.:(1982): *A Temporal Logic for Reasoning about Processes and Plans.* In: Cognitive Science 6 (2) 1982

[27] Moens,M.,Steedman,M.(1986): *The Temporal Dimension in Information Modelling and Natural Language Processing,* Acord Deliverable 2.5, Edinburgh,1986

[28] Partee,B.(1984): *Nominal and Temporal Anaphora,* in: Linguistics and Philosophy Vol.7, No.3 (1984) pp.243-287

[29] Schank, R.,Abelson, R.(1977): *Scripts, Plans, Goals and Understanding.* Hillsdale, N.J. LEA.

[30] Shoham, Y.: *Temporal Logics in AI: Semantical and Ontological Considerations.* In: Artificial Intelligence 33(1987), pp.89-104

[31] Smolka, G.(1988): *A Feature Logic with Subsorts.* LILOG-Report 33. IBM Deutschland, WT LILOG, Stuttgart

[32] Vendler,Z.(1967): *Linguistics in Philosophy,* Cornell University Press, Ithaca, New York

What KL-ONE Lookalikes Need to Cope with Natural Language

– Scope and Aspect of Plural Noun Phrases –

Jürgen Allgayer, Carola Reddig-Siekmann
FR 10.2 Dept. of Computer Science IV
University of Saarbrücken
Im Stadtwald 15
6600 Saarbrücken 11
West Germany

CSnet: {allgayer, reddig}%sbsvax.uucp@germany.csnet

Abstract

One of the major drawbacks of current NL processing systems is the lack of an adequate representation of plurals and of the means to reason about them. On one hand, this is due to the fact that current knowledge representation languages like KL-ONE lookalikes do not provide well-suited representational means either to describe sets, subsets, and elements or to deal with the respective relations or use them in specially tailored inference systems.

On the other hand, workers in linguistics provide some (although conflicting) theories about the referential, cardinal, and quantificational aspects of the meaning of plural noun phrases, namely Discourse Semantics Theory, Generalized Quantifier Theory, and Referential Net Theory.

Our goal in the present work is to extend KL-ONE lookalikes by set maintenance and to support the use of sets in an NL system with a well-suited linguistic theory.

The work presented here is being supported by the German Science Foundation (DFG) in its Special Collaborative Program on AI and Knowledge-Based Systems (SFB 314), project N1 (XTRA).

Chapter 1

Background

1.1 Thematical background: A Short Look at the XTRA System

The XTRA system is a natural language (NL) interface between a user and various expert systems (XPS) which uses different highly interacting knowledge sources for domain independent linguistic knowledge and for domain specific world knowledge. **SB-ONE**, a knowledge representation formalism which is a descendant of KL-ONE and especially tailored for natural language processing, deals with the declarative amount of these types of knowledge. The inferential aspects of knowledge that are not covered by the expressive power of **SB-ONE** – both the representation of it and its processing – is dealt with in $\mathcal{SBTWO}$, a logic-oriented embedding of **SB-ONE** (cf. Section 3).

Figure 1.1 shows the architecture of the XTRA[1] system. As can be seen, $\mathcal{SBTWO}$ and **SB-ONE** form the representational basis for the analysis and generation components as well as for the internal processing modules. Two knowledge bases are represented in the **SB-ONE** formalism: the so-called Functional Semantic Structure (FSS), and the Conceptual Knowledge Base (CKB). While the FSS contains a structured inheritance network of domain-independent, linguistic-oriented knowledge about how to combine instances of semantic classes of expressions in order to get semantically well-formed larger expressions, the CKB expresses the system's domain-specific knowledge about the world.[2]

Beyond this declarative knowledge about language and the world, $\mathcal{SBTWO}$ maintains those pieces of knowledge that express the procedural aspects of meaning.

1.2 Knowledge Representational Background: Expressive Power of SB-ONE

SB-ONE is a knowledge representation formalism (and also system) for the construction of conceptual knowledge bases to be used in NL processing systems. Compared

---

[1] For further details, see [Allgayer et al. 89].

[2] In addition, XTRA is provided with two discourse-dynamic knowledge bases: the linguistic dialog memory LDM and the belief, goal, and plan maintenance system BGP-MS. But these components are of minor interest here.

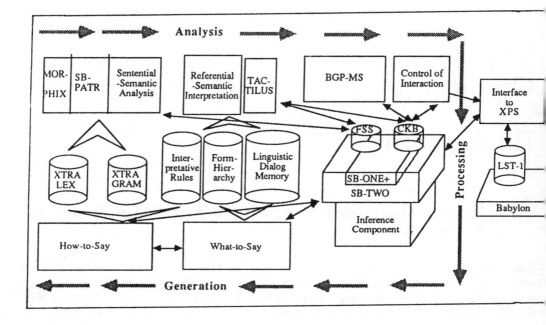

Figure 1.1: The architecture of the XTRA system

with other KR languages of the KL-ONE paradigm, it shares the following language structures with them:

1. **Primitive and defined concepts.** It is possible to define a concept by giving both necessary and sufficient descriptions for it. This can be compared to constructing a primitive concept by giving only necessary descriptions, thereby allowing for a construction of an individual of that class, but not for the decision of whether a given instance belongs to the primitive concept or not.

2. **Roles, number restriction, value restriction.** Roles can be used to define two-placed relations between concepts. Number restrictions can be added to their descriptions in order to specify how many extensions of a particular role are needed to describe a certain concept. Moreover, value restrictions, e.g. the specialization of the destination of such a two-placed predicate, can be used to define new concepts. The collection of all concept and role definitions (both defined and primitive) is usually referred to as the TBox part of such a system in which the terminology of the system is defined.

3. **Individualizations.** On the individualized level, usually referred to as the ABox in these systems, individualizations of concepts form the representational means to assert knowledge about actual situations.

The following Figure 1.2 shows the features common in KL-ONE lookalikes with the respective depictions usually used for them.

However, there are some features of **SB-ONE** not present in other systems:

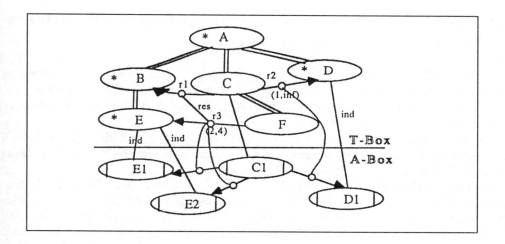

Figure 1.2: Depictions for common KL-ONE features

1. **Singleton concepts.** In addition to primitive and defined concepts, **SB-ONE** offers the possibility to define so-called singleton concepts. These are primitive or defined concepts with the additional restriction that there should be only one element in their extension.

2. **Modality.** With respect to roles, a modality can be defined in order to express whether a role has to be present for a given individualization to belong to a certain concept description or not (i.e., whether the role is **necessary** or **optional**). In other languages, this feature is covered by a number restriction ranging for example from 0 to 5, thereby expressing the possibility that there might be no role at all. The advantage of our solution is that it allows a number restriction of a role to range from 3 to 5 and nevertheless be only optional, i.e. there might be no role or, if any, then at least 3.

3. **Default NR, VR.** In addition to the features above, **SB-ONE** allows for the description of a concept by using default values for both, value restriction and number restriction. This information is used during the so-called 'default individualization'.

4. **Classifier, realizer.** The classifier – an algorithm to detect the most specific superconcept for a given concept description – takes all the new and extended language constructs into account, as does the realizer – the algorithm that determines for a given structure in the ABox the most specific general concept it is an indivdualization of.

1.3 Linguistic Background: Supporting NL Processing with Linguistic Theory

As [Shieber 87] pointed out, the goals of a theory in linguistics can be described as *completeness* and *restrictiveness* (meaning it they should characterize all and only the possible NLs), *simplicity* and *elegance*, and *declarativeness* (which means it should be order-independent or order-free).

The main goal of NL processing systems (and also of their underlying formalisms), however, is to characterize one language computationally, where the chosen formalisms should facilitate such a computational characterization.

Considering the state of the art in NL processing (see, for example, Discourse Representation Theory ([Kamp 81]), File Change Semantics ([Heim 82]), Situation Semantics ([Barwise & Perry 83]) or the Generalized Quantifier Theory ([Barwise & Cooper 81])), there is a clear tendency in NLP either to base the processing directly on a formalism taken from linguistics, or to define a well-founded underlying theory according to linguistic criteria, in order to get a system which behaves in a well-defined manner, and which is extendable for dealing with new phenomena.

Therefore, in most cases the NLP formalisms are based on first order predicate logic; but this leads to a mismatch between the properties of the natural language to be characterized and those of the knowledge representation language that should describe them.

> *Notions like quantifier, variable, sense and reference, intension and extension, ... are all technical (...) notions introduced by philosophers and logicians. They are not part of the data of natural language. It just might be that some or all of them cut across the grain of the phenomena in unnatural ways, generating artificial problems and constraining the space of possible solutions to the genuine puzzles that language presents.* ([Barwise & Perry 83, p.xi f.])

On the other hand, notions like these are very useful – and, as we believe, are indispensable – with respect to an internal formal knowledge representation language. Therefore, we look for an application of the above-mentioned notions that fits "the grain of the NL phenomena" in the best possible way.

And this, in fact, is the intention of the Generalized Quantifier Theory (GQT, [Barwise & Cooper 81]): to provide a notion of (formal) quantifiers that describes quantifiers as they occur in natural language.

If for instance B be the property of "being a Münchner (native of Munich)" and if C be the property of "drinking beer", then

Example 1.3-1
$$\forall x \ (\ B(x) \rightarrow C(x) \ )$$

may formally express the meaning of the NL sentence "All Münchner (natives of Munich) drink beer."

Looking at the constituent structure of that sentence, however, we don't find an isolated quantifier ALL which ranges over the whole sentence, but instead, there is a noun phrase (NP) "all Münchner" which in combination with the verb phrase (VP) "drink beer" forms the sentence. In Example 1.3-1, there isn't any formal correspondence to this very important notion of an NP, so one can easily imagine the problems one gets into trying to explain the very complex meaning of NPs in terms of structures like Example 1.3-1.

Even if we provide our internal representation language with a many- (or order) sorted logic, as in

Example 1.3-2

$$\forall x_{Muenchner} \; drink\text{-}beer(x_{Muenchner})$$

the problem still occurs because the Basic predicate [3] might become arbitrarily more complex than simply "Münchner"; consider e.g.

Example 1.3-3

"*Nearly every second Münchner who sits in a beer garden on a warm summer night ...* "

where nobody would suggest a sort like "Münchner who sits in a beer garden on a warm summer night".[4]

The basic idea of GQT is to regard the whole NP as the quantifier, as is shown in

Example 1.3-4

(DET Basic) Central

Example 1.3-5

(ALL Münchner) drink-beer

(where DET stand for a determiner like "all", "a", "the", "two or three", "the best", ...). Figure 1.3 (according to [Heinz & Matiasek 89]) shows the types of the constituents of Example 1.3-4.

The Basic and the Central predicate are of type P (for property); the whole quantifier (DET Basic) maps properties into truth values (type <P, t>); according to compositional semantics, determiners must be of type <P, <P, t>>. That means that they are functions from properties into sets of properties. In GQT, these determiners have been well investigated in order to discover their formal properties, and it is these very concepts we are interested in, because they are directly interpretable as inference rules for structural inheritance networks.

GQT provides an appropriate explanation for quantificational NPs and their scope, but on the other hand, determiners can be viewed as expressing a relation between a description (the NP) and the entity the speaker actually intends to refer to (see e.g. [Croft 85]). Thus, we follow [Loenning 87] in his distinction of three types of NPs

---

[3] The terms 'Basic' and 'Central' correspond to the logical terms 'restrictor' and 'scope'.

[4] As Christopher Habel pointed out during a workshop discussion, "sorts are a very precious good that we should handle with care." ...

Det	Basic	Central
< P, < P, t >>	P	
	< P, t >	P
	t	

Figure 1.3: Types of constituents

according to the respective DET: (1) indefinite noun phrases, (2) definite noun phrases, and (3) quantificational noun phrases (in the narrow sense).

Therefore, our linguistic background has to be extended by a theory concerned with the main task of indefinite NPs, namely to introduce new referents/objects into the discourse. Discourse semantics theories, especially DRT, File Change Semantics ([Heim 82]) and Situation Semantics ([Barwise & Perry 83]) fulfull this task in a manner which has been thoroughly investigated.

However, none of the above-mentioned theories deals with plural NPs in an over-all manner. In classical GQT, while considering solely individuals in the domain of application, sentences like

Example 1.3-6
"Three men lifted a piano."

cannot be interpreted *collectively* (in the sense of describing a situation where a single piano has been lifted by a team of three men together). [Heim 82] and [Kamp 81] leave aside plural NPs, too. But interesting linguistic approaches are to be found in [Link 87] who extends the GQT by introducing so-called **i-sums**. σ x Px denotes the i-sum (that is individual sum) of all individuals that are Ps.

[Habel 86] in his theory of referential nets introduced special descriptive operators (All-t,...) for plural referents which operate according to Russell's ι-operator ([Russell 05]). In contrast to GQT, but in parallel to Discourse Semantics, and, for instance, Loebner's GQT extensions ([Loebner 87]), descriptive operators build *terms* as the corresponding structures for noun phrases, which may have some additional properties not yet mentioned but which must be dealt with in NL processing systems.

Among these, the most important feature is known as Donnellan's distinction ([Donnellan 66, Kronfeld 86]): : NPs may either be referentially used (in terms of Situation Semantics: the created object is part of the situation described) or attributively described (the description conditions are part of the described situation). Hardly any of the current NL systems are concerned with this difference, which in one sense crosses the conventional extension / intension distinction.

Chapter 2

Problems

2.1 Problems with NL

The aim of this chapter is to show how we adopt (parts of) these theories in order to represent and reason upon the different aspects of the meaning of quantificational NPs.

2.1.1 The complex structure of DETs

Determiners (DETs) (or prenouns) play an important role in determining the meaning of a noun phrase and its relation to the sentence's assertion, whereas this meaning-relation of course changes from determiner to determiner.

According to [Reddig 88] we differentiate between 9 types of DET in German, which may form sequences of determiners like

Example 2.1-1
  "All die vielen tausend Zuschauer..."
  (in English:) *"All the many thousand visitors..."*

where the meaning of the DET sequence should compositionally depend on the single determiners in the sequence.

Furthermore, GQT counts as members of DET even more complex structures. Some of them indicate explicitly a *partitive* interpretation (which is the genuine quantificational reading) by including the word "of" ("three of the Basic"). Then the NP following "the", denotes the (definite) domain of quantification. Other DETs describe an interval with fixed borderlines ("between three and five") or a vague relative cardinality ("nearly half of all"). DET itself may be modified by an adverb ("nearly" in the last example) and may also contain conjunctors ("four or five").

In GQT, the complex structure of the determiner phrase leads to the distinction between basic ("every", cardinals, and possessive determiners) and composed DETs (all others), and we'll adopt this distinction to some extent in our analysis.

One of the most important results of GQT is the classification of DET according to their formal properties concerning the relation between the two predicates involved:

Example 2.1-2
    (DET Basic) Central.

Interesting special classes of determiners are those for which monotonicity ([Barwise & Cooper 81]) holds true:

Example 2.1-3
    upward: (DET Basic) Central, Central $\subseteq$ Central' $\Rightarrow$ (DET Basic) Central'
    downward: (DET Basic) Central, Central' $\subseteq$ Central $\Rightarrow$ (DET Basic) Central'

Structural inheritance networks express the relation $\subseteq$ with an ISA-link which makes it possible to map the constraint 'Central $\subseteq$ Central' onto the subsumption test. Determiner dependent inference rules (see Section 3.2.3) combine the subsumption relation and the knowledge about determiner classes so that only valid statements are infered. Membership in the class of monotonic descending determiners, for instance, (see Example 2.1-3a) implies the *specialization* of the Central predicate. This is in contrast with the 'standard' inference of generalization of the Central predicate in order to yield a new valid statement. As another example, the *stability* property of a determiner class – which holds e.g. for monotonic determiners ([van Benthem 83]) – implies that an instantiation of Example 2.1-2 will remain valid even as more members of the denotation of Central are descovered. This is especially important within systems based on the open world assumption.

However, as we have seen in Chapter 1 and as will be explained in the following sections, we shall go beyond the GQT approach in some other aspects.[1]

2.1.2   Different aspects of NP interpretation

NPs provide descriptions of objects to be talked about, and there has been a lot of work – in linguistics and philosophy as well as in NL-oriented AI – on the different aspects such NPs have. In most aspects, determiners are involved, and therefore these aspects influence the classification of determiners and determiner phrases as well as the demands on the representational and inferential capabilities of the knowledge representation language that deals internally with these expressions.

2.1.2.1 Definite and indefinite descriptions

One of the best-investigated problems is the functional distinction between definite and indefinite (singular) descriptions (see for example [Heim 82,Habel 86]). What is needed are representational means for these two distinct functions (like the ι- and the η-operators), and a procedural semantics for their interpretation.

Unfortunately, most of the investigations concerning the given/new problem only take singular noun phrases into account ([Heim 82,Kamp 81,Barwise & Perry 83]). [Habel 86] suggests an approach that introduces two more descriptive operators for collective expressions: the ALL-t operator as corresponding to ι, and the SOME-t as a plural equivalence to the function of η, but he cannot count with them in the sense of subset and element relations. So we shall have to extend the above mentioned suggestions

---

[1]A problem that remains unsolved is that of second order DETs, denoting sets of DET-denotations.

to an approach that deals with plurals as well, because even if we consider NP's as quantifiers, we have to decide whether the corresponding internal representation has to be found or created. Consider, for example,

Example 2.1-4
> *"Four children play in the sand."*

Example 2.1-5
> *"The two girls ... while the boys ... "*

The definite determiner "the" in Example 2.1-4 indicates that the respective NP describes – with uniqueness assumption – an already known object, be it a singular or, as in this case, a plural entity. Therefore, the same procedure for finding the intended referent (within the already mentioned, explicit objects, or within the wider context) has to be triggered in the singular as well as the plural case. For Example 2.1-5, the set that represents the set of four children of sentence 2.1-4 has to be divided into two subsets with cardinality 2 and an attribute describing the sex of their respective members.

The same, of course, holds true for the indefinite case: The indefinite article ("a" – in German "ein" – in the singular case, the null article in the plural case) indicates that a new structure representing the intended referent has to be introduced into the knowledge base, be it an individual or a collection of objects, regardless of its cardinality.

We shall not deal here with the problem of under which conditions a referent should be introduced at all (see, e.g., [Heim 82] for further details), but instead we argue that because there are cases in which the given/new distinction **is** the main point of the definite/indefinite problem, a system should provide means for

- explicitly marking the respective case, and

- drawing the intended inferences with respect to the maintenance of the knowledge base (which in fact is the procedural meaning of these determiner aspects)

independently of the cardinality properties of the entity described by the NP.

2.1.2.2 Referential and attributive descriptions

Another aspect that influences the meaning of NPs has been investigated both in AI and in linguistics ([Donnellan 66,Appelt 87,Kronfeld 86,Barwise & Perry 83]) and is known as the 'referential vs. attributive' use of NPs. To see the difference, consider the following little joke:

Example 2.1-6
> (1) *"Why are these guys running so fast?"*
> (2) *"The winner will get $500.-!"*
> (3) *"Ah. – And why do the others run?"*

Why is this funny? The speaker of 2.1-6(3) interpreted the NP "the winner" of sentence 2.1-6(2) referentially (that means, he has a special person in mind who wins the race), while the speaker of 2.1-6(1) used it attributively (in the sense of "the winner – whoever it will be"). Of course, the same NP could be used referentially, as in

Example 2.1-7
> *"Max, the winner of the race, splashed a bottle of champagne over the spectators."*

It is important to distinguish between these two readings, because the existence and uniqueness assumptions that hold true in the referential case don't hold true for the attributive case. The main point is that in the referential case, the **described individual** is a constituent of the interpretation of the NP (this can be seen as the **extensional** interpretation), while in the attributive reading, the **describing condition** is a constituent of its interpretation (this is called the intensional interpretation; we'll come back to these different readings in Section 2.2.2.)

Again, both readings can be found in the plural case as well:

Example 2.1-8
> *"I need two guys to help me carry the piano!"*
> *(attributive use of "two guys")*

Example 2.1-9
> *"I saw two guys carrying the piano."*
> *(referential use, although with a quantificational property)*

As in the definite vs. indefinite case, we argue that the referential/attributive distinction is an aspect *accompanying* properties of cardinality and quantification and should be handled separately from these, but in the same way for singular and for plural NPs.

2.1.2.3 Generic, individual, and prototypical descriptions

One of the most important questions concerning NPs' meanings is whether they refer to a class of objects, to a single (or a smaller set of) member(s) of that class, or to its prototypical member. In other words: Does a term correspond to a whole extension of a predicate, or to one instance of the universe to which this predicate can be applied, or to the kind expressed by the predicate? In natural language (at least in NLs like German), again, the same NP can be used to designate either:

Example 2.1-10
> *"A cat is a mammal."*

Interpretation: Every entity that is in the class of cat is a mammal, too. We'll call this the **generic interpretation. Example 2.1-11**

> *"A cat ran across the street."*

Interpretation: An (arbitrary, but fixed) entity that is a member of the class of all cats, ran ... We'll call this the **individual interpretation**.

This distinction leads to the separation of the knowledge base described in KL-ONE lookalikes into TBox and ABox, and is also reflected in most knowledge representation formalisms beyond this paradigm.

In fact, there is a third reading of the NP above (beyond the attributive reading already mentioned in Section 2.1.2.2):

Example 2.1-12
"A cat loves to catch mice."

Interpretation: Normally, cats ... or: The typical cat (which needn't exist!) ... We'll call this the **prototypical interpretation**.

The difference between the generic and the prototypical interpretation is that assertions about generic NPs hold true for all members of the respective class (in other words, they are necessary conditions for membership, and therefore part of the terminological knowledge), while assertions over prototypically interpreted NPs constitute something like a default rule and may be rejected for a specific instance of the respective class.

Again, each of these interpretations can be found to be correct for plural NPs as well (we may replace the NP "a cat" by the plural "cats" in Examples 2.1-10 – 2.1-12 without losing the interpretation). Therefore, in order to avoid redundancy, an NLP system should handle (i.e., represent and reason about) these NP aspects separately from the cardinality and the quantificational properties of the respective NP.

2.1.3 Distributive, collective, and cumulative readings. Or: Representing differences in scope

The original version of the Generalized Quantifier approach in [Barwise & Cooper 81] does not provide the intended interpretation of sentences like

Example 2.1-13
"Peter, Paul and Mary lifted a piano."

because the GQ always quantifies over the individual domain and that leads to the distributive reading

Example 2.1-14
"Peter lifted a piano and Paul lifted a piano and Mary lifted a piano."

Since then, some work has been done (e.g. by [Loenning 87]) in order to overcome this fault within the formalism of GQT. We shall adopt Loenning's classification of NPs, which provides criteria for the decision of which reading to choose.

For an NL analyzing system, four steps of reasoning seem to be necessary when dealing with possible scopings:

1. Scope readings in general: Which kinds of scope readings are possible in the respective NL?

It is common to differentiate between the **collective** reading (as informally introduced through Example 2.1-13) where a collection of things (here: persons) function as the **agent**-role filler of a verb which allows for that (so-called collective verbs); this corresponds to the narrow scope interpretation of a quantifier (if one wants to look at the NP as a quantifier at all). Loenning calls NPs that may be interpreted in this way simply **plural NPs**.

In contrast, distributive readings are forced by – what Loenning calls – **quantificational NPs**. They may, but need not be, plural on the syntactic surface:

Example 2.1-15
> "*Both children found a mushroom in the forest.*"

is quantificational (BOTH always forces the distributive reading), but

Example 2.1-16
> "*Every child found a mushroom.*"

is quantificational as well (although singular in syntax).

NPs which denote a single object (**singular NPs**) naturally cannot be read collectively.

In addition, we want to consider two more readings which in one sense are different kinds of mixtures between distributive and collective readings.

An example of a so-called **cumulative reading** is

Example 2.1-17
> "*The two brothers (together) own eleven houses.*"

which doesn't say anything about who is the owner of, say, house-4: is it brother-1, brother-2, or both of them?

[Scha & Stallard 88] deal with the problem of conjunctive NPs (or multi-level plurals) which have a partially distributive reading:

Example 2.1-18
> "*The boys and the girls meet at different places.*"

where the intended (or at least one very natural) meaning is to split the complex object of
$$\text{(boys and girls)}$$
into two groups
$$\text{(boys) and (girls)}$$
which both function as agents of instances of the (collective) verb MEET.

2. Discovering the intended reading of an actual input sentence with the help of linguistic knowledge (FSS), world (domain) knowledge, dialog knowledge. We don't want to deal with this problem in this paper. For some NPs it is unclear whether they should be classified as (definite) plurals or quantificational NPs ("all" in the individual reading). But as an example for the complexity of the problem and the area in which solutions can be found, it is interesting that determiners classified as monotone increasing may have collective reading ("many", "at least three"), while collective readings seems odd for their monotone decreasing counterparts ("few", "at most three").

3. Representing these differences is obviously the next step, and

4. Inferencing from these different structures in a way that reflects the differences in meaning should cover the non-declarative part of the respective sentence meaning.

2.2 Problems with the Knowledge Representation Language

Noun phrases with all these different readings and aspects occur frequently in NL dialogs. The value of an NL system's internal KR language should be measured according to the degree to which it reflects and maintains these phenomena (among others).

2.2.1 Sets: where to put them

Thus, one of the demands we have to face when designing our target KR language as appropriately as possible to process natural language is that of sets as representational means together with specific processing capabilities evoked by these sets.

At first glance, the natural solution to this problem seems to be to introduce the concept SET in our KL-ONE knowledge base. Then, this concept is equipped with two roles, namely *has-elements* and *cardinality*, so as to be able to speak of the cardinality of a set and its elements (cf. Figure 2.1).

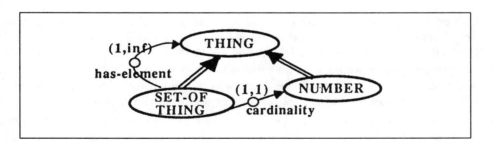

Figure 2.1: First attempt to represent sets in a TBox

Obviously, this description of sets is very natural, and if the need of our system is to *describe* sets, this is a perfect solution. However, if we want to *reason* about sets represented in this way, we have to be aware of several problems. No special processing abilities are provided in order to deal with individualizations of this 'Set of Thing' concept. For example, it is necessary in the ABox to keep the filler of the *cardinality*-role consistent with the number restriction of *has-element* – a property for which we have no computational means within the ABox. Next, we must reason about individualizations as being elements of a set with certain properties (expressed by roles). To cope with this, we have to give one role, namely the *has-element*-role, specific semantics – but that would go beyond the scope of KL-ONE lookalikes. What remains as an open problem is how in this approach we could represent the *subset*-relationship properly.

An alternative approach could be to encode the type information – i.e. the information necessary to distinguish between entities and sets of entities – by means of special concepts: one concept for the class of entities, another for the class of sets of entities. For example, we could have a concept 'Person' and another concept 'Set of Person' (cf. Figure 2.2).

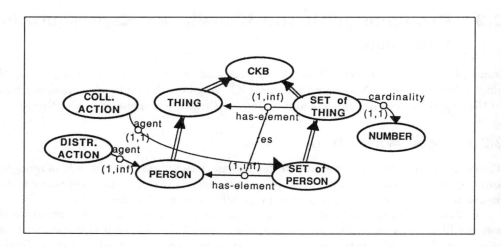

Figure 2.2: Second attempt to represent sets in a TBox

One problem which arises immediately is the impossibility of seeing any interrelationship between the two concepts! A simple solution – and quite natural in the field of KL-ONE – is to introduce an artificial concept (let's call it 'P/SoP' as in Figure 2.3) which encloses both concepts as a superconcept. Again, we are confronted with a bunch of unpleasant questions. How do we describe, e.g. which roles we should define for, P/SoP? And how can we control the obvious explosion in the number of concepts?

The basic problem seems to be the need not to represent (because this is possible as we described above) but to work with the notion of sets. The only way to do this in KL-ONE lookalikes seems to be to lower this special interpretation – i.e. their *meaning* – to the epistemological level. This, in turn, means defining new epistemological primitives for our target KR language. These primitives should have well-defined semantics and a bunch of algorithms realizing this semantics in the representation system. Our solution – in which we realize exactly this idea – is the content of Chapter 3.1.

2.2.2 Intension and extension in KL-ONE lookalikes. Or: Different aspects of a concept

As we have seen in the previous chapter, different aspects of a concept can be considered within a natural language expression. In addition to the distinction between generic and

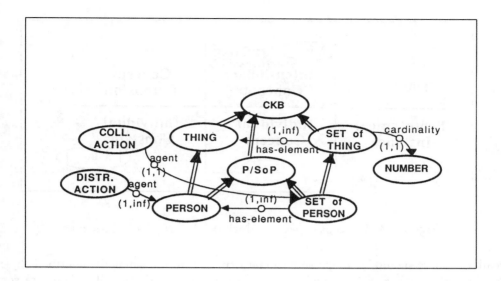

Figure 2.3: Third attempt to represent sets in a TBox

individual levels, and the attributive vs. referential reading, there are different possibilities to define classes of objects, namely intensionally and extensionally. Together, these aspects classify the notion of concept into various distinct classes.

The general aspect of a notion can be described intensionally by giving necessary and sufficient conditions of application. This is the way KL-ONE lookalikes usually define concepts in the TBox.

However, general concepts also may be defined extensionally through the explicit enumeration of their instances. In standard KL-ONE, there is no language feature to represent this kind of general information.[2]

Research in philosophical linguistics (e.g. [Putnam 75]) shows that for specific classes, especially for natural kinds, none of these possibilities can used to define the respective notion, because on the one hand there is no fixed set of common properties for all members of these classes, and on the other hand, it is impossible to enumerate the instances of those classes. Such a class may be described by means of a virtual, so-called 'prototypical' instance together with a function that collects together those instances whose properties are sufficiently near to the prototypical instance. Up till now, this notion cannot be found in any of the theories or systems grouped around the KL-ONE paradigm.

On the level of instances, too, this separation can be found. Corresponding to the extensional descriptions, instances (as well as sets of instances) of classes represent distinct objects in the world. This is just what you get when you individualize a general concept in KL-ONE.

Besides this, representatives for the two other classes on the individual level seem to be necessary as well. Nevertheless, none of the KL-ONE lookalikes provide representational means well-suited to cope with these needs. In order to deal with attributive

---

[2]First attempts in this direction were made in the LOOM system ([MacGregor & Brill 88]).

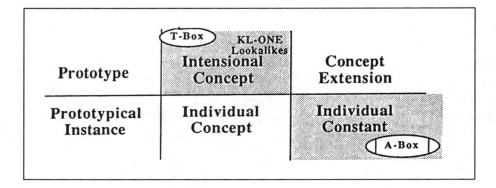

Figure 2.4: Concept aspects and what KL-ONE lookalikes cover

readings, there should be a representational structure for intensionally described objects on the individual level, so-called 'individual concepts'. Additionally, there are some kinds of prototypical reasoning on the individual level as well, for which prototypical instances are needed.

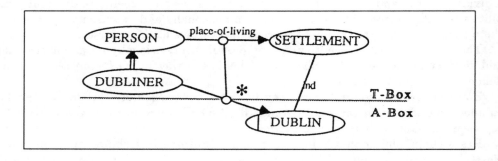

Figure 2.5: Illegal structure for DUBLINER

Another problem with KL-ONE lookalikes is the impossibility of expressing structures connecting objects of the different levels (T- and ABox) together. See for example the structure in Figure 2.5 where a TBox concept (Dubliner) is defined by means of an ABox expression (the town Dublin) or that in Figure 2.6 where an ABox assertion ranges over the extension of a TBox concept.

In the next chapters, we will show how these needs are dealt with in the $\mathcal{SBTWO}$ language. The extensional description of a concept as well as individual concepts and sets of individuals are integrated into the network, resulting in an extension of **SB-ONE** called **SB-ONE⁺**, while the structures for prototypical reasoning (on both levels) and the connections between the levels will find their correspondents in the embedding language $\mathcal{SBTWO}$.

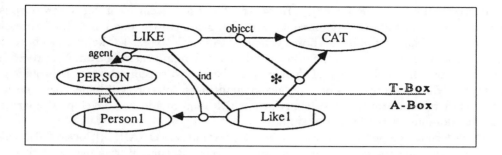

Figure 2.6: The illegal structure for "Fritz likes all cats..."

2.2.3 Maintaining assertional knowledge

In the early stages of the development of KL-ONE lookalikes, the main emphasis was laid on the terminological part of the representation formalisms and systems.[3] Up till now, there is no commonly accepted set of primitives for representing assertional knowledge. Different proposals have been made to construct systems able to handle the requirements of the embedding systems of which the KR system was part.

In the simplest approaches to the assertional level, *nexuses* were introduced as primitives to represent assertions ([Schmolze & Brachman 81]). These were connected to TBox concepts by means of *description wires*, which - together with a nexus node - denote the existence of an object satisfying that description. Nexuses themselves were collected in *contexts* which gave the ability to express 'different worlds' in the formalism. This formalism never turned into a system.

KL-TWO ([Vilain 85]) and KRYPTON ([Brachman et al. 83]) were the first systems offering a sophisticated assertional component. In KL-TWO, NIKL ([Moser 83]) – the terminological component – was combined with PENNI, a propositional inference engine based on RUP ([McAllester 82]). KRYPTON combined a terminological formalism (with less expressive power than NIKL) with a theorem prover for first order predicate logic ([Brachman et al. 85,Stickel 85].

Both systems showed the combinability of terminological languages and reasoning components in systems known as 'hybrid representation systems'. But while KL-TWO was restricted to the expressive power of propositional logic – which in the case of NL processing systems is not strong enough – the Stickel theorem prover of KRYPTON offered machinery which was much too 'expensive'. In both systems (but mainly in the second) the possible interconnection between and cooperation of TBox and ABox were not fully exploited in order to yield the balanced expressiveness between TBox and ABox aimed at in the BACK system ([von Luck et al. 87, p. 19]).

This combination of terminological and assertional components proved more useful to support pragmatical aspects ofNL processing than KR schemes developed exclusively for the purpose of handling dialog phenomena (e.g. DRT, [Kamp 81], or Situation

---

[3]We follow [von Luck 89] in distinguishing between a KR *formalism*, which is a formal language with given syntax and semantics, and a KR *system*, naming the materialization which interprets the well-formed formulae of the respective KR formalism.

Semantics [Barwise & Perry 83]). However, it doesn't seem clear for what purposes these different assertional components were designed. On the one hand, the expressive power and reasoning ability of those systems is too weak to (e.g.) take them as bases for building expert systems. On the other hand, many demands arising from NL processing systems don't lie in the scope of their power, either. No specific support is provided to analyze definite descriptions or generalized quantifiers, to represent beliefs or intentions in the assertional components themselves, which require add-ons developed separately (as was done in [Nebel 86] for example).

A system developed with these drawbacks in mind was QUARK ([Poesio 88]), which was not designed for general-purpose applications. Rather, it was designed to cover phenomena appearing in understanding dialogs, to represent linguistic and real-world contexts, and intentions and beliefs of agents. The formalism actually used to represent sentences' meaning has been IRS ([Bergmann et al. 87]) of which QUARK is a subset.

However, there remain problems in the expressive power of an assertional component one might want to have as representational bases. For one thing, the theoretical background of the findings integrated in such an ABox language should be linguistically based and well-founded. The approach should be general enough to cover a sufficiently large class of phenomena arising in NL dialogs. As a result, we will present in the next chapters our approach to handling assertional knowledge in a KL-ONE based paradigm. We do this by mixing advantages of the Generalized Quantifier Theory, Discourse Semantics, and Theory of Semantic Nets in order to have a broad base for building the representation system $SBTWO$. This system is especially tailored to handle knowledg expressed in plural NPs containing both negative and incomplete knowledge, and is intended to be flexible enough to admit extension for different operators representing modality or time.

Chapter 3

Methods: The new Representational Foundation

When looking for possible solutions to the problems described in Chapter 2 and for an adaptation of GQT, we had to keep in mind the already existing framework of the XTRA system. So, our methods for problem solving were two-fold:

1. Relating parts of the theory to existing representation structures and processing components:

   - Distinction between generic and individual descriptions, which corresponds (to a large extent) to the difference between T- and ABox.

   - Distinction between description of given versus new objects, which corresponds to·finding the intended referent in the system's knowledge base versus introducing new structures into it.

   - Classification of determiners according to their theoretical properties by means of a structured inheritance network expressed in an **SB-ONE** TBox.

2. Specifying the extensions necessary to handle the remaining part of the theory:

   - Extension of **SB-ONE** into **SB-ONE⁺** in order to deal with the descriptive aspects of sets and the subset and element-of relations introduced thereby.

   - Separating the theoretic properties of quantifiers into two groups (according to the distinction described in [Loebner 87]). From the first group, determiner based inference rules could be deduced. Properties of the second group found their correspondents in a set of so-called 'descriptive operators' ([Habel 86,Link 87]) of $\mathcal{SBTWO}$, which in some cases can be fully interpreted as **SB-ONE⁺** structures.

However, some problems dealt with in GQT are at present not fully integrated into the XTRA system. Among these are intensional quantifiers, second order quantifiers, and some special interpretations of fuzzy quantifiers. But see Chapter 5 for some suggestions on how to integrate them into the present approach.

To summarize the investigations in linguistics and logic concerning determiners, determiners carry three different kinds of information that have to be explicitly represented and reasoned upon in order to cover the meaning of an NP (cf. 3.1): **Cardinality infor-mation** determines the number of entities in question; **quantificational information** (in the narrower sense) forces the NP to have a scope; and **aspectual information** determines specific relations between the NP's denotation and the assertion made by the whole sentence.

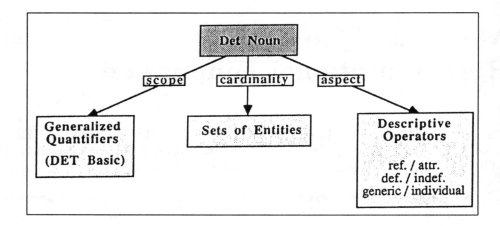

Figure 3.1: Determiner Information

In the following we present the representational and inferential means by which these different kinds of information (and all possible combinations between them) can be applied to a system based on KL-ONE.

As we have seen, there are three requirements for the new representation language: first, we want to have new representational means to deal with sets in order to meet the need to express cardinality information. We fulfill this demand with the KR language **SB-ONE⁺** which is able to deal with sets as epistemological primitives. Second and third, available scope and aspect information has to be dealt with. Besides this, there are demands to express and process a kind of general knowledge which we are not able to represent in the TBox of **SB-ONE⁺** and wich describes information about how to process certain representational structures. This is done within $SBTWO$, a language embedding **SB-ONE⁺** and enhancing it with different kinds of operators. Scope is expressed with the ordinary logical quantifers ∀ and ∃, and aspect is expressed with several so-called descriptive operators.[1]

In addition to purely representational tasks, $SBTWO$ is equipped with an interpreter which works more or less like a standard PROLOG interpreter. However, in addition to a PROLOG interpreter, there is a requirement for dealing with disjunctive, negative, inconsistent, or non-standard quantified information to be met. Another basic difference from the standard PROLOG interpreter is the need to distinguish between assertional and terminological knowledge. As we shall see, the terminological knowledge is covered by **SB-ONE⁺**, whereas the assertional part and, most important, the interactions and

---

[1]Besides these operators, others could (and will) be introduced e.g. to express modality.

interconnections between the two parts lie within the scope of $SBTWO$.[2]

The system is tailored for use in an NL processing system. Out of this, at least one additional important request arises. It is no longer satisfactory to answer questions simply with 'yes/fail' nor is it sufficient to answer 'yes / false / unknown / inconsistent' if one chooses to equip the system with a four-valued semantics à la Belnap ([Belnap 77]) (as e.g. [Levesque 84,Patel-Schneider 89,Ginsberg 86], and [Poesio 88] did in their systems). In addition, information about (positive or negative) justifications has to be given in order to support requests from the enclosing NL system components or indeed for $SBTWO$-internal reasoning processes themselves.

The system described here is built on a Horn clause compiler written in Common LISP ([Jansen-Winkeln 89]) which is extended to interact with **SB-ONE⁺** ([Allgayer 89]). This 'dump' reasoner is controlled by a meta-interpreter (called MOTHOLOG) able to prove disjunctive and negative goals ([Bartsch et al. 89]). MOTHOLOG in turn is taken as a basis and enhanced to handle non-standard quantification and aspect information.

3.1 SB-ONE⁺: Introducing Sets as Epistemological Primitives for the ABox

SB-ONE⁺ takes **SB-ONE** as the kernel and extends it with syntactic structures necessary to be able to introduce sets as epistemological primitives – primarily for the ABox. The extension is split into two parts: two additional language elements are needed to enable the TBox language to express new role descriptions within the TBox, and further additional structures enlarge the ABox language to cope with descriptions of sets on the individualized level.

3.1.1 TBox extension of SB-ONE

Provided the semantics of the **SB-ONE⁺** language assures that concept extensions contain both elements (as it is common in standard semantics for KL-ONE lookalikes) and set of elements, the new language structures to be introduced are:

TBox level:

- A **cardinality restriction**, short **CR**, is introduced to specify the type of filler – i.e. *set* or *element*.
- The notion of 'role' is split into two more refined types of relationship, namely **set-roles** and **element-roles**, reflecting the fact that the source must be of type *set* or *element*, respectively.

ABox level:

- Two representatives **IE** and **IS** for individual elements and individual sets, respectively.

---

[2]Henceforth, $SBTWO$ will stand for both the declarative part – the pure syntax – and the procedural part – the interpreter that processes this syntax. However, it should be clear from the context which aspect is referred to.

- A collection of link-types representing the various possible relationships between sets or an **IS** and a set, respectively. Here, the possible relations are: $\subset$, $\not\subset$, $\supset$, $\not\supset$, $\parallel$, $\nparallel$, $\in$, $\notin$, $\ni$, and $\not\ni$ with their obvious interpretations.

The next sections give a detailed description of how these new structures are integrated in the existing language and how the semantics for **SB-ONE$^+$** has to be adjusted in order to allow it to be an extension of the semantics for **SB-ONE**.

3.1.2 The new interpretation of a concept

In **SB-ONE$^+$**, a concept is seen as a 'black box' able to supply the ABox with elements as well as sets. The interpretations of root sets and primitive sets, from which defined concepts are constructed, therefore read:

Root Sets:
$$\mathrm{RS}_i \subset \mathcal{D} \cup \mathcal{P}(\mathcal{D}) \wedge \forall\, i, j, i \neq j\,.\ \mathrm{RS}_i \cap \mathrm{RS}_j = \emptyset$$

Primitive Sets:
$$\mathrm{PS}_j \subset \mathcal{D} \cup \mathcal{P}(\mathcal{D})$$

where $\mathcal{D}$ is the interpretation domain of concepts and $\mathcal{P}(\mathcal{D})$ is the powerset of $\mathcal{D}$.

3.1.3 The cardinality restriction CR

Within the TBox, the **CR** of a role description decides whether an (individualized) filler should be in $\mathcal{P}(\mathcal{D})$ or in $\mathcal{D}$. Figure 3.2 illustrates how this could be expressed and used in an **SB-ONE$^+$** TBox.

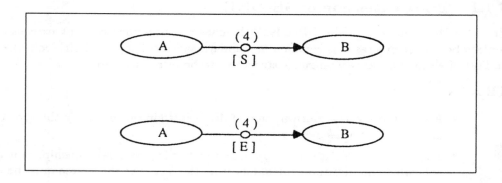

Figure 3.2: Example use of **CR** in a **SB-ONE$^+$** TBox

It shows a graphical notation of **SB-ONE$^+$** where concepts are depicted as ovals and roles as circles. Above, the concept A has 4 fillers (defined by number restriction (**NR**) (4)), all of which are defined to be sets (**CR** (S)), where below A is defined to have 4 fillers (again **NR** (4)), each of which has to be a single element (now **CR** [E]).

Table 3.1 lists the symbols allowed as **CR** and their respective meaning inside the TBox.

Symbol	Interpretation
E	Filler F in $\mathcal{D}$
S	Filler F in $\mathcal{P}(\mathcal{D})$
C	C subsumes cases E and S; i.e. nothing is said about F being in $\mathcal{P}(\mathcal{D})$ or $\mathcal{D}$.

Table 3.1: **CR** symbols and their interpetation

For the classification algorithm, Table 3.2 determines the truth value of the decision as to whether the **CR** of **r** subsumes that of **s**. It is assumed that the classification process would be successful in **SB-ONE** if there were no **CR**, i.e. it is assumed that **r** subsumes **s** ($\mathbf{r} \succeq \mathbf{s}$) neglecting the **CR** description.

CR of role r	CR of role s		
$\succeq$	C	S	E
C	1	1	1
S	0	1	0
E	0	0	1

Table 3.2: Classification of **CR**

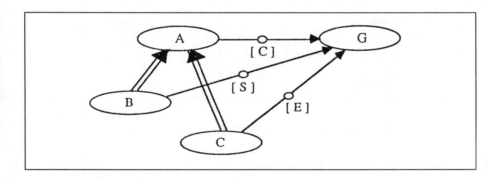

Figure 3.3: Classification of **CR** in an **SB-ONE⁺** TBox

Table 3.2 is to be read as follows: if role **r** had the **CR** [S], it would subsume role **s** with **CR** [S] and would be subsumed by role **q** having **CR** [C]. Figure 3.3 shows a hierarchy where all types of **CR** are classified according to Table 3.2.

3.1.4 Element-roles and set-roles

Assume that we define a role **r** to have fillers of concept **A**. Looking into the ABox, the question now is: How can we describe the filler of an individualized **r** (let's call it **A**) using the roles defined at the concept **A**? Here, two cases can be distinguished:

1. **A** is of type 'set', i.e. $\mathbf{A} \in \mathcal{P}(\mathcal{D})$.

2. **A** is of type 'element', i.e. $\mathbf{A} \in \mathcal{D}$.

This distinction suggests having two types of roles, one for each case. **Element-roles** have sources that are subsets of $\mathcal{P}(\mathcal{D})$, while **set-roles** have sources that are subsets of $\mathcal{D}$. This split-up allows for a clean semantics for roles with respect to the fact that concepts 'include' both sets and elements. Furthermore, it extends the expressive power of **SB-ONE$^+$** in such a way that it is now able to describe attributes of sets of things in the TBox.

role type	source type	number of extensions allowed per source individualization	filler
primsetrole, defsetrole	a set of elements of extension of A (an **individual set**)	`i..j`	for each extension of **r** an **individual set** with its cardinality not yet determined or an **individual element**
primelemrole, defelemrole	an element of extension of **A** (an **individual element**)	`i..j`	as above
	a set of elements AS of extension of A (an **individual set**)	`i..j` for each element of AS	as above

Table 3.3: Interpretation of different role types

To sum up, for the possible role descriptions

```
primelemrole (R
              domain-range (A B))
```

```
primsetrole (R
             domain-range (A B))
```

the two new role types have the semantics shown in Table 3.3.

Note the special reading for role types `primelemrole` and `defelemrole` in cases where the source is a set. We interpret this as a shorthand for: each element of the set is equipped with the description stated for the set.

Figure 3.4 depicts an example for the use of set-roles and element-roles. It shows individual elements (`P1 ...P5`) connected via ∈links to the individual set `PS1`, where

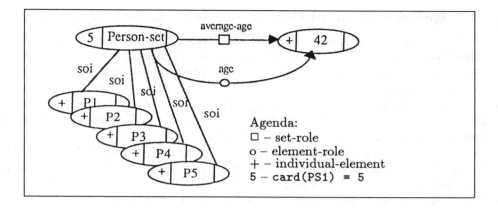

Figure 3.4: Example use of set-roles and element-roles

the set-role **average-age** was stated to hold for PS1. Furthermore, the element-role **age** was plugged at PS1, but has to be read as 'for every element of PS1...'. This means that P1 ...P5 are described with the **age** role and *not* with an **average-age** role.

3.1.5 ABox Extensions for SB-ONE

Within the ABox various new link types have to be defined to express the relationships among sets and between sets and elements.

1. Links between individual sets:
   Two sets a_1 and a_2 can be connected via the particular $\subset$-link expressing the subset relationship between a_1 and a_2. The set a_1 'inherits' the element-role descriptions of a_2 (because they were said to be valid for *all* elements of a_2), but doesn't 'inherit' the set-role descriptions: they were said to be valid only for a_2. Besides $\subset$-links, there are links for each of the relations $\subset$, $\not\subset$, $\supset$, $\not\supset$, $\parallel$, $\not\parallel$.

2. Links between individual elements and individual sets:
   An individual element a_1 and an individual set a_2 can be connected via the $\in$-link expressing the element-of relationship beween a_1 and a_2. As in case 1, only element roles are 'inherited' for a_1.

'a_1 inherits from a_2' at ABox level means 'a_1 has access to descriptions of a_2', i.e. the element-role descriptions for the elements of a_2 – which are noted in shorthand at a_2 – can be passed over to a_1, where they are again noted as a shorthand for 'for all elements of a_1...'.

Providing the system with information about set relationships results in establishing the respective links between the individuals effected. However, cases might occur where the requested relationship is not allowed or is already established implicitly. **SB-ONE⁺** keeps track of these possibilities, computes on demand all relationships that lie inside the transitive closure of the relations provided explicitly, and supports queries about allied relationships like e.g. superset, subset, or disjointness.

3.1.6 Keeping track of the various set relationships

As we have seen, a collection of various set relationships can be expressed within an **SB-ONE$^+$** ABox. We will show now how relationships between sets which exist implicitly due to explicitly expressed relationships are computed and maintained (cf. [Wellman & Simmons 88]).

It is easy to see from the facts $(\subset a_1\ a_2)$ and $(\subset a_2\ a_3)$ that the new fact $(\subset a_1\ a_3)$ could be derived because of the transitivity property of $\subset$. This composition of two relations can be generalized to the following notion:

Let a_i, $i = 1,\ldots,k$, be individualizations of concepts, so that $(\tau_i\ a_i\ a_{i+1})$ hold, $i = 1,\ldots,k-1$. Then, $(\sigma\ a_1\ a_k)$ holds with $\sigma := \tau_1 \circ \cdots \circ \tau_{k-1}$. We call σ the *implicit relation* between a_1 and a_{k-1}. The composition $\circ$ is associative and defined according to Table 3.4. This table is to be read as follows: given that $(\sigma\ a_1\ a_2)$ and $(\tau\ a_2\ a_3)$ hold, $(\sigma\circ\tau\ a_1\ a_3)$ holds (in two cases dependent on an additional constraint). No entry in the table stands for the empty relation $\square$, which underlies the property $\sigma\circ\square = \square\circ\tau = \square$.

$\sigma\circ\tau$	$\subset$	$\not\subset$	$\supset$	$\not\supset$	$\parallel$	$\nparallel$	$\in$	$\notin$	$\ni$	$\not\ni$
$\subset$	$\subset$		$\nparallel^a$	$\not\supset$	$\parallel$					$\not\supset$
$\not\subset$			$\not\subset$							
$\supset$	$\nparallel^b$	$\not\subset$	$\supset$		$\not\supset$	$\nparallel$			$\ni$	
$\not\supset$	$\not\supset$									
$\parallel$	$\not\supset$		$\parallel$				$\not\supset$	$\not\ni$		
$\nparallel$	$\nparallel$				$\not\supset$					
$\in$	$\in$				$\notin$					
$\notin$	$\notin$									
$\ni$							$\nparallel$			
$\not\ni$										

^{a}card(a_1)+card(a_3) >card(a_2)
^{b}card(a_2)>0

Table 3.4: Composition tableau

The computation of an implicit relation between two individualizations a_1 and a_2 yields a net entry of a so-called *implicit link* between these two objects. These links might have to be rejected in case of deletion or introduction of an explicit relation between objects of the net. The maintenance of these dependencies are taken care of in an efficient way in **SB-ONE$^+$**.[3]

---

[3]Depending on the implementation of the underlying network – which is based upon an associative network implemented in ASCON ([Bosch & Wellner 89]) – the algorithm that maintains the deletion or introduction of new links can be handled in time linear in the number individualizations in the net ([Allgayer & Fabri 89]).

Descriptive Operators	Task	Example
γ	Generic Description: description or definition of a general concept, using an A-Box expression (if necessary)	"Dubliner": γ [**supers** (PERSON), **VR** (place-of-living, γ [Dublin])] (where 'Dublin' is a pointer to the resp. A-Box individualization)
α	Individual Concept: describing conditions of an A-Box element (intensional interpretation)	"the winner of the race" α [**ONE** (**isa** (PERSON), **irole** (is-winner, Race1))] (where 'Race1' is a pointer to the resp. A-Box individualization)
ι	Definite (singular or plural) description, referring to a given individualized structure	"the two girls" ι [**TWO** (**isa** (GIRL))]
η	Indefinite (singular or plural) description of a new individualized structure	"two girls" (non-quant.reading) η [**TWO** (**isa** (GIRL))]
π	Prototypical instance: description of the (non-existing) typical member of a class of objects	"the (typical) lion" π [**supers** (LION)]

Table 3.5: Descriptive operators in $\mathcal{SBTWO}$

3.2 $\mathcal{SBTWO}$: Expressing Scope and Aspect

So far, **SB-ONE⁺** provides – on the TBox level – concepts which are to be interpreted as classes of individuals as well as sets of individuals of a certain type, and – on the ABox level – representational structures for plural entities (sets), partitive constructions between them ($\subset$, $\in$), and other relationships.

However, **SB-ONE⁺** keeps both the strict distinction and the representation of the respective contents. In order to overcome this borderline in the sense described in Section 2.2.2, specific term-forming operators, the so-called descriptive operators[4], are to be defined as elements of the embedding representation language $\mathcal{SBTWO}$.

3.2.1 Descriptive operators

$\mathcal{SBTWO}$-expressions may, but need not, be fully mapped into **SB-ONE⁺** net structures, as the above Table 3.5 shows.

Besides these descriptive operators, $\mathcal{SBTWO}$ provides operators to express plural entities of a certain cardinality. These cardinalities may be fixed or vague. Examples for cardinality operators with fixed cardinality (ONE, TWO, ...) can be found in the same Table 3.5 as arguments of α, ι and η operators.

Vague cardinality operators (SOME, FEW, ...) show the same syntactic structure as fixed ones. Their interpretation, however, may be complex, as described in Section 4.3.

---

[4]They are called descriptive operators and are designated with characters of the Greek alphabet in order to emphasize the parallel with Russel's well-known ι- and η-operators ([Russell 05]).

3.2.2 Representing Scope

In $SBTWO$ the natural language quantifiers find their correspondents in the descriptive operators (e.g. two, some ...). Thus the logical quantifiers $\forall$ and $\exists$ are available to represent the scope aspects of these NLQs. The process of making scope ambiguities explicit follows well-known algorithms to be found in [Hobbs & Shieber 87,Allen 87] for example.

Consider for instance the following $SBTWO$-expression representing the NL sentence "Two Dubliners drank three whiskies."

Example 3.2-1
$\exists u \in TWO(Dubliner) \; \forall d \in u \; \exists w \in THREE(Whisky)$
  $agent(Drink, d) \wedge object(Drink, w)$

The task now is to interpret this description of scoped expressions as **SB-ONE⁺** net structures. In the case of narrow scope, the corresponding $SBTWO$-expression contains a sequence of logical quantifiers of the same type (i.e. $\forall$ or $\exists$). Wide scope (as in the example above), then, is reflected in sequences of logical quantifiers of alternating type. These scope types correspond to the different readings mentioned in Section 2.1.3, where distributive reading is expressed with a quantifier of wide scope and cumulative/collective reading with a quantifier of narrow scope.

One argument for introducing sets as epistemological primitives into our representation language **SB-ONE⁺** was to be able to distinguish among these three readings. We now show how an $SBTWO$-expression with logical quantifiers corresponds to an **SB-ONE⁺** structure.

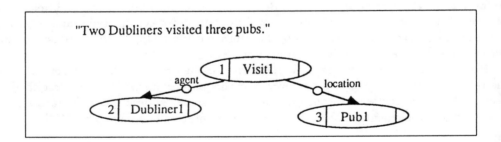

"Two Dubliners visited three pubs."

Figure 3.5: Net translation for distributive reading

In contrast, the sentence "Two Dubliners visited three pubs." will preferentially be represented as in

Example 3.2-2
$\exists u \in TWO(Dubliner) \; \exists \; THREE(Pubs)$
  $agent(Visit, u) \wedge object(Visit, THREE(Pubs))$

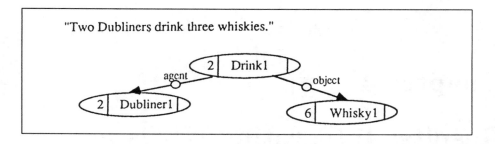

Figure 3.6: The net translation for collective reading

3.2.3 Determiner-dependent inference rules

Knowing about properties of determiners, we are able to classify our set of determiners according to them. Thus, each determiner inherits as a member of a special class of determiners their intrinsic behavior in regard to their processing.

We express this knowledge about determiner-dependent properties by means of inference rules for the different determiner classes. So, we describe the property of a determiner to be monotonic in the following form:

```
gen-quant(?DET ?BASIC ?CENTRAL) :-
        detclass(?DET persistent/raising);
        substitutable(?DET-S ?DET);
        subsumesp(?SubC ?BASIC);
        gen-quant(?DET-S ?SubC ?CENTRAL).


gen-quant(?DET ?BASIC ?CENTRAL) :-
        detclass(?DET monoton/raising);
        substitutable(?DET-S ?DET);
        subsumesp(?SubC ?CENTRAL);
        gen-quant(?DET-S ?BASIC ?SubC).
```

We hereby take advantage of having represented the classified determiner classes by means of an **SB-ONE**[+] TBox. Therefore, the predicate detclass(?DET ?PROP) maps to the predicate supers*(?DET ?PROP), which is already part of the TELL operation. The predicate substitutable expresses the possibility to replace one determiner by another according to a given semi-ordering: "at-least-two x" in this ordering is more special than "at-least-three x" and therefore can be replaced inside an ASK-expression.

In addition, the processing of an inference rule of this kind involves search processes in order to find the corresponding translation for ?BASIC and ?CENTRAL, as we described earlier. These processes are only triggered within gen-quant rules. Thus, these rule structures can also be used to express general knowledge of a rule-like nature within $\mathcal{SBTWO}$.

Chapter 4

Results: Integrating NLQs in XTRA

In this chapter, we will describe how the methods presented in Chapter 3 are integrated in the NL interface system XTRA. We show the representation of determiners at the level of Functional Semantic Structures (Chapter 4.1). Their processing with respect to cardinality of sets is described in Section 4.3[1]. While these processes work solely on the network structure of **SB-ONE+**, the processing of the formal properties of quantifiers and descriptive operators lies within $\mathcal{SBTWO}$ (Chapter 1.1).

As we have seen in Chapter 1.1, knowledge is classified according to the linguistic vs. world knowledge distinction on one side and according to the declarative vs. procedural distinction on the other. We now have to assign the determiner relevant knowledge to the appropriate knowledge sources.

1. GQT classification of determiners defines a structural inheritance network which is obviously domain-independent and is therefore encoded in the TBox of the Functional Semantical Structure.

2. The description of the world knowledge, however, does not contain any explicit representation about determiners. Instead, a part of the information expressed by determiners is interpreted and stored in the Conceptual Knowledge Base ABox using elements, sets, and their relationships as representational means.

3. The remaining part of the meaning of determiners is given through interpretation rules expressing their procedural semantics within an NL processing system.

The solution chosen in our system brings along the problem that the uniform structure of GQ formulae (DET Basic)(Central) doesn't appear on the $\mathcal{SBTWO}$-level. This is because of the fact that an arbitrary property (that describes the Basic and Central predicates in GQT) cannot be expressed (in all cases) with a given KL-ONE TBox (cf. our example in Chapter 1.3, "Münchner who sits in a beer garden on a warm summer night"). Therefore, we have to find the correspondents to the (DET Basic)(Central) structures still present in the syntax and the intermediate functional semantic structure.

---

[1]This section shows the processing of determiners from the analysis point of view; but as in the XTRA system all knowledge sources are used bidirectionally (see [Allgayer et al. 89]); the respective reverse processing holds true for NL generation as well.

These correspondents may be arbitrarily complex $\mathcal{SBTWO}$-structures. The $\mathcal{SBTWO}$-translation for the Basic of the above example phrase reads:

$$\gamma \; [\; \textbf{isa}(\text{Muenchner}),$$
$$\textbf{irole}(\text{is-agent-of},$$
$$\textbf{vr}(\text{To-Sit})),$$
$$\textbf{irole}(\text{time},$$
$$\textbf{vr}(\text{Summer-Night})),$$
$$\textbf{irole}(\text{location},$$
$$\textbf{vr}(\text{Beergarden})) \;]$$

The determiner-dependent inference rules – which are expressed in an elegant way in terms of DET, Basic, and Central – are in fact applied to the respective $\mathcal{SBTWO}$-translation thereof.

The operations required to apply the DDI rules (e.g. generalization, restriction, intersection) are mapped onto the basic **SB-ONE**[+] operation 'subsumption' described in Sections 1.2 and 3.1.

As it is possible in $\mathcal{SBTWO}$ to combine general and individual structures within one expression, it is also possible to express both generic and referential quantification. Therefore, the problems of GQT as described e.g. in [Loebner 87] don't arise in our system.

4.1 Classification of Extensional Determiners / Quantifiers

As we have seen in Section 1, determiners take a predicate and define interpretational rules for the respective NLQ built up out of the NP. In XTRA, the processing of these interpretational rules take place during the transformation of the functional semantic structure (FSS) of an input sentence into a structure of the conceptual knowledge base (CKB), a process which is called the referential-semantic interpretation (RSI, see Figure 1.1)

The FSS – while containing linguistics-oriented knowledge about the semantic structure of NL expressions – still entails explicit representatives corresponding to determiners, i.e. the class DET, and its subclasses. On the CKB level, however, DET no longer occurs explicitly, but is already interpreted with regard to cardinality, scope, and aspectual properties.

Note that the FSS does not deprive sentences of their ambiguity with respect to these properties. The set of possible readings for a sentence is the union of possible readings over the set of its well-formed FSSs. But a sentence's FSS must carry all information necessary for the next analyzing step.

In the FSS TBox, structures corresponding to (classes of) noun phrases are distinguished from others (e.g. verb correspondences) by a role *det* which relates them to the class DET of determiners and its subclasses. Figure 4.1 shows a detail of the FSS TBox.[2]

---

[2]This conceptualization reflects our investigations so far concerning the representational and inferential realizability of properties of DETs. It might be extended in future work.

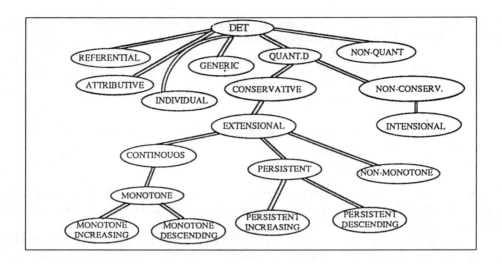

Figure 4.1: Detail of the FSS TBox

Determiners, as they are defined in the semantic lexicon, are thereafter integrated as leaves, that is as subconcepts (of one or more) of these FSS concepts, and are individualized as they occur in the surface structure of an actual input.

Each subconcept of DET represents a type of inference rule that holds true for the respective NLQ. Some of the criteria according to which DET is divided into its subclasses have already been explained in previous sections (quantificational +/-, referential vs. attributive, definite vs. indefinite, generic vs. individual, monotone +/-, with subclasses increasing vs. descending, and persistent +/-).

The *conservativity* feature may be characterized ([van Benthem 83]) by

Example 4.1-1
   (DET Basic) Central ≡ (DET Basic) (Basic ∩ Central)

and holds true for exactly the extensional determiners, e.g.

Example 4.1-2
   "All Dubliners drink whisky." ≡ "All Dubliners are Dubliners and drink whisky."

but not for intensional determiners, e.g.   .

Example 4.1-3
> *"Some alleged murderers are innocent."* ≢ *"Some alleged murderers are*
> *murderers and innocent."*

Continuity is a principle somewhat weaker than monotonicity:

Example 4.1-4
> $[(DET\ Basic)\ \text{Central}_1] \wedge [(DET\ Basic)\ \text{Central}_2] \wedge$
> $\text{Central}_1 \subseteq Central \subseteq \text{Central}_2$
> $\Rightarrow (DET\ Basic)\ Central$

[Thijsse 82] has shown that all simplex determiners denote continuous relations. But MANY in the *relative frequency* reading[3] ([Westerstahl 85]) may fail to fulfill this principle for example.

Extendable determiners guarantee "stability under growth of the universe" [vanBenthem 83, p. 435]:

Example 4.1-5
> $Basic, Central \subseteq U \subseteq U' \Rightarrow (\text{DET}_E\ Basic)\ Central \equiv (\text{DET}_{E'}\ Basic)\ Central$

The principle of extension describes just the context-independent expressions, while MANY in the relative frequency reading just mentioned, for example, may fail to pass this test.

In order to classify determiners, we used the terminological part of our KR formalism, **SB-ONE**. The properties underlying this classification correspond to different kinds of inference rules, the so-called *determiner dependent inference rules* (DDI), which form the interpretational basis for $\mathcal{SBTWO}$.

4.2 Processing the Formal Properties of Determiners

The strategy of how to apply the DDIs to a given query is one of the tasks of the ASK component of $\mathcal{SBTWO}$. There are several ways to process the query. The first possibility is a check whether the information asked for is already part of the system's knowledge base. Second, a test whether standard procedures of the underlying net formalism (like classifier, realizer, and matcher) succeed in the deduction of the respective information. But in some cases, in fact in those where quantified expressions are involved, these deductions have to be suppressed. Instead, and third, DDI might open up new possibilities in deriving the required information.

We now give two examples of situations where the standard deductions are to be replaced by activation of DDI in order to yield the correct answer and avoid incorrect inferences.

---

[3]The relative frequency of ⟦ Central⟧ in ⟦ Basic⟧ exceeds that of ⟦ Central⟧ in the whole universe.

Example 4.2-1
   "Exactly two Münchner go by bus to the Wies'n."

Imagine the following query:

```
ASK [ ∃(m ∈ E-TWO(isa(Münchner)))
      (agens(Drive, m) ∧ destination(Drive,Wies'n))]
```
which represents the NL sentence "Are there two Münchner who go to the Wies'n?".

The normal **SB-ONE$^+$** formalism would answer this query positively, because the information known is more specific than that asked for. But, because a generalized quantifier is involved, this inference has to be blocked. Instead, the properties of this GQ have to be taken into account. Because of the non-monotonicity of E-TWO (standing for 'exactly two'), we can infer that a generalization of the Central predicate ('going to the Wies'n by bus' into 'going to the Wies'n')[4] is not possible. Therefore, the system's answer is NO, together with some additional information to be used e.g. for overanswering.
   Suppose for example the data base contains the input

Example 4.2-2
   "Few Münchner go to the Wies'n."

and the following query is to be answered:

```
ASK [ ∃(m ∈ FEW(isa(Münchner)))
      (agens (Drive, m) ∧ object(Drive, bus)
      ∧ destination(Drive, Wiesn)) ]
```

being the representation of "Are a few Münchner going by bus to the Wies'n?".
   The standard inference procedures of the net formalism yields the answer unknown, because this would require 'descending' in the data base which is not supported. By means of the respective DDI rule, we find out that FEW is monotone-descending, i.e. a structure more general than the equivalence of the Central predicate in the query is sufficient to answer positively.

4.3   Interpretation of Cardinality Operators

As the second kind of information carried by determiners, cardinality has to be interpreted. We distinguish between determiners with fixed and those with vague cardinality information. Determiners with fixed cardinalities (as "two", "the three tall", "exactly five") are directly expressible as an **SB-ONE$^+$** set with the respective cardinality. Determiners with vague cardinality have to be divided further into two subgroups: those describing an interval with fixed boundaries ("between two and four", "more than five"[5])

---

[4]This inference is based on (1) the specification of the translation of the Central predicate, and (2) the decision whether the Central predicate of the first phrase is more specific than that of the second. The later is done by the classifier or realizer, respectively.

[5]This is possible because **SB-ONE$^+$** supports 'inf' as upper boundary expressing infinitely many.

find their correspondents in the net structure as well. Still open are determiners expressing fuzzy cardinality information ("few", "some"). The problem is that these determiners are context-dependent in the sense that their interpretation depends on the cardinality of the denotation of the Basic as well as the Central structure. The first solution just ignores these dependencies and takes a standard interpretation of each of these fuzzy determiners in order to compute the respective cardinalities (e.g., the standard interpretation for "some" is "more than one"). The second solution tries to take into account the context dependency by defining rule bases which involve the cardinality of the concept denotation – if known. $\mathcal{SBTWO}$'s solution provides means to do both.

4.4 Interpretation of Descriptive Operators

Descriptive operators as defined in Section 3.2.1 may but need not be fully mapped into **SB-ONE**+ net structures. Figure 4.1 shows the processing of descriptive operators during the referential semantic interpretation.

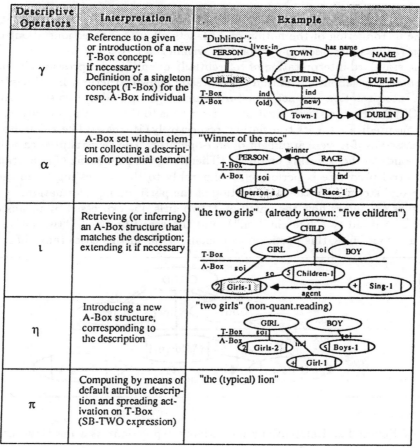

Table 4.1: Interpretation of descriptive operators

In this process, **SB-ONE** tools like the **SB-ONE** matcher ([Aue et al. 89]) and the spreading activation component (which is still under development) are involved.

4.5   Interpretation of Quantificational NPs

The last source of determiner information is concerned with quantification. During the
analysis so far, the XTRA system has differentiated between generic and referentially
quantifying expressions. The interpretation has to reflect this distinction.

In the case of a generic quantifying expression, we introduce a representative for
the extension of the respective concept into the **SB-ONE$^+$** ABox. Together with this
so-called ALL-set, processing algorithms are provided that ensure that

- all individual sets of this concept form subsets of the ALL-set;

- all individual elements are elements of the ALL-set;

- therefore, all properties provided for the ALL-set hold true for each subset or
  element, respectively.

The introduction of the notion of ALL-sets allows for assertions about concept extension (as was discussed in Section 2.2) within $SBTWO$.[6]

The standard interpretation of referentially quantifying expressions is that they describe assertions about a partition of an already known structure. In principle, the result
of this interpretation is a net structure which is computed in two steps. According to
the GQT theory, the Basic predicate corresponds to the counting predicate. Therefore,
the Basic predicate (or its translation, respectively) is treated like a definite NP; i.e., it
is processed as if it were argument of an ι-operator. The Central predicate, in contrast,
corresponds to the counting predicate. Therefore, the processing of (the translation of)
the Central predicate is performed comparably to the processing of an η-expression.
The second step is now the introduction of the partitioning information.

Figure 4.2 shows how the partition ("two girls") is expressed by means of a subset
of the known entity ("five children"), and the assertion itself ("two girls sing") by a
structure containing this partition as domain or (as in this case) range of roles.

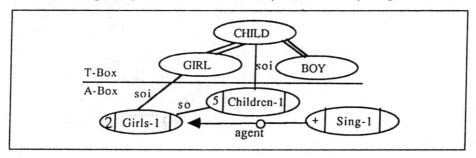

Figure 4.2: Referentially quantifying expressions as a net structure

---

[6]Another type of generic quantifying expressions, MOST, seems to express default information. Because the problem of default reasoning in KL-ONE lookalikes is still open, we are able to trigger the
respective $SBTWO$ inference procedure, but the processing itself is still under consideration.

Chapter 5

Summary and Future Work

In the work presented here, we have developed a representational and inferential framework for treating of the various aspects of (especially plural) NPs. The determiner classes are an example of how the TBox can be used to structure inferential knowledge. The inference concepts obtained in this manner are connected to processing procedures which realize them. Corresponding inferences for certain NL expressions are triggered by syntactical symbols of the KR language, thereby enlarging the inferential power of the KR system with reasoning capabilities additional to those offered by standard subsumption hierarchy sytems.

SB-ONE (as an example for KL-ONE lookalikes) has been extended by structures representing sets of entities as well as subset- and element-relations. By means of the embedding ABox language $SBTWO$ a number of other shortcomings of KL-ONE lookalikes concerning the representation of a variety of concept aspects – such as the use of ABox elements to define TBox entities, structures representing the extension of concepts, or structures reflecting attributively described entities – can be overcome.

Our future work will be concerned with the investigation of the expressive and inferential power of this framework:

- the semantics of sequencing and nesting of descriptive and cardinal operators – although syntactically possible – is not yet quite clear;

- second order quantifiers – ranging over sets of sets of individuals – need to be integrated;

- intensional determiners which result from regarding preposed intensional adjectives as part of DET should be processed within the same framework;

- studies in the kinds of prototype reasoning on the basis of **SB-ONE**$^+$ and $SBTWO$ will follow;

- the principal approach will be adapted for treating modality; and

- the integration of XTRA's NL analysis and generation component will be completed, resulting in an NL system that can properly process noun phrases with respect to the properties described here.

A   The language $\mathcal{L}_{\mathrm{SB-ONE^+}}$: a term language for defining T- and ABox of SB-ONE$^+$

A.1   The Syntax for $\mathcal{L}_{\mathrm{SB-ONE^+}}$

TELL structure:

TBoxExpression ::=	TBoxDefinition |
	TBoxRestriction |
	TBoxExtension
	TBoxRelaxation
TBoxDefinition ::=	Concept | Role

Concept definition

DomainConcept ::=	Concept
RangeConcept ::=	Concept
DefaultVRConcept ::=	Concept
Concept ::=	PrimConcept |
	DefConcept |
	SingConcept
PrimConcept ::=	**primconcept**(CName, CSpecList) |
	Name
DefConcept ::=	**defconcept**(CName, CSpecList) |
	Name
SingConcept ::=	**singleton**(CName, CSpecList) |
	Name
ConceptList ::=	Concept | Concept, ConceptList
CSpecList ::=	**supers**(ConceptList) |
	CSpec |
	CSpec, CSpecList
CSpec ::=	RoleInhRes | CSpec1
CSpec1 ::=	**nr**(RName, NRTriple) |
	cr(RName, CRSpec) |
	necres(RName, NecSpec)
CSpec2 ::=	RoleRaiseRes | CSpec1
RoleRaiseRes ::=	**raise-inh**(RName, RoleRes, RoleRange)
RoleInhRes ::=	**restrict-inh**(RName, RoleRes, RoleRange)
RoleRange ::=	**range**(RangeConcept, DefaultVRConcept)
NRTriple ::=	(MinNR, MaxNR, DefNR)
CRSpec ::=	**C** | **S** | **E**
NecSpec ::=	**opt** | **nec**
MinNR ::=	0 | PosInteger
MaxNR ::=	PosInteger | **inf**
DefNR ::=	PosInteger
Number ::=	0 | PosInteger | **inf**

Role definition

Role ::=	PrimRole |
	DefRole |
	vrdiff(Role, DSpecList) |
	Name
PrimRole ::=	**primelemrole**(PrimElemRoleName, PrimRoleSpec) |
	primsetrole(PrimSetRoleName, PrimRoleSpec)
PrimRoleSpec ::=	**domain-range**(DomainConcept, RangeConcept,
	DefaultVRConcept)
DefRole ::=	**defelemrole**(ElemRoleName, DefRoleSpec) |
	defsetrole(SetRoleName, DefRoleSpec)

DefRoleSpec ::=	RoleRes, RoleDomain
RoleRes ::=	**restricts**(Role)
DSpecList ::=	DSpec, DSpec | DSpec, DSpeclist
DSpec ::=	(DRoleName, NRTriple, RangeConcept DefaultVRConcept)

Attribute definitions

TBoxRestriction ::=	DisjointnessRestriction | AttributeRestriction
DisjointnessRestriction ::=	**disjoint**(PrimConceptList)
PrimConceptList ::=	PrimConcept | PrimConcept, PrimConceptList
AttributeRestriction ::=	**attributes**(Concept, AttributeList)
AttributeList ::=	AttributeElement | AttributeList
AttributeElement ::=	(AttributeName AttributeValue)

Concept expansion

TBoxExtension ::=	**expand**(CName, CSpecList)

Relaxation of roles and concepts

TBoxRelaxation ::=	RoleRelax | ConceptRelax
RoleRelax ::=	**relaxrole**(RName, RoleRange)
ConceptRelax ::=	**relaxconcept**(CName, CSpecList1)
CSpecList1 ::=	CSpec2 | CSpec2, CSpecList1

Individualization

ABoxExpression ::=	ABoxIndividualization
ABoxIndividualization ::=	IConcept | IRole

Concept Individualization

IConcept ::=	**definst**(CName, ICSpecList) | Name
ICSpecList ::=	ICSpecList1, ICSpecList2
ICSpecList1 ::=	**isa**(CName), **card**(CardSpec)
ICSpecList2 ::=	ICSpec2 | ICSpec2, ICSpecList2
ICSpec2 ::=	**irole**(RName, IRSpecList2)
CardSpec ::=	PosInteger | (MinNR, MaxNR)

Role Individualization

IRole ::=	**defirole** (RName, IRSpecList)
IRSpecList ::=	IRSpecList1, IRSpecList2
IRSpecList1 ::=	**at**(CName)
IRSpecList2 ::=	IRSpec2, IRSpec3 |
	IRSpec1, IRSpec2, IRSpec3
IRSpec1 ::=	**name**(IRole)
IRSpec2 ::=	**nr**(NRTriple)
IRSpec3 ::=	**vr**(IConcept)

A.2 The Semantics for $\mathcal{L}_{SB-ONE^+}$

Following [Schmolze 87], the semantics for the TBox part of TELL is developed in two steps. First, a structure $S = <D, M_p>$ is defined with

D = domain, here $D_S \cup \mathcal{P}(D_S)$, where D_S is the standard domain, and

M_P : Primitives $\mapsto D \cup (D \times D)$

This defines the primitive part of the net: all primitives are asserted to be a subset of the interpretational domain. Henceforth, these can be intersected with defined concepts (i.e. $M[(C1....Cn)]$). Next, the interpretation M_P is extended on M:

$$M : \text{terms} \mapsto D \cup (D \times D)$$

This defines the semantics of defined terms. $\preceq$ stands for the subsumption relation, i.e. $C_1 \preceq C_2 \Longleftrightarrow$ supers(C_1, C_2).

Thus, the semantics for the TBox part of $\mathcal{L}_{SB\text{-}ONE^+}$ can be given as follows:

$$M[\text{supers}(C)] = M[C]$$

$$M[C_1, \ldots, C_n] = \bigcap_i M[C_i]$$

$M[\text{supers}(C_1, \ldots, C_n)]$ yields the intersection of $C_1, \ldots, C_n$.

$$M[\text{primconcept}(C_1, \text{supers}(C_2))] = PS \cap M[C_2]$$

primconcept yields a subset of the extension of the given superconcept; which subset is described by the information 'primitive'.

$$M[\text{defconcept}(C_1, \text{supers}(C_2))] = M[C_2]$$

$$M[\text{singleton}(C_1, \text{supers}(C_2))] \Longleftrightarrow M[C_1] = M[C_2] \wedge \text{card}(M[C_1]) = 1 \wedge M[C_1] \in D$$

The extension of a singleton concept contains only one element.

$$M[\text{nr}(\text{SetRole}, (i, j, k))] =$$
$$\{x \in \mathcal{P}(D) \mid i \leq \textstyle\sum_l \text{card}(e_l) \leq j, \text{ with } e_l \in \{y \in D \times \mathcal{P}(D) \mid (x,y) \in M[\text{SetRole}]\}$$
$$\wedge$$
$$y \in D \vee y \in \mathcal{P}(D)\}$$

$$M[\text{nr}(\text{ElemRole}, (i, j, k))] =$$
$$\{x \in D \mid i \leq \textstyle\sum_l \text{card}(e_l) \leq j, \text{ with } e_l \in \{y \in D \times 2^D \mid (x,y) \in M[\text{ElemRole}]\}$$
$$\wedge$$
$$y \in D \vee y \in \mathcal{P}(D)\}$$

$$M[\text{cr}(\text{Role}, C)] = M[\text{Role}]$$

$$M[\text{cr}(\text{Role}, S)] = M[\text{Role}] \cup (D \times \mathcal{P}(D_S))$$

$$M[\text{cr}(\text{Role}, E)] = M[\text{Role}] \cup (D \times D_S)$$

$M[\text{primelemrole}(R, \text{domain-range}(C_1, C_2, C_3))]:$
$$R \subseteq (C_1 \cup D) \times C_2$$
$$\wedge$$
$$C_3 \preceq C_2 \}$$

For any 'element role', the source has to be $\subseteq D_S$.

$M[\text{primsetrole}(R, \text{domain-range}(C_1, C_2, C_3))]:$
$$R \subseteq (C_1 \cup \mathcal{P}(D)) \times C_2$$
$$\wedge$$
$$C_3 \preceq C_2 \}$$

Here the source is a set, i.e. $\in \mathcal{P}(D_S)$.

$$M[\text{restricts}(R)] = M[R]$$

The role hierarchy. **restricts** corresponds to **supers** when defining a concept.

$M[\text{restricts-inh}(R_1, \text{restricts}(R_2), \text{range}(C_1, C_2))] =$
$$M[R_1] = M[R_2] \cap (D \cup \mathcal{P}(D)) \times C_1$$

Specializing the VR.

$$M[\textbf{raise-inh}(R_1, \textbf{restricts}(R_2), \textbf{range}(C_1, C_2))] =$$
$$M[R_1] = M[R_2] \cap (D \cup \mathcal{P}(D)) \times C_1$$

Generalization of a role definition.

The semantics for the ABox part is given according to [von Luck et al. 87] by naming a translation procedure that transforms ABox expression into logical formulars.

The general form of an ABox expression

$$\textbf{definst}(a_1, \textbf{isa}(A), \textbf{card}(C), \textbf{irspec}_1(a_1), \ldots, \textbf{irspec}_n(a_1))$$

is transformed into:

$$C = E \Rightarrow a_1 \in M[C] \cap D_S$$
$$\wedge A(a_1)$$
$$\wedge \textbf{irspec}_1(a_1)$$
$$\wedge \vdots$$
$$\wedge \textbf{irspec}_n(a_1)$$
$$\vee$$
$$C = S \Rightarrow$$
$$a_1 \in M[C] \cap \mathcal{P}(D_S)$$
$$\wedge A(a_1)$$
$$\wedge \textbf{irspec}_1(a_1)$$
$$\wedge \vdots$$
$$\wedge \textbf{irspec}_n(a_1)$$

The form

$$\textbf{irspec}_i(a_1) = \textbf{irole}(R, \textbf{name}(r_i, \textbf{nr}((1, r)), \textbf{vr}(B))$$

can be transformed as follows:

$$R(r_i) \wedge r(a_1, b_1) \wedge \exists_l x : ir(a_1, x) \wedge \neg \exists_{k+1} x : ir(a_1, x)$$

The default descriptions which might be part of the TBox definition of a concept are translated into default rules à la Reiter ([Reiter 80]) when instantiating a concept. The theory defined with the logical formulae which so far constite the ABox, together with the default rules, can be pointwise circumscribed ([Lifschitz 86]) yielding a set of logical formulae without defaults ([Bauer & Merziger 89]). The successful test of these formulae guarante the correct individualization.

Bibliography

[Allen 87] J. **Allen**. *Understanding Natural Language.* 1987.

[Allgayer & Fabri 89] J. **Allgayer** and A. **Fabri**. *Efficient maintenance of set relationships in* **SB-ONE**+. Memo, Univ. des Saarlandes, 1989.

[Allgayer 89] J. **Allgayer**. $\mathcal{L}_{SB-ONE^+}$: *Die Zugangssprache zu* **SB-ONE**+. Memo, Univ. des Saarlandes, 1989.

[Allgayer et al. 89] J. **Allgayer**, R. M. **Jansen-Winkeln**, C. **Reddig** and **N. Reithinger**. Bidirectional Use of Knowledge in the Multi-Modal Natural-Language Access System XTRA. In: *Proceedings of the 11th International Joint Conference on Artificial Intelligence, Detroit, USA*, 1989.

[Appelt 87] D. **Appelt**. Reference and Pragmatic Identification. In: *Proc. of the TINLAP-3*, pp. 128–132, 1987.

[Aue et al. 89] D. **Aue**, S. **Heib** and A. **Ndiaye**. *SB-ONE Matcher: Systembeschreibung und Benutzeranleitung.* Memo Nr. 32, Univ. des Saarlandes, 1989.

[Bartsch et al. 89] W. **Bartsch**, A. D. **Kader**, I. **Raasch** and R. **Schmitt**. *MOTHOLOG: ein Prolog Meta-Interpreter zum erklärbaren Beweisen disjunktiver und negativer Ziele.* Memo, Univ. des Saarlandes, 1989.

[Barwise & Cooper 81] J. **Barwise** and R. **Cooper**. Generalized Quantifiers and Natural Language. *Linguistics and Philosophy*, 4:159–219, 1981.

[Barwise & Perry 83] J. **Barwise** and J. **Perry**. *Situations and Attitudes.* Bradford Books. Cambridge, Mass.: MIT Press, 1983.

[Bauer & Merziger 89] M. **Bauer** and G. **Merziger**. *Conditioned Circumscription: Translating Defaults to Circumscription.* Memo 34, Univ. des Saarlandes, 1989.

[Belnap 77] N. D. **Belnap**. A Useful Four-Valued Logic. In: J. Dunn and G. Epstein (eds.), *Modern Uses of Multiple-Valued Logic*, pp. 8–40, 1977.

[Bergmann et al. 87] H. **Bergmann**, M. **Fliegner**, M. **Gerlach**, H. **Marburger** and M. **Poesio**. *IRS – The Internal Representation Language.* WISBER Memo 14, Univ. of Hamburg, 1987.

[Bosch & Wellner 89] G. **Bosch** and I. **Wellner**. *ASCON: a reimplementation of the FUZZY associative network.* Memo, Univ. des Saarlandes, 1989.

[Brachman et al. 83] R. J. **Brachman**, R. F. **Fikes** and H. J. **Levesque**. KRYPTON: Integration Terminology and Assertion. In: *Proceedings of the 3rd National Conference of the American Association for Artificial Intelligence, Washington, DC*, 1983.

[Brachman et al. 85] R. J. **Brachman**, V. P. **Gilbert** and H. J. **Levesque**. An Essential Hybrid Reasoning System: Knowledge and Symbol Level Accounts of KRYPTON. In: *Proceedings of the 9th International Joint Conference on Artificial Intelligence, Los Angeles, CA*, 1985.

[Croft 85] B. **Croft**. Determiners and Specification. In: J. Hobbs et al. (ed.), *Commonsense Summer: Final Report*, Chapter 7, Leland Stanford Junior University: Center for the Study of Language and Information, October 1985.

[Donnellan 66] K. **Donnellan**. Reference and Definite Description. *Philosophical Review*, 75:281–304, 1966.

[Ginsberg 86] M. L. **Ginsberg**. Multi-valued logic. In: *Proceedings of the 5th National Conference of the American Association for Artificial Intelligence, Philadelphia, PA*, pp. 243–247, 1986.

[Habel 86] Ch. **Habel**. *Prinzipien der Referentialität*. Informatik-Fachberichte, 122. Berlin, Heidelberg, New York, London, Paris, Tokyo, Hong Kong: Springer, 1986.

[Heim 82] I. **Heim**. *The Semantics of Definite and Indefinite Noun Phrases*. PhD thesis, Univ. of Massachusetts, 1982.

[Heinz & Matiasek 89] W. **Heinz** and J. **Matiasek**. Die Anwendung Generalisierter Quantoren in einem natürlichprachigen Datenbank-Interface. In: *Proc. of the Proceedings der 4. Österreichischen Artifical Intelligence Tagung*, p. ??, Berlin, Heidelberg, New York, London, Paris, Tokyo, Hong Kong, 1989.

[Hobbs & Shieber 87] J. R. **Hobbs** and S. M. **Shieber**. An Algorithm for Generating Quantifier Scopings. *Computational Linguistics*, 13(1-2):47–63, January-June 1987.

[Jansen-Winkeln 89] R. M. **Jansen-Winkeln**. *HC2LC: a PROLOG to Lisp Compiler*. Memo, Univ. des Saarlandes, 1989.

[Kamp 81] H. **Kamp**. A Theory of Truth and Semantic Representation. In: J. A. G. Groenendijk, T. M. V. Janssen, and M. B. J. Stokhof (eds.), *Formal Methods in the Study of Language*, pp. 277–322, Amsterdam: Mathematical Centre, 1981.

[Kronfeld 86] A. **Kronfeld**. Donnallan's Distinction and a Computational Model of Reference. In: *Proceedings of the 24th Annual Meeting of the ACL, Columbia University, New York, NY*, pp. 186–191, 1986.

[Levesque 84] H. **Levesque**. A Logic of Implicit and Explicit Belief. In: *Proceedings of the 4th National Conference of the American Association for Artificial Intelligence, Austin, TX*, pp. 198–202, 1984.

[Lifschitz 86] V. **Lifschitz**. Pointwise Circumscription: Preliminary Report. In: *Proceedings of the 5th National Conference of the American Association for Artificial Intelligence, Philadelphia, PA*, pp. 406–410, 1986.

[Link 87] G. **Link**. Generalized Quantifiers and Plurals. In: P. Gärdenfors (ed.), *Generalized Quantifiers. Linguistical and Logical Approaches*, pp. 151–180, Dordrecht, Boston, Lancaster, Tokyo: D. Reidel, 1987.

[Loebner 87] S. **Loebner**. Natural Language and Generalized Quantifier Theory. In: P. Gärdenfors (ed.), *Generalized Quantifiers. Linguistical and Logical Approaches*, pp. 181–201, Dordrecht, Boston, Lancaster, Tokyo: D. Reidel, 1987.

[Loenning 87] J. T. **Loenning**. Mass Terms and Quantification. *Linguistics and Philosophy*, 10:??, 1987.

[MacGregor & Brill 88] R. **MacGregor** and D. **Brill**. *LOOM User Manual*. DRAFT, ISI, 1988.

[McAllester 82] D. A. **McAllester**. *Reasoning Utility Package User Manual*. AI Memo 667, MIT, April 1982.

[Moser 83] M. G. **Moser**. *An overview of NIKL, the New Implementation of KL-ONE*. Annual Report Nr. 5421, BBN, 1983.

[Nebel 86] B. **Nebel**. NIGEL Gets To Know Logic. In: *Proceedings des GWAI-86 und der 2. Österreichischen Artificial-Intelligence-Tagung, Ottenstein/Niederösterreich*, pp. 75–86, Berlin, Heidelberg, New York, London, Paris, Tokyo, Hong Kong, 1986.

[Patel-Schneider 89] P. F. **Patel-Schneider**. A Four-Valued Semantics for Terminological Logics. *Artificial Intelligence*, 38(3):319–152, 1989.

[Poesio 88] M. **Poesio**. *The QUARK Reference Manual*. Memo Nr. 22, Univ. Hamburg, 1988.

[Putnam 75] H. **Putnam**. The Meaning of 'Meaning'. In: K. Gunderson (ed.), *Language, Mind, and Knowledge*, Minneapolis: University of Minnesota Press, 1975.

[Reddig 88] C. **Reddig**. 3D in NLP: Determiners, Descriptions, and the Dialog Memory in the XTRA System. In: *Proceedings des GWAI-88, Geseke, West Germany*, p. ??, Berlin, Heidelberg, New York, London, Paris, Tokyo, Hong Kong, 1988.

[Reiter 80] R. **Reiter**. A Logic for Default Reasoning. *Artificial Intelligence*, 13(2):81–132, 1980.

[Russell 05] B. **Russell**. On Denoting. *Mind*, 14:479–493, 1905.

[Scha & Stallard 88] R. **Scha** and D. **Stallard**. Multi-level plurals and distributivity. In: *Proceedings of the 26th Annual Meeting of the ACL, State University of New York, Buffalo, NY*, pp. 17–24, 1988.

[Schmolze & Brachman 81] J. G. **Schmolze** and R. J. **Brachman**. *Proc. of the 1981 KL-ONE Workshop*. FLAIR Technical Report No. 4, Fairchild Laboratory for AI Research, 1981.

[Schmolze 87] **Schmolze**. *Syntax and Semantics of NIKL*. Technical Report, ??, 1987.

[Shieber 87] S. M. **Shieber**. Separating Linguistic Analysis from Linguistic Theories. In: P. Whitelock, M. M. Wood, H. L. Somers, R. Johnson, and P. Bennett (eds.), *Linguistic Theory & Computer Applications*, pp. 1–36, London, San Diego, New York, Berkeley, Boston, Sydney, Tokyo, Toronto: Academic Press, 1987.

[Stickel 85] M. A. **Stickel**. Automated Deduction by Theory Resolution. In: *Proceedings of the 9th International Joint Conference on Artificial Intelligence, Los Angeles, CA*, pp. 1181–1186, 1985. (Also in: Journal of Automated Reasoning, Vol. 1, No. 4, 1985, pp.333-355.).

[Thijsse 82] E. **Thijsse**. On some Proposed Universals of Natural Language. In: A. ter Meulen (ed.), *Studies in Modeltheoretic Semantics*, Dordrecht: Foris, 1982.

[van Benthem 83] J. **van Benthem**. Determiners and Logic. *Linguistics and Philosophy*, 6:447–478, 1983.

[Vilain 85] M. **Vilain**. The restricted language architecture of a hybrid representation system. In: *Proceedings of the 9th International Joint Conference on Artificial Intelligence, Los Angeles, CA*, pp. 547–551, 1985.

[von Luck 89] K. **von Luck**. *Räpresentation assertionalen Wissens im BACK-System – eine Fallstudie –*. KIT Report 72, TU Berlin, 1989.

[von Luck et al. 87] K. **von Luck**, B. **Nebel**, Ch. **Peltason** and A. **Schmiedel**. *The Anatomy of the BACK-System*. KIT-BACK Report, TU-Berlin, 1987.

[Wellman & Simmons 88] M. P. **Wellman** and R. G. **Simmons**. Mechanisms for Reasoning about Sets. In: *Proceedings of the 7th National Conference of the American Association for Artificial Intelligence, Saint Paul, Minnesota*, pp. 398–402, 1988.

[Westerstahl 85] D. **Westerstahl**. Logical Constants in Quantifier Languages. *Linguistics and Philosophy*, 8:387–413, 1985.

Functor-Argument Structures
for the Meaning of Natural Language Sentences
and Their Formal Interpretation

Bernd Mahr and Carla Umbach

Technische Universität Berlin
Institut für Software und Theoretische Informatik
Franklinstraße 28/29
D-1000 Berlin 12

Introduction

FAS (Functor Argument Structure) is the sentence semantic level of representation and transfer that is presently being developed in the project KIT-FAST[1]. KIT-FAST is a research project concerned with the machine translation (MT) of natural languages.

FAS is the second of three transfer levels foreseen in the machine translation model used in KIT-FAST. The FAS component is the project's main concern in this phase, where we work on translation between German and English. This includes the definition of FAS, the development of transfer rules in both directions and the implementation of procedures for analysis, transfer and generation.

In this paper we propose a method of associating model-theoretic semantics with FAS expressions which is based on an approach of the ADJ-group (see [Goguen/Thatcher/Wagner, 76]). The syntax of FAS is given in a context free grammar with some additional characteristics. According to the ADJ method a context free grammar is translated into a signature and the corresponding term algebra is defined as initial semantics of the context free language.

Due to the additional characteristics of FAS syntax -non-atomic categories and multiple use of functional features- the ADJ procedure has to be modified, since these characteristics cannot be adequately described by classical signatures. However, the problems with applying the ADJ approach are not specific to FAS: Non-atomic categories are well-known as

---

[1]The project KIT-FAST constitutes one of the complementary research projects of Eurotra-D and receives grants by the Federal Minister for Research and Technology under contract 1013211. (KIT=Artificial Intelligence and Text Understanding, FAST=Functor Argument Structure for Translation). This paper was presented at the workshop "GPSG and Semantics", cf. [Busemann et al. 89].

structured or polymorphic types; multiple use of functional features leads to the problem of overloading.

In order to overcome these problems the concept of polymorphic signatures is introduced and suitable models are declared. A polymorphic signature allows overloading operators and comprises a type signature, which makes it possible to describe polymorphic types.
On this basis, a procedure can be defined which translates an FAS syntax into a signature. Thus FAS expressions can be interpreted as terms.

Expanding the traditional signature into the polymorphic signature has wide-reaching consequences for the concept of model theory corresponding to signatures. This expansion, however, seems to be justified, as both overloading and structured types are well-known and widespread phenomena. Examples can be found in programing languages, in syntactic descriptions of natural languages and in various languages of predicate logic for the representation of knowledge. Thus our approach goes beyond an FAS interpretation.

Due to the expansion to polymorphic signatures a large amount of the theoretical work in the field of term interpretation needs to be revised. This enterprise is still in the initial stages. This paper is limited to the following points:
- outlining the concept and introducing the syntax of FAS,
- showing the problems of a term interpretation of FAS,
- gradually expanding the concept of signatures to include overloading operators and polymorphic types,
- defining the interpretation of the extended signature,
- specifying a procedure for translating FAS syntax into polymorphic signatures.

1. The concept of FAS

In the KIT-FAST project we start from a model of machine translation which comprises several levels of representation and transfer. In this model a natural language text is represented at four different levels. The first three also form transfer levels (see Figure 1).
- The syntactic level represents the surface syntax of sentences; the formalism is Generalized Phrase Structure Grammar (GPSG) following [GKPS, 85] and [Hauen-schild/Busemann, 88].
- The sentence semantic level is realized by FAS. Its main features are the logical relationships between the elements in a sentence, the semantic valency frames of predicates and the thematic structuring of sentences.
- The textual level of representation is to make explicit referential relations in the text. We plan to use some version of a formalism developed in the field of AI.
- The deepest level of representation contains the argumentative structure of the text and is conceived as invariant during translation. The development of a corresponding formalism is one of the future goals of the project.

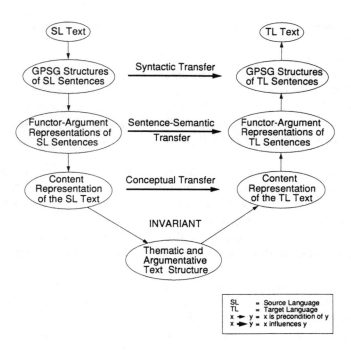

Figure 1: Levels of Representation and Transfer in the Machine Translation Model

The concept of FAS originates in two very different linguistic approaches: on one hand the proposal of semantic interpretation made for GPSG in [GKPS, 85], and on the other the Prague School's theory of Functional Sentences Perspective. For details see [Hauenschild/ Umbach, 88] and [Busemann/Hauenschild, 89].

Two aspects can be distinguished within the information represented in FAS: distribution and function. The distributional aspect refers to what is typically represented by categories in context free syntax, whereas logical relations in a sentence are an example of the functional aspect. In our view, both aspects are of equal importance, which means that the FAS formalism has to be able to cope with both of them.

We have chosen a context-free grammar with complex categories, where a distinct feature (op) must be present in every category. (For the concept of complex categories see [Kratzer/Pause/ Stechow, 73]). On the one hand complex categories permit a differentiated representation of distributional information, on the other, they make it possible to label the nodes with functional relations too. This is done with the help of the distinct feature "op" (operation).

2. The Syntax Formalism for FAS

Definition: FAS Syntax
An FAS syntax is given by:

    CAT   - a set of categories,
    LEX   - a set of lexicals,
    PROD - a set of production rules,
              the rules have the form $c ==> w$ or $c ==> a$
              where $c \in CAT$, $w \in CAT^+$, $a \in LEX$

such that

(1) The categories of an FAS syntax are given by:

    H - a set of 'main categories'
    M - a set of features
    W - a set of values
    $(V_m)_{m \in M}$ - value variables for each feature $m \in M$
    $f: H \dashrightarrow 2^M$ - assigning a set of features to each 'main category'
    $wb: M \dashrightarrow 2^W$ - assigning a value domain to each feature

Thus the set CAT of categories can be defined by:
$$CAT = \{\ (h, (<m,v>,...))\ |\ h \in H,\ m \in f(h),\ v \in wb(m) \cup V_m\ \}$$

(2) Every FAS syntax contains the distinct feature op:

    - $op \in M$
    - $op \in f(h)$ for each $h \in H$

Remarks:
- The categories of an FAS syntax are called 'complex categories' according to their internal structure.
- The set of features assigned to a main category is assumed to be ordered.

Definition: FAS Expression, FAS Language
- An FAS expression is a terminated derivation tree.
- An FAS language consists of all FAS expressions licenced by an FAS syntax.

3. Term Interpretation of FAS

There are different reasons for defining formal semantics for FAS. One of these is for FAS to be comparable with other sentence-semantic approaches, in particular those used in the machine translation field. Another reason can be found in the transition from the FAS level to a textual level, which implies the necessity of inference processes. FAS semantics is not designed as just an intermediate level but as a formal basis for transition.

The idea of FAS semantics can be formulated in the classical framework of linguistic formal semantics, i.e. denotational model-theoretic semantics associated with a compositional interpretation function. (see e.g. Montague's Universal Grammar, [Montague, 70]). In this framework we chose term interpretation, because this implies the concept of initial semantics, which is able to serve as a basis of comparison and further interpretation.

In 1976 the ADJ-group developed a procedure for defining initial semantics for contex-free (cf) languages. (See [Goguen/Thatcher/ Wagner/Wright, 77]). The cf-grammar is translated into a signature where the sorts are given by the nonterminals and for each rule there is an operation declaration consisting of the nonterminals of the right hand side as argument sorts and the nonterminal of the left hand side as range sort.
Each terminated derivation tree of the cf-grammar corresponds to exactly one term of the sort <start>. The term algebra of this signature is conceived as the initial semantics of the cf-language.

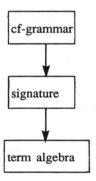

In adapting this procedure to FAS syntax, two problems emerge which are caused by the differences between conventional cf-grammars and FAS:
- the name problem: how to choose the names of the signature's operations?
- the sort problem: how to construct the sorts of the signature?

Both problems are connected with interesting aspects of signatures and their interpretation. Neither problem is specific for FAS syntaxes but they are well known e.g. in the field of programming languages. The name problem refers to overloading operators, the sort problem refers to polymorphic typing.

When translating a cf-grammar into a signature there are various possibilities to choose the names of the operators. For this reason the translation procedure is now formulated without a special naming convention but leaves it open as a parameter.
In the next section four different name mechanisms and their effects on the signature will be discussed.

<u>Translation of a cf-grammar into a signature:</u>
Given a context free grammar CF = (NT, T, P, S) where
   NT: set of nonterminals,
   T: set of terminals,
   P: set of production rules, each rule has the form 'n ==> w' with $n \in NT$, $w \in (NT \cup T)^+$,
   S: the startsymbol, $S \in NT$.
The cf-grammar is translated into a signature SIG(CF) = (S, OP) where
     - sorts S = NT,
     - for each rule 'n ==> w' $\in P$ there is an operation declaration op that has the form:
     name('n ==> w'): nt(w) --> n
     where nt(w) provides the string of nonterminals in w, and name ('n ==> w') is a
     naming mechanism (see next section).

4. The Name Problem

Various mechanisms can be conceived to select the operator names in translating a cf-grammar into a signature. Both the ADJ mechanism and the numbering mechanism are independent of the cf- grammar in that for each rule a new operator name is introduced. The third mechanism has been developed for the demands of a specification language. The FAS mechanism takes into account that FAS syntax contains explicit information about the functional aspect of phrases, this being the value of the op-feature.

<u>Name mechanism:</u>
name: cf production --> operation symbol
 (1) According to the ADJ procedure
   The operation name is created by conceiving the production rule as a string:
   name ('n ==> w') = "n==>w"
 (2) Numbering
   The production rules are assumed to be numbered. The respective number of a rule gets
   the operator name: name ('n ==> w') = #('n ==> w')
 (3) According to [Ehrig/Mahr, 85]
   The terminals of the right side of a rule are used as the operation name:
   name ('n ==> w') = t(w) (t(w) provides string of the terminals in w)
   (This procedure is restricted to grammars with a 'fixed word order', i.e. if there are
   rules p1, p2 with t(w1)=t(w2)≠λ, then nt(w1)≠nt(w2)).
 (4) Designed for FAS
   The production rules are assumed to be in Chomsky normal form. In terminal rules, the
   terminal is conceived as operation name. In nonterminal rules, the value of the op-feature
   in the left hand side category constitutes the operation name:
   name ('n ==> w') = if 'n ==> a' then a
                       else   val(op,n)
   (val(op,n) yields the value of the op-feature of category n)

When name mechanisms 1 and 2 are applied the translation yields classical signatures. Using mechanisms 3 or 4 operator names may occur several times in the signature. Such operators are called overloading operators. Classical signatures do not admit overloading operators.

Example: Translation of a cf-grammar
       cf = { A,B,C },
          { a, f },
          { A ==> f B,
            A ==> f C,
            B ==> a,
            C ==> a }

with Name-Mechanism 2:	with Name-Mechanism 3:
sig(cf) = <u>sorts</u> A, B, C	sig(cf) = <u>sorts</u> A, B, C
<u>opns</u> #1: B --> A	<u>opns</u> f: B --> A
#2: C --> A	? f: C --> A
#3: --> B	a: --> B
#4: --> C	? a: --> C

Certainly, overloading operators can be avoided by indexing. But then generalizations present in the FAS syntax were no longer expressed in the signature. Moreover, overloading is a well known phenomenon. Programming languages usually provide operators for addition and multiplication which are defined on integers and real numbers as well. It generally appears convenient to admit overloading operators in signatures. In Section 6 a signature with overloading is defined and its effects on the evaluation of terms is examined.

5. The Sort Problem

Types, sorts, categories and nonterminals are notions from different formalisms denoting the same fact; that being the distribution of objects, terms, phrases etc. Hence nonterminals in a context free syntax are translated as sorts of the corresponding signature.

In contrast to classical cf-grammars the categories in an FAS syntax are not atomic but structured and, moreover, polymorphic. There are main categories, features and values, and rules as to how to construct FAS categories. Moreover, the FAS categories can contain variables for feature values. In translating FAS syntax into a signature we encounter the problem of how FAS categories can be interpreted as sorts.

FAS categories are not recursively constructed and the set of feature values is finite. Thus another auxiliary construction is conceivable: each 'ground instance' of a complex category can be taken as a particular sort. But again generalizations present in the FAS syntax are no longer expressed in the signature.

Polymorphic types are no peculiarity of FAS syntax. Such type concepts can be found in various different areas. Structured types such as lists and records are used in programing languages. For example, the programing language ML provides polymorphic types, see [Milner, 84]. Syntax formalisms such as GPSG and categorial grammar handle complex categories, that are structured types. Languages of logic, e.g. intensional logic IL (Montague), make use of structured types.

Example: structured types in different languages (simplified)

Pascal, integer list:	ARRAY [1..n] OF integer
ML, any list:	'a list
GPSG, verbal phrase:	[N:-, V:+, ... SLASH:[..]]
Categorial Grammar, transitive verb:	(n\s)/n
typed logic, 2nd order predicate:	<<e,⊳>, ⊳
FAS-D, nomen in any configuration:	n_pred(... conf: A, ...)

Although the types may appear different their concepts are similar. There are atomic elements, constructors and syntactic rules as how to construct a type. In some of the type concepts variables are allowed. Hence it is possible to define operations on types including variables. Such types are called polymorphic types.

In more recent developments in categorial grammar also axioms on types are used, i.e. 'type lifting rule' [Partee, 87] or 'Geach rule' [Geach, 71]. In this context the linguistic aspects of a type calculus called Lambek calculus, [Lambek, 58], are examined. So the question emerges whether such a type concept can be transfered to the sorts of a signature to cope with polymorphic types in an algebraic framework.

These considerations together with those concerning overloading operators lead to an expansion of the concept of signatures and the corresponding model theory with applications beyond an FAS interpretion. The ideas of this expansion are presented in the following sections; their full elaboration would be beyond the scope of this paper.

6. Signature with Overloading

Classical signatures entail the restriction that the sets of operation symbols have to be disjunct. Hence overloading operators cannot appear in classical signatures. Here, the definition of signatures is revised to allow overloading operators. The consequences as to the termalgebra and the evaluation function are discussed. A characterization is given for signatures that preserve a unique evaluation function.

Definition: Signature with Overloading
A signature with overloading SIG = (S, OP) consists of

 S: set of sort symbols,

 $OP = (OP_{w,s})$ $w \in S^*$, $s \in S$

  $OP_{w,s}$ set of operation symbols with argument sorts w and range sort s.

(Contrary to the classical definition the condition of disjoint sets of operation symbols is missing.)

Overloading creates a problem with the evaluation function *eval* from the signature's term algebra to each sig-algebra. (For a definition of *eval* see [Ehrig/Mahr, 85]). In the case of classical signatures the evaluation function is unique as a homomorphism. Term algebras referring to signatures with overloading may, however, contain the same terms in different carriers, which is not possible in classical term algebras. Hence, the uniqueness of the evaluation function may get lost.

Example: eval: T_{sig_1} --> A is not unique:

sig_1 =		T_{sig_1} =	A =
sorts	A	{f(a)}	Z
	B	{a}	Z^+
	C	{a}	Z^-
opns	f: B --> A	f^T: {a} --> {f(a)}	*2: Z^+ --> Z
	f: C --> A	f^T: {a} --> {f(a)}	*2: Z^- --> Z
	a: --> B	a^T: --> {a}	+1: --> Z^+
	a: --> C	a^T: --> {a}	-1: --> Z^-

with Z denoting the set of integers and accordingly defined subsets Z^+, Z^- and operations.

eval: T_{sig_1} --> A is a family of functions:
 $eval_A$: T_{SIG-A} --> Z
 $eval_B$: T_{SIG-B} --> Z^+
 $eval_C$: T_{SIG-C} --> Z^-

term evaluation:
 $eval_B$ (a) = +1
 $eval_C$ (a) = -1
?? $eval_A$ (f(a)) = *2($eval_B$(a)) = +2
or $eval_A$ (f(a)) = *2($eval_C$(a)) = -2

As shown in this example there are several possibilities to evaluate the term f(a), either +2 or -2. Hence, there are several evaluation functions $eval_A$ with respect to the same carrier.

It is not the case that the evaluation function is necessarily ambiguous in connection with overloading operators. Several constellations preserve the uniqueness of *eval*. This concerns term algebras in which for each term the types of its subterms are fixed, at least in the respective context. The context of a subterm is determined by several components: on the one hand by the operator and the type of the superterm and on the other by the additional subterms and their possible types. That leads to a characterization of signatures which preserve unique evaluation:

Definitions:
 (1) t:p is is called <u>typing of term t</u> if t is a term of sort p, i.e. $t \in T_p$.
 (2) A declaration $f:s_1..s_i..s_n$ --> r is <u>consistent</u> with a typed term $f(t_1,..,t_i,..,t_n)$:r if for each subterm t_i there is a typing $t_i:s_i$. (n≥0)

Fact: Characterization of Signatures Preserving Unique Evaluation
A signature with overloading preserves the unique evaluation function eval: $T_{SIG} \to A$ into each sig-algebra A iff for each typed term $t = f(t_1,..,t_n):r$ there is exactly one consistent declaration.

Remark:
Surprisingly, this characterization allows various combinations of overloading operators (see example sig_2 below).

Proof:
==>: Given a typed term $f(t_1,...,t_n):r$ $(n>0)$. The term evaluation is defined by
$$eval_r(f(t_1,...,t_n)) = f^A(eval_{typ(t1)}(t_1),...,eval_{typ(t2)}(t_n))$$
If $eval_r$ is unique, the domain of the eval functions in the recursion must be fixed. Hence the types of the subterms are fixed as soon as the type of the superterm is fixed.
This, however, means that there is exactly one consistent declaration for a typed term:
$f: typ(t_1) ... typ(t_n) \to r$
Given a typed constant f:r, then $eval_r(f) = f^A$ is unique and there is exactly one consistent declaration, anyway.
<==: Given a typed term $f(t_1,...,t_n):r$ $(n \geq 0)$ and its unique consistent declaration
$f:s_1...s_n \to r$. Then eval(t) can be defined in a unique way:
$$eval_r(f(t_1,...,t_n)) = f^A(eval_{s1}(t_1),...,eval_{sn}(t_n))$$

Remark:
Definition and characterization are restricted to terms without variables. Since the types of variables are unique in a signature with overloading as well, both definition and characterization can easily be extended to terms with variables. Given a variable assignment
ß: X $\to$ A, the evaluation function has to expand this assignment: $eval_ß: T_{SIG}(X) \to A$
with $eval_ß(x) = ß(x)$.

Examples:
(1) sig_1 = <u>sorts</u> A, B, C
          <u>opns</u> f: B --> A
                f: C --> A
                a: --> B
                a: --> C
sig_1 does not preserve unique term evaluation, since the typed term f(a):A has two consistent declarations: f: B --> A is consistent with f(a):A on account of a:B, and
f: C --> A is consistent with f(a):A on account of a:C.

(2) sig_2 = <u>sorts</u> A, B, C, D
          <u>opns</u> f: B --> A
                f: C --> D
                a: --> B
                a: --> C

sig_2 preserves unique term evaluation, since for each typed term there is exactly one consistent declaration: for f(a):A the declaration f: B --> A; for f(a):D the declaration f: C --> D; for the typed constants there is exactly one consistent declaration in any case.

(3) sig_3 = <u>sorts</u> A, B, C, D, E, G
         <u>opns</u> f: A D --> C
                 f: B E --> C
                 f: G D --> C
                 a: --> A
                 a: --> B
                 d: --> D
                 d: --> G
                 e: --> E

sig_3 preserves unique term evaluation, since for each typed term there is exactly one consistent declaration: for f(a,d):C the declaration f: A D --> C; for f(a,e):C the declaration f: B E --> C; for f(d,d):C the declaration f: G D --> C; (the same for the constants).

Now it has to be considered which consequences result from signatures not satisfying the above characterization. Such signatures imply several evaluation functions due to the different possibilities of evaluating a term. The evaluation can then be defined as a set of functions:

<u>Definition:</u>
EVAL is the smallest set of functions eval: T_{SIG} --> A, where
(1) For each constant declaration k:r there is $g \in$ EVAL with $g(k)=k^A$.
(2) Given $g_1...g_n \in$ EVAL. For all typed terms $t=f(t_1,...,t_n):r$ (n>0) and for each consistent declaration of a typed term there is a mapping $g \in$ EVAL with
   $g(f(t_1,...,t_n))=f^A(g_1(t_1),...,g_n(t_n))$.

In case a signature preserves unique evaluation EVAL is a singleton. Otherwise the evaluation is at least not arbitrarily ambiguous since the set EVAL of evaluation functions can be constructed according to the signature and each mapping of EVAL is still homomorphic.

Non-unique evaluation raises many further questions which cannot be dealt with here. Their linguistic interpretation in connection with FAS, however, seems to be very interesting.

7. The Polymorphic Signature

In the translation from an FAS syntax into a signature there is a problem in that contrary to the sorts of a signature the categories of an FAS syntax are structured and also polymorphic. Since such a type concept is used in various fields it seems to be reasonable to transfer this type concept to the sorts of a signature to cope with polymorphic types in an algebraic framework.

The type concept of classical signatures is restricted to atomic sorts. There is an extension of signatures where the type concept includes subsorting. Thus a partial order on sorts can be defined. But the sorts remain atomic. Further extensions may be found in [Poigné, 86] and [Möller/Tarlecki/Wirsing, 88] but are not yet fully elaborated.

A type concept comprising the types shown in Section 5 entails basic expressions, type constructors, variables, predicates on types and axioms. Moreover the type constructors are restricted to specific types in their 'argument places'. Thus the types or sorts of the types themselves have to be taken into account.

It is convenient to describe such a type concept itself by a signature. The constants of the signature form the basic type expressions and the operators form the type constructors. The sorts of types can be expressed and the constructors can be limited to certain sorts and numbers of arguments. Moreover it is possible to define predicates on types to stipulate certain type axioms.

Terms to a given type signature represent type expressions. These type expressions serve as sorts of a signature on the object level.
The main point, however, in a concept of polymorphic types is the presence of variables: if type expressions contain variables one can define polymorphic operations in the object signature.

These observations lead to the definitions of type signature, type theory and object signature, and by combination to the definition of a polymorphic signature.

Definition: Type Signature
   A <u>type signature</u> is a tuple $\tau = (K, C, P, X)$ where
    K - a set of kinds
    $C = (C_{w,k}) w \in K^*, k \in K$
        $C_{w,k}$ a set of type constructors with argument kinds w and range kind k,
    $P = (P_w) w \in K^+$
        P_w a set of type predicates with argument kinds w,
    $X = (X_k) k \in K,$
        X_k a set of type variables of kind k,

Definition: Type Expressions
$T_C(X)$ is the set of <u>type expressions</u>:
   - type variables are type expressions,
   - type constructors with arity zero are type expressions,
   - if c is a type constructor of the form c: $k_1...k_n$ --> k and $t_1,..,t_n$ are type expressions of
     suitable kind, then $c(t_1,..,t_n)$ is a type expression.
T_C is the set of <u>ground type expressions</u>, i.e. type expressions without variables.

Given a type signature $\tau = (K, C, P, X)$, we define <u>type axioms</u> to be closed first order

τ-formulas build up from atomic τ-formulas by logical constants and quantification. Atomic τ-formulas may be of the form $P(t_1,...,t_n)$, $t_1=t_2$, or $t_1 \leq t_2$.

Using this concept of type axioms we define the notion of type theory as follows:

Definition: Type Theory

A <u>type theory</u> is a pair (τ, Ax) consisting of a type signature τ and a set of type axioms Ax with respect to τ.

(we will use the symbol τ for denoting type signatures as well as type theories).

Definition: Object Signature

Given a type theory τ, an <u>object signature</u> is a tuple $\sigma = (S, OP)$:

    S - a set of type expressions, $S \leq T_C(X)$,

    $OP = (OP_{w,t})$ $w=t_1..t_n$, $t_1,..,t_n,t \in S$, $n \geq 0$

        $OP_{w,t}$ a set of operation symbols with   argument sorts $t_1,..,t_n$ and range sort t.

Definition: Polymorphic Signature

A <u>polymorphic signature</u> is a pair (τ, σ) consisting of a type theory τ and an object signature σ with respect to τ.

<u>Examples of Type Theories</u>:

(1) $\tau 1 = $ <u>kinds</u> sort

        <u>constr</u> s_1: --> sort

                :

             s_n: --> sort

$\tau 1$ produces type expressions $s_1...s_n$. Since variables are missing one cannot express polymorphic types. A polymorphic signature based upon $\tau 1$ corresponds to a classical many sorted signature.

(2) $\tau 2 = $ <u>kinds</u> sort

        <u>constr</u> s_1: --> sort

                :

             s_n: --> sort

        <u>prds</u> $\leq$: sort sort

        <u>axms</u> $s_1 \leq s_2$

Likewise, type signature $\tau 2$ cannot express polymorphic types. The axiom determines the inclusion s_1 in s_2. A polymorphic signature based upon $\tau 2$ corresponds to a signature with subsorting.

(3) $\tau 3 = $ <u>kinds</u> α, ß

        <u>constr</u> M: α α --> ß

               D: ß --> α

               C: ß --> α

        <u>vars</u> X,Y: α

        <u>axms</u>   D(M(X,Y)) = X

               C(M(X,Y)) = Y

Type signature $\tau 3$ contains variables. Hence it is possible to express polymorphic types.

There are various possibilities to interpret τ3, e.g. maps, domain and codomain, or product, first and second projection.

Examples of Polymorphic Signatures:

(4) τ = <u>kinds</u> sort
          <u>constr</u> bool: --> sort
          <u>vars</u> X: sort

     σ = <u>sorts</u> X, bool
          <u>opns</u> if_then_else: bool X X --> X
                  true: --> bool
                  false: --> bool
          <u>eqns</u> if true then x else y = x
                  if false then x else y = y

In (4) the polymorphic if_then_else is defined which is frequently used in classical signatures without being explicitly defined.

(5) τ = <u>kinds</u> type
          <u>constr</u> e: --> type
                  t: --> type
                  <>: type type --> type
          <u>vars</u> X, Y: type

     σ = <u>sorts</u> e, t, X, Y, <X,Y>, <X,t>
          <u>opns</u> fkt_App: <X,Y> X --> Y
                  neg: t --> t
                  and: t t --> t
                  join: <X,t> <X,t> --> <X,t>

(5) shows part of a typed logical language. Functional application is defined as polymorphic operation fkt_App. The operation *join* represents 'polymorphic and'. (This definition of *join* does not cover all "conjoinable types" as defined by [Partee/Rooth, 83]).

8. Interpretation of Type Theories and Polymorphic Signatures

<u>Definition</u>: Interpretation of a Type Theory

A type theory τ = (K, C, P, X, Ax) is interpreted by a <u>type model</u> A with
- carriers k_i^A for each $k \in K$, all carriers k_i^A are sets of sets;
- operations c^A: --> k^A for each constant constructor c:-->k in C,
$$c^A: k_1^A \text{ x...x } k_n^A \text{ --> } k^A \text{ for each constructor } c:k_1..k_n\text{-->}k \text{ in } C,(n{\geq}1);$$
- predicates $p^A \subseteq k_1^A \text{ x...x } k_n^A$ for each type predicate $p:k_1...k_n$ in P (n≥1).
- all axioms are valid in A.

<u>Remark</u>:

Validity of type axioms is based upon variable assignments ß: X --> A and evaluation functions $\text{eval}_ß$: $T_τ(x)$ --> A and is defined as standard first order validity where equality and inclusion have the fixed interpretation as equality and inclusion of sets.

Definition: Ground Type Instance
Given a polymorphic signature (τ, σ) with $\sigma = (S, OP)$ we define for $t \in S$ the sets of <u>ground type instances</u> GS(t) with respect to σ as follows:
- $t \in$ GS(t) if t is a ground type expression
- $t[x_1|t_1,...,x_n|t_n] \in$ GS(t) if t contains type variables $x_1,...,x_n$ and $t_1,...,t_n$ are ground type instances of some type expression $s_1,...,s_n \in S$ of appropriate kind.
  (Here square brackets denote substitutions in the usual sense.)
By $GS(\sigma)$ we denote the set of ground type instances with respect to σ.

Definition: Interpretation of a Polymorphic Signature
A polymorphic signature (τ, σ) is interpreted by a polymorphic structure P.
A <u>polymorphic structure</u> is a pair (A, B) consisting of a type model A with respect to the type theory τ and an <u>object structure</u> B with carriers and operations, that refer to sets provided by the type model A as follows:
- carriers t^A for each ground type instances t with respect to σ, when t^A denotes the value of t in A;
- operations - $(f^B: \rightarrow t^A)$ $_{t \in GS(s)}$ for each constant operation symbol f: $\rightarrow$ s in OP,
  - $(f^B: t_1^A \times ... \times t_n^A \rightarrow t_{n+1}^A)$ $_{ti \in GS(si)}$ for each operation symbol f: $s_1..s_n \rightarrow s_{n+1}$ in OP, $(n \geq 1)$

Remarks:
- Type expressions containing variables represent sets of types, that are the sets of types that can be obtained by substitution with ground type instances from the sort list.
- Only ground type instances denote sets in the object structure.
- Operations represent families of operations indexed by all ground type instances that yield ground type instances rather than just single functions.
- If there are no suitable ground type instances, then the indexing set is empty.

Restriction:
Neither type constructor symbols in a type theory nor operation symbols in an object signature have to be disjunct. Hence overloading is permitted in the type theory as well as in the object signature. But there is a restriction in that type theory and object signature have to preserve unique evaluation according to the characterization in Section 6.

Example:

polymorphic signature (τ, σ)	polymorphic structure (τ^*, σ^*)
type theory	type model
τ = <u>kinds</u> s1	τ^* = <u>doms</u> {B, N ...}
<u>constr</u> bool: $\rightarrow$ s1	<u>constr</u> B: $\rightarrow$ {B, N,...}
nat: $\rightarrow$ s1	N: $\rightarrow$ {B, N,...}
<u>vars</u> X: s1	

object signature	object structure
σ = <u>sorts</u> X, bool	σ* = <u>carrs</u> B
<u>opns</u>	<u>opns</u>
if_then_else: bool X X --> X	if_then_else*: B B --> B
true: --> bool	T: --> B
false: --> bool	F: --> B

In this example there is one substitution for the type variable X that yields a ground type. Hence the if_then_else family of operations consists of just one operation. As soon as further ground type expressions (with suitable kinds) are added to the sort list, e.g. kind nat, the if_then_else is defined on them, as well.

9. Procedure to Translate FAS Syntaxes in Polymorphic Signatures

The starting point of the considerations leading to the concept of polymorphic signatures was the need to translate FAS languages into term algebras and thereby declare the semantics of FAS. Now, a procedure is defined to translate arbitrary FAS syntaxes into polymorphic signatures. This procedure is based on the ADJ approach of cf language interpretation. But instead of classical signatures polymorphic signatures are available now. So complex categories and functional features can be appropriately handled.

The procedure consists of two steps:
(1) Translating the FAS system of complex categories into a type theory,
(2) Translating the FAS production rules into operations of an object signature with respect to the above type theory.

Given an FAS syntax (CAT, LEX, PROD).
(1) The <u>FAS categories</u> CAT = (H, M, W, V, f, wb) are translated into a type theory
$\tau_{FAS} = (K, C, P, X, Ax)$ where
- K = M + {cat}
- C = { w:-->m | for all w∈wb(m), for each m∈M }
  +
  { h:m_1...m_j-->cat | where f(h)={m_1,...m_j}, for each h∈H }
- X = { x:m | for each x∈V_m, for each m∈M }
- P = φ
- Ax = φ

This procedure yields a mapping from FAS categories into type expressions of kind *cat*:
typ: CAT --> $T_C(X)_{cat}$ where typ([h,[<m_1:v_1>,...<m_j:v_j>]]) = h(v_1,...v_j).

<u>Remarks:</u>
- The features and the distinct symbol *cat* constitute the kinds of the type theory.
- The feature values constitute the constant constructors, their kinds being fixed by the respective feature to which they belong.

- The main categories form the other constructors. Their argument kinds are defined by their adjoining features, their range kind is represented by the distinct kind *cat*.
- There are no predicates and no axioms, i.e. P and A are empty.

Restriction:

According to the FAS definition the domains of features need not be disjoint. Thus there may be overloading constant constructors in the type theory.

The interpretation of polymorphic signatures has been restricted to those preserving unique evaluation. For this reason it is required that feature domains in an FAS syntax are disjoint.

(2) The <u>production rules</u> of an FAS syntax are translated into an object signature
$\sigma_{FAS} = (S, OP)$ with respect to the type theory τ_{FAS}:
- $S = T_C(X)_{cat}$
- for each production rule '$C_0==>C_1..C_n$' $\in$ PROD there is an operation declaration
name('$C_0==>C_1..C_n$'): $typ(C_1)...typ(C_n)$ --> $typ(C_0)$
with name('$C_0==>C_1..C_n$') = if '$C_0==>C_1$' and $C_1 \in$ LEX then C_1 else val(op,C_0)
(where val(op,n) provides the value of the op-feature of catgorie n)

The type theory τ_{FAS} and the object signature σ_{FAS} constitute the polymorphic signature $(\tau, \sigma)_{FAS}$ of an FAS syntax. Thus, a procedure for term interpretation of FAS expressions is defined.

Conclusion

The polymorphic signature introduced here is not supposed to be a detailed concept. It demands elaboration in different areas. As to the field of algebraic theory, the results concerning the classical concept of signatures have to be revised in the light of polymorphic signatures. In addition, a comparison with other type theories is required. On the other hand, the application to linguistic problems has to be examined. This will raise interesting questions concerning linguistic theory.

Besides, there is a series of questions to be considered, e.g.
- How to extend the concept of polymorphic signatures to include equations?
- What consequences arise when signatures do not preserve unique evaluation but induce a set of evaluation functions? (e.g. the term algebra is no longer initial object in the respective class).
- The concept of parameterized specifications, [Ehrig/Mahr, 85], seems to be related to that of polymorphic signatures. What do they have in common?
- Which phenomena are appropriately represented by overloading and which by polymorphic typed operators? How do the models differ?

With regard to linguistic aspects the concept of polymorphic signatures offers a new perspective. Linguistic phenomena that are regarded as polymorphic can be handled within the framework of denotational semantics. Here a comparison should be made with the linguistic applications of Lambek calculus (see e.g. [Moortgat, 88]). Prior to this there are detailed problems that have to be considered such as:

- Which natural language phenomena should be regarded as overloading and which as polymorphic typing?
- It is an essential task to construct linguistically motivated models.
- Concerning the transition from FAS into a textual representation one must take into account how to base knowledge representation and inferencing on such models.
- It may be a further interesting application of the concept presented here to translate other natural language representations into polymorphic signatures, e.g. our syntactic representation GPSG or the Eurotra-D semantic representation IS. As to LFG the above approach would present interesting problems due to LFG's two levels.
- Finally an implementation of the procedure translating FAS syntaxes into polymorphic signatures would be useful.

Acknowledgement

We thank Christa Hauenschild and all other members of the KIT-FAST group for critical and encouraging comments and Lucy Wilson for her help in translation.

References

[Busemann/Hauenschild, 89]
    S.Busemann, Ch.Hauenschild: From FAS Representations to GPSG Structures. In: [Busemann et al. 89].
[Busemann et al. 89]
    S.Busemann, Ch.Hauenschild, C.Umbach (eds): Views of the Syntax Semantics Interface, TU Berlin KIT-Report 74, 1989.
[Ehrig/Mahr, 85]
    H.Ehrig, B.Mahr: Fundamentals of Algebraic Specification 1, Springer Verlag, Berlin, 1985.
[Geach, 71]
    P.Geach: A Program for Syntax, Synthese 22, 1971, 3-17.
[GKPS, 85]
    G.Gazdar, E.Klein, G.Pullum, I.Sag: Generalized Phrase Structure Grammar, Blackwell, Oxford, 1985.
[Goguen/Thatcher/Wagner, 76]
    J.A.Goguen, J.W.Thatcher, E.G.Wagner: An Initial Algebra Approach to the Specification, Correctness and Implementation of Abstract Data Types, IBM Research Report RC 6487, 1976.
[Goguen/Thatcher/Wagner/Wright, 77]
    J.A.Goguen, J.W.Thatcher, E.G.Wagner, J.B.Wright: Initial Algebra Semantics and Continuous Algebras, Journal of the ACM 24, 1977.

[Hauenschild, 86]
Ch.Hauenschild: KIT-NASEV oder die Problematik des Transfers bei der maschinellen Übersetzung. In: I.Batori, H.J.Weber (eds.): Neue Ansätze in maschineller Sprach-übersetzung: Wissensrepräsentation und Textbezug, Niemeyer, Tübingen, 1986, 167-195.

[Hauenschild/Busemann, 88]
Ch.Hauenschild, S.Busemann: A Constructive Version of GPSG for Machine Translation. In: E.Steiner, P.Schmidt, C.Zelinsky-Wibbelt (eds.): From Syntax to Semantics, Pinter Publishers, London, 1988.

[Hauenschild/Umbach, 88]
Ch. Hauenschild, C. Umbach: Funktor-Argument-Struktur. Die satzsemantische Repräsentations- und Transferebene im Projekt KIT-FAST. In: J.Schütz (ed.): Workshop "Semantik und Transfer". IAI Working Papers Nr. 6, Saarbrücken, 1988.

[Kratzer/Pause/Stechow, 73]
A.Kratzer, E.Pause, A.v.Stechow: Einführung in Theorie und Anwendung der generativen Syntax, Athenäum, Frankfurt/M, 1973.

[Lambek, 58]
J.Lambek: The Mathematics of Sentence Structure, In: American Mathematican Monthly 65, 1958, 154-169.

[Milner, 84]
R.Milner: The Standard ML Core Language, Edinburgh University Internal Report CSR-168-84, 1984.

[Möller/Tarlecki/Wirsing, 88]
B.Möller, A.Tarlecki, M.Wirsing: Algebraic Specification with Buildt-In Domain Constructions. Lecture Notes in Computer Science 295, Springer Verlag, 1988.

[Montague, 70]
R.Montague: Universal Grammar, in: Theoria 36, 1970, 373-398; Reprinted in: H.Thomason (ed.): Formal Philosophy, Selected Papers of Richard Montague, Yale University Press, New Haven, 1974, 222-246.

[Moortgat, 88]
M.Moortgat: Categorial Investigations, Logical and Linguistical Aspects of the Lambek Calculus, Dissertation, Amsterdam, 1988

[Partee/Rooth, 83]
B.Partee, M.Rooth: Generalized Conjunctions and Type Ambiguity. In: R.Bäuerle, Ch. Schwarze, A.v.Stechow (eds.), Meaning, Use and Interpretation of Language, de Gruyter, Berlin, 1983, 361-383.

[Partee, 87]
B.Partee: Noun Phrase Interpretation and Type-Shifting Principles. In: J.Groenendijk, D.de Jongh, M.Stockhof (eds.), Studies in Discourse Representation Theory and the Theory of Generalized Quantifiers, Foris, Dordrecht 1987.

[Poigné, 86]
A.Poigné: On Specifications, Theories, and Models with Higher Type. Information and Control 68, 1986, 1-46.

List of Contributors

J. Allgayer
Department of Computer Science
Universität des Saarlandes
Im Stadtwald 15
D 6600 Saarbrücken

C. Beierle
IBM Germany
Scientific Center
Institute for Knowledge Based Systems
P.O. Box 80 08 80
D 7000 Stuttgart 80

K.H. Bläsius
Fachhochschule Dortmund
Fachbereich Informatik
Sonnenstr. 96-100
D 4600 Dortmund

W. Dilger
Fraunhofer-Institut für Informations- und Datenverarbeitung
Fraunhoferstr. 1
D 7500 Karlsruhe

K. Eberle
Institut für Maschinelle Sprachverarbeitung
Universität Stuttgart
Keplerstr. 17
D 7000 Stuttgart 1

U. Hedtstück
IBM Germany Scientific Center
Institute for Knowledge Based Systems
P.O. Box 80 08 80
D 7000 Stuttgart 80

K. von Luck
IBM Germany Scientific Center
Institute for Knowledge Based Systems
.O. Box 80 08 80
D 7000 Stuttgart 80

B. Mahr
Technische Universität Berlin
FB 20 - Institut für Software und Theoretische Informatik
Franklinstr. 28/29
D 1000 Berlin 12

B. Nebel
IBM Germany Scientific Center
Institute for Knowledge Based Systems
P.O. Box 80 08 80
D 7000 Stuttgart 80

A. Oberschelp
Universität Kiel
Abteilung Logik
Olshausenstr. 40-60
D 2300 Kiel

B. Owsnicki-Klewe
PHILIPS Research Laboratories Hamburg
P.O. Box 540 840
D 2000 Hamburg 54

U. Pletat
IBM Germany Scientific Center
Institute for Knowledge Based Systems
P.O. Box 80 08 80
D 7000 Stuttgart 80

C. Reddig-Siekmann
Department of Computer Science
Universität des Saarlandes
Im Stadtwald 15
D 6600 Saarbrücken 11

C.-R. Rollinger
IBM Germany Scientific Center
Institute for Knowledge Based Systems
P.O. Box 80 08 80
D 7000 Stuttgart 80

P.H. Schmitt
Institute for Logic, Complexity, and Deduction Systems
University of Karlsruhe
D 7500 Karlsruhe 1

J.H. Siekmann
FB Informatik
Universität Kaiserslautern
Postfach 2080
D 6750 Kaiserslautern

G. Smolka
IBM Germany Scientific Center
Institute for Knowledge Based Systems
P.O. Box 80 08 80
D 7000 Stuttgart 80

C. Umbach
Technische Universität Berlin
Institut für Software und Theoretische Informatik
Franklinstr. 28/29
D 1000 Berlin 12

H. Voß
Gesellschaft für Mathematik und Datenverarbeitung
Schloß Birlinghoven
D 5205 St. Augustin

Ch. Walther
Institute for Logic, Complexity, and Deduction Systems
University of Karlsruhe
D 7500 Karlsruhe

W. Wernecke
IBM Germany Scientific Center
Institute for Knowledge Based Systems
Wilckensstr. 1a
D 6900 Heidelberg 1